Cancer Sensitizing Agents for Chemotherapy
THERAPEUTIC STRATEGIES TO OVERCOME ALK RESISTANCE IN CANCER

VOLUME 13

Cancer Sensitizing Agents for Chemotherapy Series

Series Editor: Benjamin Bonavida, PhD

Cancer Sensitizing Agents for Chemotherapy

THERAPEUTIC STRATEGIES TO OVERCOME ALK RESISTANCE IN CANCER

VOLUME 13

Edited by

LUC FRIBOULET, PhD

Predictive Biomarkers and Novel Therapeutic Strategies in Oncology, INSERM U981,
Gustave Roussy Cancer Campus, Paris-Saclay University, Villejuif, France

ELSEVIER

ACADEMIC PRESS
An imprint of Elsevier

Academic Press is an imprint of Elsevier
125 London Wall, London EC2Y 5AS, United Kingdom
525 B Street, Suite 1650, San Diego, CA 92101, United States
50 Hampshire Street, 5th Floor, Cambridge, MA 02139, United States
The Boulevard, Langford Lane, Kidlington, Oxford OX5 1GB, United Kingdom

Notices
Knowledge and best practice in this field are constantly changing. As new research and experience broaden our
understanding, changes in research methods, professional practices, or medical treatment may become
necessary.

Practitioners and researchers must always rely on their own experience and knowledge in evaluating and using
any information, methods, compounds, or experiments described herein. In using such information or methods
they should be mindful of their own safety and the safety of others, including parties for whom they have a
professional responsibility.

To the fullest extent of the law, neither the Publisher nor the authors, contributors, or editors, assume any liability
for any injury and/or damage to persons or property as a matter of products liability, negligence or otherwise, or
from any use or operation of any methods, products, instructions, or ideas contained in the material herein.

Library of Congress Cataloging-in-Publication Data
A catalog record for this book is available from the Library of Congress

British Library Cataloguing-in-Publication Data
A catalogue record for this book is available from the British Library

ISSN: 2468-3183

ISBN: 978-0-12-821774-0

For information on all Academic Press publications
visit our website at https://www.elsevier.com/books-and-journals

Publisher: Stacy Masucci
Senior Acquisitions Editor: Rafael Teixeira
Editorial Project Manager: Samantha Allard
Production Project Manager: Punithavathy Govindaradjane
Cover Designer: Greg Harris

Typeset by SPi Global, India

Cover Image Insert

The cover image illustrates the current situation in oncology precision medicine aiming at selecting the appropriate targeting agent for each individual cancer patient. This is particularly true for ALK-driven cancers.

Aims and Scope for Series "Cancer Sensitizing Agents for Chemotherapy"

Current cancer management strategies fail to adequately treat malignancies with chemotherapy, with multivariable dose-restrictive factors such as systemic toxicity and multidrug resistance; hence, limiting therapeutic benefits, quality of life and complete long-term remission rates. The resistance of cancer cells to anticancer drugs is one of the major reasons for the failure of traditional cancer treatments. Cellular components and dysregulation of signaling pathways contribute to drug resistance. If modulated, such perturbations may restore the drug response and its efficacy. The recent understanding of the molecular mechanisms and targets that are implicated for cancer chemoresistance have paved the way to develop a large battery of small molecules (sensitizing agents) that can target resistance and, thus, allowing their combination with chemotherapeutic drugs to be effective and to reverse chemoresistance. A large variety of chemotherapy-sensitizing agents has been developed and several have been shown to be effective in experimental models and in cancer patients.

The main objective of the proposed series "Cancer Sensitizing Agents for Chemotherapy" is to publish individualized and focused volumes whereby each volume is edited by an invited expert Editor(s). Each volume will dwell on specific chemo-sensitizing agents with similar targeting activities. The combination treatment of the sensitizing agent with chemotherapy may result in a synergistic/additive activity and the reversal of tumor cells resistance to drugs.

The Editor(s) will compile nonoverlapping review chapters on reported findings in various cancers, both experimentally and clinically, with particular emphases on underlying biochemical, genetic, and molecular mechanisms of the sensitizing agent and the combination treatment.

The scope of the Series is to provide scientists and clinicians with updated and clinical information that will be valuable in their quest to investigate, develop, and apply novel combination therapies to reverse drug resistance and, thereby, prolonging survival and even cure in cancer patients.

Dr. Benjamin Bonavida, PhD (Series Editor)

About the Series Editor

Dr. Benjamin Bonavida, PhD, (Series Editor), is currently Distinguished Research Professor at the University of California, Los Angeles (UCLA). He is affiliated with the Department of Microbiology, Immunology and Molecular Genetics, UCLA David Geffen School of Medicine. His research career, thus far, has focused on investigations in the fields of basic immunochemistry and cancer immunobiology. His research investigations have ranged from the biochemical, molecular, and genetic mechanisms of cell-mediated killing and tumor cell resistance to chemo-immuno cytotoxic drugs. The reversal of tumor cell resistance was investigated by the use of various selected sensitizing agents based on molecular mechanisms of resistance. In these investigations, there was the newly characterized dysregulated NF-κB/Snail/YY1/RKIP/PTEN loop in many cancers that was reported to regulate cell survival, proliferation, invasion, metastasis, and resistance. Emphasis was focused on the roles of the tumor suppressor Raf Kinase Inhibitor Protein (RKIP), the tumor promoter Yin Yang 1 (YY1), and the role of nitric oxide as a chemo-immuno-sensitizing factor. Many of the earlier mentioned studies are centered on the clinical challenging features of cancer patients' failure to respond to both conventional and targeted therapies.

The Editor has been active in the organization of regular sequential international miniconferences that are highly focused on the roles of YY1, RKIP, and nitric oxide in cancer and their potential therapeutic applications. Several books edited or coedited by the Editor have been published. In addition, the Editor has been the Series Editor of books (over 23) published by Springer on *Resistance to Anti-Cancer Targeted Therapeutics*. In addition, the Editor is presently the Series Editor of three series published by Elsevier/Academic Press on "*Cancer Sensitizing Agents for Chemotherapy,*" *Sensitizing Agents for Cancer Resistance to Antibodies*, and *Breaking Tolerance to Anti-Cancer Immunotherapy*. Lastly, the Editor is the Editor-in-Chief of the Journal *Critical Reviews in Oncogenesis*. The Editor has published over 500 research publications and reviews in various scientific journals of high impact.

Acknowledgments: The Editor wishes to acknowledge the excellent editorial assistance of Ms. Inesa Navasardyan who has worked diligently in the completion of this volume, namely, in both the editing and formatting of the various contributions of this volume.

The Editor acknowledges the Department of Microbiology, Immunology and Molecular Genetics and the UCLA David Geffen School of Medicine for their continuous support. The Editor also acknowledges the assistance of Mr. Rafael Teixeira, Acquisitions Editor for Elsevier/Academic Press, and the excellent assistance of Ms. Samantha Allard, Editorial Project Manager for Elsevier/Academic Press, for their continuous cooperation throughout the development of this book.

Aims and Scope of Volume

This volume aims to provide the necessary knowledge on the biology of ALK RTK (receptor Tyrosine Kinase), focusing on its tissue expression, structure, and function. It also gives a comprehensive overview of the current treatment options and the benefits of ALK tyrosine kinase inhibitors in lung and other cancer types, such as ALCL, neuroblastoma, and inflammatory myofibroblastic tumors. Most importantly, the volume encompasses evidence on systemic treatments other than TKI, including chemotherapy, immunotherapy, and antiangiogenic agents in ALK-driven cancers.

Dr. Luc Friboulet, PhD

About the Volume Editor

Team leader and Deputy Director at the INSERM U981 research unit which is hosted by Gustave Roussy Cancer Center in France, Dr. Luc Friboulet (PhD) is a Principal Investigator funded by the European Research Council since 2016. After his PhD degree on apoptotic regulation in cancer from the University Paris-Saclay in 2009, Luc Friboulet joined Pr. Jean-Charles Soria's team to study how the DNA repair protein ERCC1 and its isoforms can predict patient response to platinum-based chemotherapy in lung cancer patients. He then went to Dr. Jeffrey A. Engelman's lab in Massachusetts General Hospital in Boston, Harvard, where he evaluated response to ALK inhibitors and the ability of next-generation ALK Tyrosine Kinase inhibitors to overcome resistance in lung cancer patients. His ongoing scientific work focuses on targeted therapies in cancer and the early detection of drug-tolerant persister cells during treatment. Linked to a clinical trial named MATCH-R allowing the collection of repeated biopsies in patients treated with ALK, ROS1, RET, EGFR, BRAF, FGFR inhibitors, he characterizes the resistance mechanisms and establishes new laboratory models. The characterizing of the drug tolerance process aims to identify adaptive therapeutic strategies that might prevent or eliminate drug-tolerant cancer cells. Cited in more than 2000 scientific articles, he has published as a first or last author in journals such as the *New England Journal of Medicine, Cancer Discovery, Clinical Cancer research*, and *Nature Reviews Clinical Oncology.*

Preface

One of the main goals in cancer research is to improve survival outcomes for patients and it frequently relies on the understanding of biologic events. ALK-driven cancers are a remarkable example of how the molecular characterization and functional study of a receptor have allowed to reach such a goal. In less than a decade, first, second, and third generation ALK tyrosine kinase inhibitors (TKIs) have dramatically raised quality of life and outcome of ALK-driven cancers. Nevertheless, resistance to ALK TKIs forces clinicians to face a therapeutic impasse after the exhaustion of multiple treatment lines. The characterization of resistant cancer samples has led to the identification of multiple resistance mechanisms and innovative strategies trying to overcome these, but they have not been exhaustively reviewed in a single document for different cancer types. Facing efficacy failure of ALK TKIs in oncology, clinicians and researchers would benefit from general guidelines on how to identify the acquired resistance mechanism and how to select optimal therapeutic strategies to overcome the specific resistance.

Dr. Luc Friboulet is a cancer scientist at the INSERM institute and the Gustave Roussy Comprehensive Cancer Center in France. He is renowned for several breakthroughs in the understanding of acquired mechanisms of ALK-driven lung cancer. Together with the best experts in the world on the topic, he has delivered a genius book containing the most comprehensive overview on ALK and cancer to date. It presents the structure and function of the ALK receptor and its role, not only in ALK-rearranged lung cancer, but also in ALK-positive pediatric cancers. It is also an extraordinary lesson and guidance into resistance mechanisms and tumor drug adaptation. The book is also a remarkable resource on drug combinations able to enhance ALK TKI antitumor efficacy in ALK-driven cancers and presents many future perspectives in the field. Among cancer researchers, clinicians, members of the biomedical community, and even students, most of them will find new inspiration and useful information in this text to face their daily fight to treat ALK-driven cancers.

Ken A. Olaussen
Université Paris-Saclay, Gif-sur-Yvette,
France

Contents

Contributors

Sylvain Baruchel Department of Pediatrics, The Hospital for Sick Children Research Institute; Institute of Medical Sciences, University of Toronto, Toronto, ON, Canada

María Castro-Henriques Department of Medical Oncology, Centro Integral Oncológico Clara Campal (HM-CIOCC), Hospital HM Delfos, HM Hospitales, Barcelona, Spain

Ana Collazo-Lorduy Department of Medical Oncology, Centro Integral Oncología Clara Campal Madrid, HM-Sanchinarro, Medical Oncology Department; Department of Medical Oncology, Hospital Universitario Puerta de Hierro, Madrid, Spain

Marie-Emilie Dourthe Department of Pediatric Hematology, Robert Debré University Hospital, AP-HP, Paris, France

Francesco Facchinetti Predictive Biomarkers and Novel Therapeutic Strategies in Oncology, INSERM U981, Gustave Roussy Cancer Campus, Paris-Saclay University, Villejuif, France

Luc Friboulet Predictive Biomarkers and Novel Therapeutic Strategies in Oncology, INSERM U981, Gustave Roussy Cancer Campus, Paris-Saclay University, Villejuif, France

Robert E. Hutchison State University of New York, Upstate Medical University, Syracuse, NY, United States

Beatriz Jiménez Department of Medical Oncology, Centro Integral Oncología Clara Campal Madrid, HM-Sanchinarro, Medical Oncology Department, Madrid, Spain

Ryohei Katayama Division of Experimental Chemotherapy, Cancer Chemotherapy Center, Japanese Foundation for Cancer Research; Department of Computational Biology and Medical Sciences, Graduate School of Frontier Sciences, The University of Tokyo, Tokyo, Japan

Alessandro Leonetti Medical Oncology Unit, University Hospital of Parma; Department of Medicine and Surgery, University of Parma, Parma, Italy

Luca Mologni School of Medicine and Surgery, University of Milano-Bicocca, Monza, Italy

Silvia Novello Department of Oncology, University of Turin, San Luigi Gonzaga Hospital, Turin, Italy

Gonzalo Recondo Medical Oncology, Center for Medical Education and Clinical Research (CEMIC), Buenos Aires, Argentina

Jordi Remon Department of Medical Oncology, Centro Integral Oncológico Clara Campal (HM-CIOCC), Hospital HM Delfos, HM Hospitales, Barcelona, Spain

Charlotte Rigaud Department of Pediatric Oncology, Gustave Roussy Cancer Campus, Villejuif, France

Luisella Righi Department of Oncology, University of Turin, San Luigi Gonzaga Hospital, Turin, Italy

Fabrizio Tabbò Department of Oncology, University of Turin, San Luigi Gonzaga Hospital, Turin, Italy

Riccardo Taulli Department of Oncology, University of Turin, San Luigi Gonzaga Hospital, Turin, Italy

Marcello Tiseo Medical Oncology Unit, University Hospital of Parma; Department of Medicine and Surgery, University of Parma, Parma, Italy

Libo Zhang Department of Molecular Medicine; Department of Anesthesia, The Hospital for Sick Children Research Institute, Toronto, ON, Canada

The ALK receptor tyrosine kinase journey: From physiological roles to pathological disruptions

Fabrizio Tabbò, Luisella Righi, Riccardo Taulli, and Silvia Novello

Department of Oncology, University of Turin, San Luigi Gonzaga Hospital, Turin, Italy

Abstract

The overwhelming path of Anaplastic Lymphoma Kinase (ALK) goes from the characterization of its physiological functions as Receptor Tyrosine Kinase (RTK) to the identification of pathogenetic mechanisms spurring neoplastic transformation. If ALK role in mammal development is partially understood, molecular disruptions have been observed in a variety of cancers either at high frequency in rare tumors or with low occurrence in more prevalent ones, such as Non-Small Cell Lung Cancer (NSCLC). Deciphering ALK perturbing effects in cancer cells gave us mechanistic insights in the comprehension of its physiological roles and the basis to develop multiple small Tyrosine Kinase Inhibitors (TKI) capable to restrain ALK-driven oncogenesis. A deep understating of ALK biology has, therefore, direct therapeutic implications: amplifications, point-mutations and translocations differently affect human cells and influence patients' response to TKI or other treatment strategies.

Abbreviations

AKT	protein kinase B
ALCL	anaplastic large cell lymphoma
ALK	anaplastic lymphoma kinase
ALKAL	ALK and LTK ligands
AP1	activator protein 1
BIM1	B cell-specific Moloney murine leukemia virus integration site 1
CAS9	CRISPR-associated protein 9
CD4	cluster of differentiation 4
CGH	comprehensive genomic hybridization
CRISPR	clustered regularly interspaced short palindromic repeats
CRKL	CRK-like protein

CTLC	clathrin heavy chain
DDR	DNA damage response
DLBCL	diffuse large B cell lymphoma
DNMT1	DNA methyltransferase 1
EML4	echinoderm microtubule-associated protein-like 4
EZH2	enhancer of zeste homolog 2
FAM	augemntor alfa/beta
FDA	Food and Drug Administration
GCS	granular cytoplasmic staining
GEM	genetically engineered model
GR	glycine-rich
HEN1	hesitation behavior 1
IMT	inflammatory myofibroblastic tumor
JAK	Janus kinase
Jeb	jelly belly
JNK	c-Jun N-terminal kinase
JUNB	transcription factor Jun-B
KD	kinase domain
KIF5B	kinesin family member 5B
LDLa	low-density lipoprotein
lncRNA	long non-coding RNA
LTK	leucocyte tyrosine kinase
MAM	meprin, A-5 protein, Mu protein
MAPK	mitogen-activated protein kinase
MCL1	induced myeloid leukemia cell differentiation protein Mcl-1
MDM2	mouse double minute 2 homolog
miRNA	micro RNA
MMR	DNA mismatch repair
MSH	MutS homolog
mTOR	mammalian target of rapamycin
MYCN	N-myc proto-oncogene protein
NB	neuroblastoma
NF1	neurofibromatosis type 1
NGS	next-generation sequencing
NHL	non-Hodgkin lymphoma
NPM1	nucleophosmin 1
p16INK4a	inhibitor of CDK4
PI3K	phosphoinositide 3 kinase
PLC-γ	phospholipase C Gamma
PRDM1	positive regulatory domain maturation 1
PTN	pleiotrphin
RAP1	Ras-related protein 1
RB	retinoblastoma
RET	Ret proto-oncogene
RPTP	receptor-like protein tyrosine phosphatase
SETD2	SET domain containing 2
SHP1	Src homology region 2 domain-containing phosphatase-1
SNP	single nucleotide polymoprhism
STAT	singal transducer and activator of transcription
TAPE	tandem atypical propeller in EMLs
TGF-β	transforming growth factor beta
TP53	tumor protein P53
TPM3	tropomyosin alpha-3 chain
TRAF1	TNF receptor associated factor 1

Conflict of interest

No potential conflicts of interest were disclosed.

Introduction

Receptors Tyrosine Kinases (RTKs) exert a crucial role in the cellular physiology orchestrating fundamental mechanisms such as cell-cycle progression, proliferation, differentiation and migration. Their ligand-dependent activation allows a temporal and spatial control of the catalytic function, whose final fate is to induce specific biological responses through a complex and redundant signaling cascade [1]. Negative regulators exist to attenuate and block intracellular firing when a new cell "status" is achieved; however, diverse disrupting events are capable to sustain RTK enzymatic activity determining an hyperactive condition [2,3]. Structural alterations, either prompting a ligand-independent constitutive activation or expressing uncontrolled kinases out of their physiological context, make RTKs well-recognized oncogenes, from which cancer cells very often depend [4,5]. The Anaplastic Lymphoma Kinase (ALK) RTK, which plays key roles in organism development, belongs to this family of growth-enhancing proteins that, if functionally or structurally altered, may promote uncontrolled cancer growth [6]. Across different tumors, multiple ALK alterations have been described: full-length receptor point mutations in neuroblastoma (NB) and gene translocations in Anaplastic Large Cell Lymphoma (ALCL) and Non-Small Cell Lung Cancer (NSCLC) are the most relevant [7]. Characterization of these molecular defects, in terms of occurrence, protein structure, signaling perturbation and cancer cell dependency had led to the understanding of their pathogenic role and development of a plethora of ALK inhibitors (ALKi) available in the clinical arena.

The ALK proto-oncogene

The Anaplastic Lymphoma Kinase (ALK) was originally identified in Anaplastic Large-Cell non-Hodgkin's Lymphoma (ALCL) cell lines as the product of a chromosomal translocation t(2;5; p23;q35) involving the Nucleophosmin (NPM1) partner [8,9]. The human ALK locus is located at the 2p23.2–p23.1 chromosome region and encodes for 26 exons that result in a 1620 amino acids protein of about 180 kDa, that is the extensively glycosylated [10,11]. At the structural level the ALK tyrosine kinase receptor, together with the leukocyte tyrosine kinase (LTK), belong to the insulin receptor superfamily due to their similarities in the kinase domain (KD). In contrast to LTK, the extracellular region of human ALK consists of a glycine-rich domain (GR), two meprin A-5 protein, receptor protein-tyrosine phosphatase µ regions (MAM) and a low-density lipoprotein motif (LDLa) (Fig. 1) [10–13]. The biological significance of LDLa is unknown, while at the structural level the MAM domain can potentially establish extracellular cell-to-cell interactions. The functional relevance of MAM and GR regions have been demonstrated in a series of Drosophila ALK mutants, in which specific amino acid substitutions resulted in a lethal gut phenotype during development [14]. At the molecular level, the activation loop includes an YxxxYY motif that is in common with the insulin

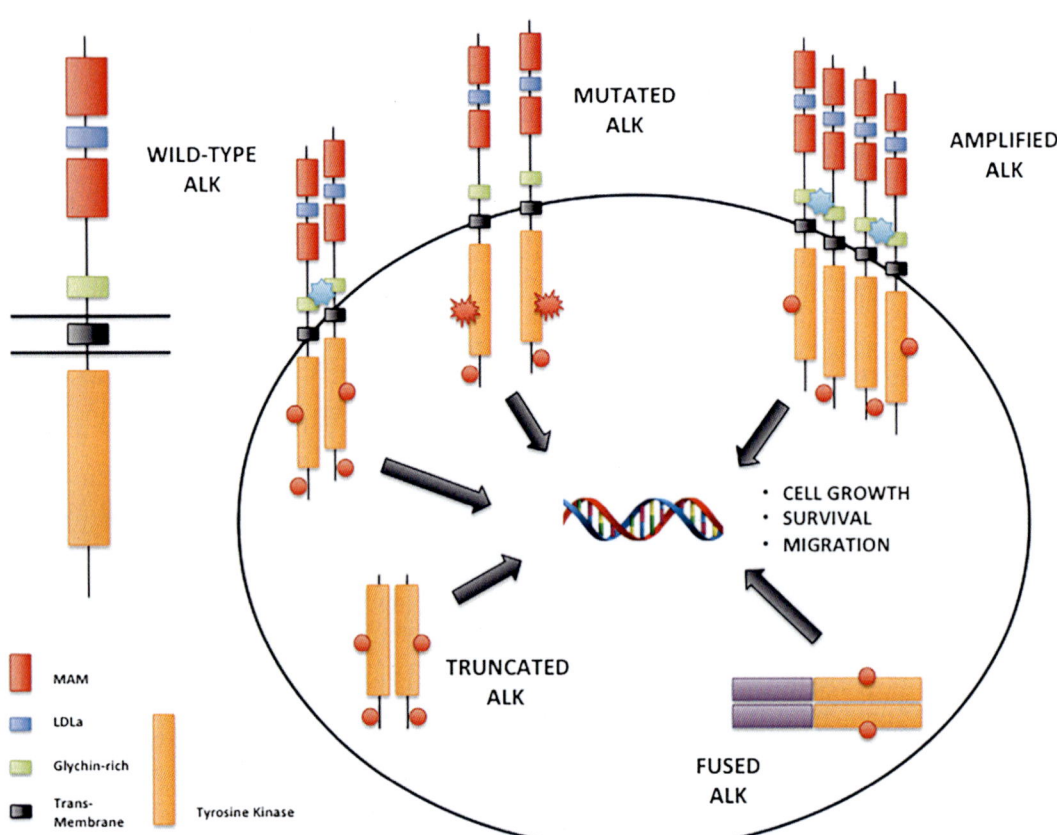

FIG. 1   The anaplastic lymphoma kinase (ALK) receptor tyrosine kinase structure and its pathological alterations. As other receptor tyrosine kinases, ALK is composed by an extracellular domain (with two MAM domains, one LDLa domain and one glychin-rich region), an intracellular domain (the tyrosine kinase domain—TKD) and a connecting trans-membrane region. Physiological activation of ALK receptor depends by ligand binding (light blue star) then requires auto- and trans-phosphorylation of specific residues (red dots) and finally promotes signaling to the nucleus. Intra-TKD mutations determine a constitutive ligand-independent activation, whereas ALK amplification sustains an increased downstream signaling in a ligand-dependent manner. Truncated ALK mutants, loosing extracellular domain, appear overexpressed within the cytoplasm and rearranged ALK proteins, with TKD juxtaposed to a fusion partner with dimerization sites, aberrantly phosphorylates protein substrates and constitutively sustain cell survival and proliferation.

receptor. Notwithstanding, a detailed molecular analysis has revealed that, specifically, the ALK Tyr1278 is critical for KD activation and for oncogenic transformation [15,16].

ALK function relies on classical receptor tyrosine kinase mechanisms: via a ligand-dependent dimerization and trans-phosphorylation of specific residues, its catalytic capabilities are activated. In addition to Y1278, Y1282 and Y1283 also appear as relevant tyrosine residues within the activation loop [13,17]. Like other tyrosine kinase receptors, phosphorylated ALK generates a downstream signaling cascade. The vast majority of insights about ALK signaling are gained by aberrant isoforms, mainly ALK fusions, where canonical and non-canonical effectors are involved; in those conditions, the fusion partner dictates lots of the ALK chimeras' biological properties, influencing spatial and temporal distributions. General signaling pathways activated are: PLCγ, JAK-STAT, PI3K-AKT-mTOR, JUNB—as part of the AP1 complex, CRKL-C3G-RAP1, MAPK signaling and a large number of other downstream targets (Fig. 2) [7,18–20].

Although an active scientific debate is ongoing to establish a unique and specific human ALK ligand [21–23], in *D. melanogaster* and in *C. elegance*, Jelly belly (Jeb) and hesitation behavior 1 (HEN-1), both characterized by a peculiar LDLa domain, have been recognized as effective ALK ligands [24–29]. More recently, in vitro and in vivo studies validated ALKALs (FAM150A and FAM150B) as ligands of ALK, since their binding to the ALK extra-cellular domain activates the receptor and its signaling cascade. Also heparin has been claimed as activator of ALK, since an heparin-binding motif is part of the ALK N-terminal region [22,30–32].

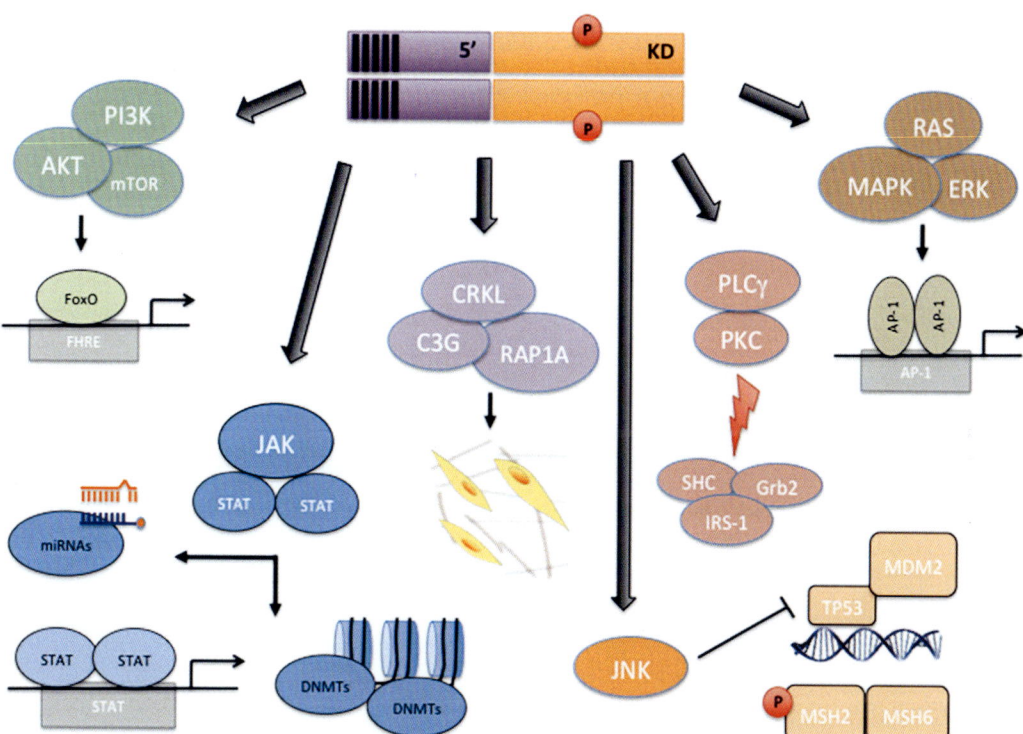

FIG. 2 Anaplastic lymphoma kinase (ALK) fusion protein and the signaling network. ALK chimeric proteins are composed of ALK tyrosine kinase domain (orange) and the fusion partner (purple), which often contain a dimerization domain (black lines). Once transactivation of catalytic domains is gained (red dots indicate phosphorylated sites) ALK fusions activate multiple canonical signaling pathways (PI3K/AKT/mTOR, JAK/STAT, PLCγ/PKC, and RAS/MAPK/ERK) modulating different downstream mediators and transcriptional factors. Multiple perturbing events are induced: epigenetic reprogramming by DNMTs expression and miRNA interference; cytoskeleton reshaping and migration induced by CRKL/C3G/RAP1A axis; genomic integrity jeopardized counteracting DNA damage response pathway and mismatch repair system. *KD*, kinase domain; *DNMTs*, DNA methyltransferases; *miRNA*, microRNA.

Defining ALK biology exploiting animal models

Pioneering studies in *Drosophila* have defined the relevance of ALK in gut development. Indeed, during embryogenesis, Jeb is involved in the specification of founder cells that are responsible of visceral musculature formation [14,24,25]. Other functional analyses have revealed a critical role of the Jeb/ALK axis in cognitive functions and in the modulation of the body size [33]. Interestingly, a recent study in Neurofibromin 1 (NF1) mouse model confirms the importance of ALK signaling in associative learning and sleep, proposing also the impairment of ALK activity as potentially beneficial in patients affected by Neurofibromatosis [34]. In *Drosophila* the Jeb/ALK pathway is also involved in synapses and neuromuscular junction formation, in visual system development, sparing also organ growth during caloric restrictions [35–38]. Moreover, recent evidences support the involvement of ALK signaling in the phenomenon of longevity and long-term memory modulation [39,40]. A functional crosstalk between ALK and Transforming Growth Factor β (TGF-β) pathway has been observed in Drosophila during gut development, an interplay then demonstrated also in *C. elegans* by an epistatic analysis [29,41]. Moreover, in olfactory adaptation of *C. elegans*, the HEN-1 signaling contributes with other pathways to the regulation of sensory responses and in particular to the proper forgetting function [42]. In Zebrafish (*Danio rerio*) the ALK family is composed of two members: Dr*Ltk* and Dr*Alk*. The DrALK exhibits a smaller extracellular domain that lacks one MAM domain, while DrLTK includes both MAM regions, as the human counterpart [13]. Interestingly, the Dr*Ltk* mutant (*shady*) exhibits profound defects in pigmentation [43]. As the mammalian ALK, Dr*Ltk* is expressed in iridophores of the neural crest [13,43,44]. Thus, Zebrafish represents a useful model to recapitulate human pathological alterations as recently demonstrated by the gain of function phenotype observed in chimeric Zebrafish models harboring human cancer mutations [45]. Recently, in vivo evidences support a functional interaction between Dr*Ltk* and *alkal1*, *alkal2a* and *alkal2b*, which represent the homologs of the human secreted small activating proteins ALKAL1 and 2 [31]. In mammalian, pioneering studies on ALK have highlighted a specific pattern of mRNA expression in the central nervous system during mouse development [10,11,46]. Notably, ALK expression decreases upon weaning, approximately at 3 weeks of age [11]. A similar modulation of ALK levels has also been confirmed in chicken, rat and humans, where additional ALK variants have been observed in other body districts as small intestine, colon, prostate and testis [8,47,48]. ALK loss of function studies reveal that, at least in mice, ALK is dispensable for survival during development; however, a deep analysis of the central nervous system has recently showed an increase of hippocampal performance and dopamine levels in these mutants [49,50].

Genetically engineered models (GEMs) of ALK pathogenesis

GEMs have actively contributed to define the oncogenic role of ALK in different tumor types. The expression of the NPM1-ALK translocation under the control of the CD4 promoter results in the spontaneous development of T-cell lymphomas and plasma cell tumors, showing that both ALK and its effector Signal Transducer and Activator of Transcription 3 (STAT3) act as critical drivers in this subset of hematological diseases [51–53]. In 2008, ALK KD mutations were reported in familiar Neuroblastoma (NB), supporting a potential oncogenic role

- Gain-of-function mutations in NB: mainly involving hotspot region **F1174**, **R1245** and **R1275**; most relevant are: F1174I, F1174L, F1245C, F1245V, R1275Q, R1275L, D1091N, G1128A, M1166R, I1171N, R1192P, and I1250T.

- L1198F and G1201E in ATC and F1174L in cSCC.

- Truncated activated ALK mutants: ALK Δ2-3, Δ1-5, Δ4-11 and Δ2-17 in NB, ALCL and SS.

- Acquired ALK mutations as mechanisms of resistance; in NSCLC most frequent are: L1196M after 1st, G1202R after 2nd and *compound mutations* after 3rd generation TKIs.

FIG. 3 Figurative summary of anaplastic lymphoma kinase (ALK) mutations. Gain-of-function mutations are mainly point mutations occurring within the kinase domain of ALK receptor. The majority has been identified in neuroblastoma (NB), either as congenital or acquired alterations, interesting, in the 80% of the cases, three hot spot regions (in bold). Mutations have been identified also in anaplastic thyroid cancer (ATC) and cutaneous squamous cell carcinoma (cSCC) cases. Truncating mutations that eliminate the extracellular domain have been described in primary samples or cell lines of NB, anaplastic large cell lymphoma (ALCL) and synovial sarcoma (SS). Lastly, in non-small cell lung cancer (NSCLC) ALK intra-kinase domain mutations are a well-known mechanism of acquired resistance to ALK inhibitors (ALKi). A large spectrum has been reported, with peculiar difference based on the type of administered ALKi. Less frequently identified after first-generation ALKi, their percentage augment after second-generation inhibitors, being G1202R the most frequent. With the sequential administration of multiple ALKi, after third-generation ALKi, *so called* compound mutations may occur.

of the receptor in this malignancy (Fig. 3) [54–58]. These mutations were also observed in commercial NB cell lines and trigger tumor growth in nude mice [55]. At the genomic level, in NB, the ALK mutation F1174L co-segregates with N-myc proto-oncogene (MYCN). Accordingly, in both mouse and zebrafish models, the expression of this variant in the neural crest, in concomitance with MYCN, results in more aggressive NB tumors in terms of development, penetrance and lethality [59–61]. Importantly, the use of ALK inhibitors (ALKi) exerts a dramatic effect on tumor growth, confirming the relevance of ALK-addiction in this disease [60,61]. Molecular analysis on knock-in mice bearing the ALK F1178L or the R1279Q mutations has also revealed that ALK is able to trigger Ret proto-oncogene (RET) up-regulation during transformation. Indeed, RET inhibition reduces tumor growth in double MYCN/ALK knock-in models, supporting also a pathogenetic role for this tyrosine kinase receptor in ALK-mutated NB tumors [62]. Recently, a novel GEM model, in which the ALK F1174L variant is temporally and conditionally controlled by the inducible Cre-loxP system, has been generated. In this more physiological setting, tumor development is observed only in association with MYCN, confirming that the two oncogenes actively cooperate in promoting NB growth [63].

The identification of the Echinoderm Microtubule-associated protein-Like 4 (EML4)-ALK translocation in non-small cell lung cancer (NSCLC) patients, has pointed out the potential oncogenic role of ALK also in this tumor type [64]. Thereby, several transgenic animals expressing the EML4-ALK fusion, or other X-ALK fusions, under the control of surfactant protein C gene, specifically in lung alveolar epithelial cells, have been generated [65–67]. These models have been extensively exploited to investigate novel therapeutic options for the treatment of ALK + lung tumors, including targeted therapies, immunological approaches or different combination regimens [67]. The improvements of next generation sequencing (NGS) methodologies have recently expanded the spectrum of ALK fusion partners in NSCLC to about ninety different variants [68]. Since the majority of these novel chimeric entities have not been investigated, but can exhibit different sensitivity to anti-ALK directed therapies, innovative approaches are urgently required for in vivo functional validation studies. Intriguingly, the use of CRISPR/Cas9 genome editing has been proposed to engineer specific chromosomal rearrangements directly in murine lungs [69]. This technology represents an extraordinary opportunity to faithfully recapitulate, in a simple and effective manner, the plethora of ALK rearrangements found in lung cancer patients [70]. Finally, this new generation of GEMs can be extremely useful to test ALKi in a personalized setting, to investigate the mechanisms of acquired resistance and to explore novel therapeutic regimens.

Aberrant forms and altered expression of ALK

ALK exists in multiple different conditions that may be physiological alternatives or, as above described, pathological disruptions. As results of alternative transcription initiation (ATI) site, three different isoforms (ALK^{ATI}) have been recognized. These ALK^{ATI} are kinase-active, possess oncogenic capacity and their expression has been identified in certain tumors: melanoma, lung adenocarcinoma and renal clear cell carcinoma [71]. Why these alternative isoforms exist and how they signal is still not elucidated.

Mechanisms that alter ALK expression are substantially three: ALK amplification, ALK mutations ad ALK fusions. Several groups have investigated ALK overexpression in neoplastic tissues (e.g., thyroid carcinoma, NSCLC, breast cancer, melanoma, tumors of the central nervous system, ewing- or rabdomyo-sarcoma and malignant fibrous histocytoma), although its significance at the molecular level remains elusive [72–76]. Evidences suggest a potential role of ALK in breast cancer due to its strong expression, associated with the Pleiothropin (PTN)/Receptor-like Protein Tyrosine Phosphatase (RPTP)β/ζ signaling, claiming an hypothetical ALK oncogenic property in a patients' subset [77]. Amplifications of chromosome 2, encoding both for ALK and MYCN, play a relevant role in NB especially when associated with ALK mutations [76,78]. For this reason, as discussed, ALK amplification and gain-of-function mutations are deeply characterized in the pathogenesis of NB [79].

Mutations within "hot spot" regions of the tyrosine KD, disrupting the auto-inhibitory mechanism, are able to determine a constitutively active full-length receptor. Almost 85% of them occupy specific positions: R1275, F1174 and F1245, which are observed in a relevant number of familial NB and also in sporadic cases [54–58,80,81]. Full-length receptor mutations, which affect tissues physiologically expressing ALK rather than cause an ectopic expression like in ALK-rearranged tumors, are observed also in other cancer types, as anaplastic thyroid cancer (mutations in L1198F and G1201E), rhabdomyosarcoma, primitive neuroectodermal tumor, osteosarcoma and cutaneous squamous cell cancer [82–84]. Mainly three different classes of ALK receptor mutations may be identified: ligand-independent gain-of-function, ligand-dependent and truncated activated ALK mutants. Among this latter category, ALK Δ2-3, Δ1-5 and Δ4-11 have been identified in neuroblastoma-derived cell lines, while other variants (ALK Δ2-17) in ALCL and synovial sarcoma cell lines as results of complex genomic rearrangements [85–88]. Lastly, intra-ALK KD mutations may occur as an acquired mechanism of resistance during treatment with ALKi: in NSCLC patients their appearance is influenced by inhibitors of different generations (Fig. 3) [89, 90].

Structure and origin of ALK chimeras

From the first identification of the NPM1-ALK fusion protein generated by the chromosomal translocation t(2;5; p23;q35) in ALCL in 1994, an overwhelming plethora of ALK chimeras has been characterized, often occurring as rare event in a wide range of different tumors. Today ALK translocations play a relevant role in the group of tyrosine kinase fusions as potential therapeutic targets [7,8,91]. Of note ALK detection and targeting is part of gold-standard diagnostic-therapeutic protocols of specific tumors, like NSCLC. Even if reported as a quite rare event (2–7%), the EML4-ALK fusion, generated by an intra-chromosomal inversion inv.(2)(p21p23), represents the most prevalent ALK chimera due to the high prevalence in NSCLC patients [64,92,93]. A relevant aspect in the generation of chimeric protein, either created by inter- or intra-chromosomal rearrangements, is the forced oligomerization caused by loss of inhibitory domains and due to the ALK fusion partner: for instance, the coiled-coil domain of EML4 induces dimerization and activation of the fused protein (Fig. 2). Breakpoints generating ALK fusions occur within exons coding for the juxta-membrane region (typically exon 20), thus creating a protein with the entire tyrosine KD. Since the transmembrane and the promoter region are lost, the fusion partner will dictate its localization and ectopically expression. In addition, some of the ALK chimeras have multiple variants for the same partner (i.e., EML4-ALK with more the 15 variants), therefore augmenting the biological complexity of the fusion protein itself [94,95].

Tyrosine kinase fusions are well-recognized oncogenes, however multiple efforts have been spent to clarify its role as simple passengers or drivers of the neoplastic transformation. Observations that NPM1-ALK fusion transcripts are detectable also in non-neoplastic healthy T lymphocytes or hematopoietic cells suggest that maybe fusions are necessary, but not sufficient, to drive oncogenesis and other alterations are required [96]. In lung adenocarcinoma, EML4-ALK fusions appear as events of early oncogenesis and the acquisition, only after long latency, of other genetic defects (i.e., SETD2 mutations) determines the emergence of tumor lesions [97]. Furthermore, a differential role in tumor development may be due to different levels of ALK protein expression, which depend on different transcriptional activity of the fusion partner and different patterns of translational regulation of the chimeric protein: these evidences shed light on the concept of near-neutrality also for ALK rearrangements during early phases of neoplastic transformation [98].

ALK fusions and their partners: Biological implications

Chromosomal rearrangements are perturbing events within the cell, whichever they represent the inception of oncogenesis or just a step forward. The generated chimeric protein alters the intracellular biology, accelerating transformation with multiple different modalities: recruits molecular adaptors, phosphorylates signaling substrates, alters transcriptional programs interacting with nuclear factors, influences epigenetic modifications and genomic stability, determines its proper spatial localization and prevents fusion protein degradation (Fig. 2). ALK fusions activate a plethora of signaling pathways, as above elucidated, and are capable to potentiate oncogenic pathways directly, like in the case of the NF-kB signaling for the TRAF1-ALK fusion [99]. NPM1-ALK tends to accumulate also into the nucleus, due to the oligomerization with wild-type NPM1, an ubiquitously expressed protein with nucleocytoplasmatic shutting role [100]. Due partially to its sub-localization, NPM1-ALK influences directly and indirectly DNA damage response (DDR) pathway and DNA mismatch repair (MMR) system. From one side, as other oncogenes, ALK fusions induce DNA damages and curtail pathways capable to induce cellular senescence: NPM1-ALK inhibits p53 via MDM2 and JNK and deregulates p16INK4a/RB. On the other side they alter function of the MMR mechanism: MSH2 is a substrate of NPM1-ALK phosphorylations, thus blocking the MSH2/MSH6 interaction and hindering DNA repair properties in presence of DNA damage (Fig. 2) [101–103]. If we add that, in ALK+ ALCL patients, microsatellite instability and loss of TP53 have been reported, we conclude that ALK fusions are able to perturb genomic stability [103–105].

Skewing of the epigenetic pathways by NPM1-ALK happens through different mechanisms; enhancement of gene silencing by promoting DNA methylation is mediated via STAT3, a central node of ALK signaling, which induces expression of DNA methyltrasferase 1 (DNMT1)—a central enzyme that maintain methylation status [52,106]. In ALK+ ALCL, different methylated genes, therefore repressed, have been identified: both positive regulators for cellular proliferation (e.g., BIM) and negative controllers of signaling (e.g., SHP1, STAT5A) that normally act as tumor suppressors [107–109]. Gene expression could also be subverted by activity of miRNAs, facilitating malignant phenotype. Different studies have tried to interrogate expression profiles of ALK+ versus ALK- ALCL, identifying key miRNAs in ALK+ cell lines and primary patients' samples: among others, miR-17-92 cluster, which is regulated by STAT3 and seems vital of ALK+ cells and miR-101, whose targets are mTOR, MCL1 and EZH2 [110]. Moreover, long non-coding RNAs (lncRNAs), having role in post-transcriptional regulation of mRNA, have been associated with ALK positivity, but their significance is still matter of investigation. Lastly, epigenetic programs are diverted by NPM1-ALK capacities to influence transcription factors: AP1 complexes and cJUN and JUNB directly interact with chimeric protein at nuclear level and facilitate cell cycle progression (Fig. 2) [111,112].

As discussed below, a vast variety of N-terminal partners has been discovered in a wide range of cancers, mainly occurring at a very low frequency. EML4, the most characterized one, with its different truncation sites and derived variants (v-1, -2, -3a/b, -5a/b) accounts for the vast majority of ALK fusions in NSCLC. Different 5′ ALK partners, and different variants of the same chimeric protein, may influence functional aspects: localization within the cell (EML4 v1 is found in the cytoplasm whereas v3a/b is more associated with microtubules), intrinsic kinase activity, protein-protein interactions (TPM3-ALK impairs co-localization of WT TPM3 influencing cytoskeleton shaping) and protein stability (presence or absence of TAPE moiety influences unstable forms recruiting heat shock proteins) [113–115]. Pre-clinical and clinical retrospective studies highlighted therapeutic implications based on different partners and variants: patients bearing EML4-ALKv1 tend to respond better to ALKi, resistance mutations to next generation inhibitors are identified preferentially in EML4-ALKv3 and KIF5B-ALK is less sensitive to crizotinib and lorlatinib. Conversely, large prospective studies are reporting equal clinical responses—in terms of objective response, progression free survival and duration of response—in ALK+ NSCLC bearing different EML4 variants [116–119].

ALK rearrangements in hematological cancers

NPM1-ALK fusion protein is expressed in the majority of ALCL, approximately 55% of adult patients and more than 90% of pediatric ones. ALCL is a rare subtype of non-Hodgkin lymphoma (NHL), most common in children and young adults, with a male predominance (male/female ratio, 3.0) and accounting for 10–15% of pediatric and adolescent NHL, while it represents approximately 3% of adult NHL [120]. ALK+ ALCL, expressing ALK protein due to translocations that involve the 2p23 locus, represents a distinct type of peripheral T-cell lymphoma in the World Health Organization (WHO) tumors' classification [121,122]. The more frequent fusion partner in ALK+ ALCL is NPM1, but an increasing number of ALK fusion proteins are being identified in hematological neoplastic diseases (Table 1).

The localization of the ALK fusion protein in the lymphoma cells detected by immunostaining differs depending on the partner gene. For the common NPM1-ALK fusion, immunohistochemical detection of the ALK antigen shows both nuclear and cytoplasmic staining due to the heterodimerization of NPM1-ALK with normal NPM1. Approximately 15% of ALK+ ALCL cases and other ALK+ NHL lack the nuclear staining pattern, indicating that aberrant ALK expression is due to a partner gene other than NPM1 (Table 1) [139,141]. Even though the ALK fusion is a key oncogenic event in the pathogenesis of ALK+ ALCL, additional genetic imbalances have been described. By comprehensive genomic hybridization (CGH) array, 58% of the ALK+ ALCL demonstrated chromosomal imbalances including loss of 4q13-q28, 6q13-q22, 11q14-q23, and 13q and also gains of 7p11-pter and chromosome 17 [142]. By single nucleotide polymorphism (SNP) array analysis, ALK+ ALCL displayed losses of 17p13.3-p12 and 6q21 in 35% of the case, and gains of 1q, 7q32.3 and 7p22.3-p21.3 were observed in 15% of samples. Positive Regulatory Domain Maturation 1 (PRDM1) gene, a well-known tumor suppressor, located on 6q21, is less frequently inactivated in ALK+ ALCL compared to ALK- ALCL [104]. The role of these additional alterations in the pathogenesis of the disease and prognosis is still unclear.

Diffuse Large B-cell lymphoma (DLBCL) is the most common type of non-Hodgkin lymphoma diagnosed in adults and accounts for 30–40% of newly diagnosed lymphomas in the United States [122]. ALK-positive large B-cell lymphoma (ALK+ DLBCL) is a rare subtype of large B cell lymphoma defined as a neoplasm of ALK-positive large immunoblast-like B-cells, resembling ALCL with occasional plasmablastic differentiation. The most frequent fusion partner in ALK+ LBCL is Clathrin heavy Chain (CLTC) gene, followed by NPM1, associated with a granular cytoplasmic and a nuclear/cytoplasmic immunostaining pattern, respectively.

TABLE 1   ALK translocations, their partners, and relative frequencies in hematological cancers.

ALK partner	Chromosomal location	Entity	Frequency (%)	ALK immunostaining	Reference
NPM1	t(2;5)(p23;q35)	ALCL, DLBCL	75–80, N/A	Nuclear and cytoplasmic	[8,123]
RNF213/ ALO17	t(2;17)(p23;q25)	ALCL	<1	Diffuse cytoplasmic	[124]
TFG	t(2;3)(p23;q21)	ALCL	2	Diffuse cytoplasmic	[125]
MSN	t(2;X)(p32;q11-12)	ALCL	<1	Membranous	[126]
TPM3	t(1;2)(q25;p23)	ALCL	12–18	Cytoplasmic and membranous	[127]
TPM4	t(2;19)(p23;p13)	ALCL	<1	Diffuse cytoplasmic	[128]
ATIC	inv(2)(p23;q35)	ALCL	2	Diffuse cytoplasmic	[129]
MYH9	t(2;22)(p23;q11.2)	ALCL	<1	Diffuse cytoplasmic	[130]
TRAF1	t(2;9)(p23;q33)	ALCL	<1	Diffuse cytoplasmic	[131]
CLTC	t(2;17)(p23;q23)	ALCL, DLBCL	2, N/A	Granular cytoplasmic	[132,133]
SQSTM1	t(2;5)(p23.1;q35.3)	DLBCL	N/A	Diffuse cytoplasmic with spots	[134]
SEC31A	t(2;4)(p24;q21)	DLBCL	N/A	Granular cytoplasmic	[135]
RANBP2	inv(2)(p23;q13)	AML, DLBCL	<1, N/A	Nuclear membrane with spots	[136,137]
EEF1G	t(2;11)(p23;q12.3)	ALCL	<1	Diffuse cytoplasmic	[138]
GORASP2	inv(2)(p23;q31.1)	DLBCL	<1	Diffuse cytoplasmic with perinuclear spots	[139]
EML4	inv(2)(p23;p21)	DLBCL	<1	Diffuse cytoplasmic	[140]

ALCL, anaplastic large cell lymphoma; *DLBCL*, diffuse large B cell lymphoma; *AML*, acute myeloid leukemia; *N/A*, not applicable.

Tumor cells of rare cases harboring other gene fusions are characterized by different ALK staining patterns (Table 1). Indeed, a prognostic role of ALK immunohistochemical pattern was suggested: segregating cases with a granular cytoplasmic staining (GCS) pattern or a non-GCS pattern showed lower survival in this latter category [140].

ALK rearrangements in solid cancers

Gene rearrangements are found in up to 3% of all human tumors [143]. Since ALK was described in ALCL in 1994, several ALK fusion proteins have been identified in the next two decades as oncogenic drivers in numerous different not-hematological malignancies. Although the most relevant is the EML4-ALK, due its frequency in NSCLC, the advent of NGS-based diagnostics led to the identification of a long list of different ALK fusion partner genes across multiple solid neoplasms (Table 2). For example, the inflammatory myofibroblastic tumor (IMT) is the first solid tumor found to harbor ALK rearrangements, which occur in up to 50% of cases; nevertheless, in ALK+ IMT EML4-ALK has not been reported, but a variety of other fusion partners, not seen in NSCLC, have instead been identified [202,203]. NSCLC is the second solid tumor in which oncogenic ALK fusions were detected, accounting for 3–7%

TABLE 2 ALK translocations, their partners, and relative frequencies in solid cancers.

ALK partner	Chromosomal location	Entity	Frequency (%)	Reference
		IMT	Up to 50	
A2M	12p13.31			[144]
ATIC	2q35			[145]
CARS	11p15.5			[124]
CLTC	17q23.1			[146]
DES	2q35			[147]
HNRNPA1	12q13.13			[148]
IGFBP5	2q35			[149]
KIF5B	10p11.22			[150]
NUMA1	11q13.4			[151]
PPFIBP1	12p11			[152]
RANBP2	2q13			[153]
SEC31A	4q21.22			[154]
SQSTM1	5q35.3			[155]
THBS1	15q14			[149]
TPM3	1q21.2			[156]
TPM4	19p13.1			[156]
		NSCLC	3–7	
BCL11A	2p16.1			[157]
CLTC	17q23.1			[158]
CMTR1	6p21.2			[159]
CRIM1	2p22.2			[158]
CUX	7q22.1			[160]
DCTN1	2p13.1			[161]
EML4	2p21			[64]

Continued

TABLE 2   ALK translocations, their partners, and relative frequencies in solid cancers—cont'd

ALK partner	Chromosomal location	Entity	Frequency (%)	Reference
GCC2	2q12.3			[162]
HIP1	7q11.2			[163]
KIF5B	10p11.22			[164]
KLC1	14q32.3			[165]
LCLAT	2p23.1			[166]
MREG	2q35			[167]
NLRC4	2p22.3			[168]
PTPN3	9q31.3			[169]
SEC31A	4q21			[170]
SOS1	2p22.1			[171]
SQSTM1	5q35.3			[161]
STRN	2p22.2			[172]
TFG	3q12.2			[173]
TNIP2	5q33.1			[174]
TPR	1q31.1			[175]
VIT	2p22.2			[176]
WDPCP	2p23.3			[177]
CLIP1	12q24.31	LUNG NET	N/A	[178]
		CRC	<1	
WDCP/C2ORF44	2p23.3			[179]
CAD	2p23.3			[180]
EML4	2p21			[181]
SMEK2	2p16.1			[182]
		RCC	<1	
EML4	2p21			[183]
HOOK1	1p32.1			[184]
STRN	2p22.2			[185]
TPM3	1q21.2			[186]
VCL	10q22.2			[187]
		LMS	<1	
ACTG2	2p13.1			[188]

TABLE 2   ALK translocations, their partners, and relative frequencies in solid cancers—cont'd

ALK partner	Chromosomal location	Entity	Frequency (%)	Reference
KANK2	19p13.2			[188]
TNS	2q35			[189]
		MMyST	N/A	
FBXO28	1q42.11			[190]
NPAS2	2q11.2			[190]
TPM3	1q21.2			[190]
PPFIBP1	12p11			[190]
		EFH	N/A	
MLPH	2q37.3			[191]
PRKAR2A	3p21.31			[191]
SQSTM1	5q35.3			[191]
VCL	10q22.2			[191]
TPM3	1q21.2			[191]
EML4	2p21			[191]
EML4	2p21	BC	N/A	[181]
TPM1	15q22.2	TCC	N/A	[182]
EML4	2p21	EC	N/A	[192]
TPM4	19p13.1	ESCC	N/A	[193]
A2M	12p13.31	FLIT	N/A	[194]
FN1	2q35	GIL	N/A	[195]
PPP1CB	2p23.2	HGG	N/A	[196]
FN1	2q35	OC	N/A	[197]
		MPM	N/A	
STRN	2p22.2			[198]
TPM1	15q22.2			[198]
ATG16L1	2q37.1			[198]
		ST	N/A	
MLPH	2q37.3			[199]
GTF3C2	2p23.3			[200]
CLIP1	12q24.31			[200]
		TC	<1%	

Continued

TABLE 2   ALK translocations, their partners, and relative frequencies in solid cancers—cont'd

ALK partner	Chromosomal location	Entity	Frequency (%)	Reference
STRN	2p22.2			[201]
GTF2IRD1	7q11.23			[182]

IMT, imflammatory myofibroblastich tumor; *NSCLC*, non-small cell lung cancer; *NET*, neuroendocrine tumors; *CRC*, colorectal cancer; *RCC*, renal cell cancer; *LMS*, leiomyosarcoma; *MMyST*, melanocytic myxoid spindle cell tumor; *EFH*, epithelioid fibrous histiocytoma; *BC*, breast cancer; *TCC*, transitional cell cancer; *EC*, endometrial cancer; *ESCC*, esophageal squamous cell cancer; *FLIT*, fetal lung interstitial tumor; *GIL*, gastro-intestinal leiomyoma; *HGG*, high-grade glioma; *OC*, ovarian cancer; *MPM*, malignant peritoneal mesothelioma; *ST*, Spitz tumor; *TC*, thyroid cancer; *N/A*, not applicable.

of lung cancers and being associated with non-smoking habit, younger age, and adenocarcinoma histology [64,173,204]. Even though the relative proportion of NSCLC bearing an ALK rearrangement is significantly lower compared to ALCL or IMT, NSCLC constitute the largest subset of tumor with an ALK rearrangement due to the high incidence of lung cancer worldwide [93]. Notably, studies in NSCLC have identified several additional ALK fusion proteins, which collectively occur less frequently than EML4-ALK (Table 2). The identification of ALK rearrangements in lung cancer patients had sparked the development of a series of ALK TKI from different companies. To date, five ALK TKI (crizotinib, ceritinib, alectinib, brigatinib and lorlatinib), with different efficacy and resistance profiles, have received approval by the Food and Drug Administration (FDA) for ALK+ NSCLC treatment [205,206].

At low frequency, ALK rearrangements have been detected in other solid cancers, including colorectal, breast, renal cell, esophageal, ovarian, and anaplastic thyroid carcinoma (Table 2). Due to the wide and routine use of NGS, more and more reports, even of single case, of new ALK fusion partners have been described, sometimes associated with different sensitivity to TKI [68]. The development of deep sequencing, coupled with immunohistochemistry and fluorescent-in-situ-hybridization, had shed light on the complexity of ALK rearrangements and how their occurrence is an heterogeneous event; multiple rearrangements within the ALK locus in the same specimen, "not-productive" translocations without a targetable protein or, even, nonreciprocal/reciprocal translocations with the retained 5′ ALK domain determine multiple phenotypes with different clinical behaviors [207,208].

Conclusion

ALK structure and signaling investigation are propaedeutic to ALK targeting in the context of human cancers. Structural alterations with pathological implications are detectable across different tumor types and the administration of ALKi demonstrated unexpected clinical impacts in the management of ALK+ cancers both in pediatric and adult patients [119,209]. The comprehension of ALK molecular biology, acquired mainly leveraging disease models driven by ALK mutations or translocations, represents the bedrock for the development of therapeutic strategies targeting ALK aberrant activation. Today multiple ALKi, with different activity profiles, are available in the clinic area and experimental or combinatorial approaches, such as antibody-drug conjugate or vaccination strategies, are under development [67,210]. Nevertheless, the management of ALK+ cancer patients still requires answers to open questions. Gene rearrangements are complex events that deeply alter cellular homeostasis and their occurrence not always translates into a targetable event; moreover, the countless list of partner genes makes ALK fusion a proteiform molecule, whose impact in the pathogenesis may differ notably from one cancer to others. Thus, correct patients' selection and identification is a first mandatory step to proper guide future decisions and NGS techniques are widen our comprehension [68,117,207]. Then, the identification of precise sequence of ALKi administration and the implementation of strategies to overcome resistance mechanisms will help us reaching a better disease control in ALK+ cancer patients. Toward this end, an extensive comprehension of ALK-related disrupting molecular events, and derived pathogenetic modifications, is a crucial requirement to do not lag behind the ALK enemy.

References

[1] Lemmon MA, Schlessinger J. Cell signaling by receptor tyrosine kinases. Cell 2010;141:1117–34. https://doi.org/10.1016/j.cell.2010.06.011.

[2] Lemmon MA, Freed DM, Schlessinger J, Kiyatkin A. The dark side of cell signaling: positive roles for negative regulators. Cell 2016;164:1172–84. https://doi.org/10.1016/j.cell.2016.02.047.

[3] Hanahan D, Weinberg RA. Hallmarks of cancer: the next generation. Cell 2011;144:646–74. https://doi.org/10.1016/j.cell.2011.02.013.

[4] Croce CM. Oncogenes and cancer. N Engl J Med 2008;358:502–11. https://doi.org/10.1056/NEJMra072367.

[5] Weinstein IB. Cancer. Addiction to oncogenes – the Achilles heal of cancer. Science 2002;297:63–4. https://doi.org/10.1126/science.1073096.

[6] Hallberg B, Palmer RH. The role of the ALK receptor in cancer biology. Ann Oncol 2016;27(Suppl 3):iii4–iii15. https://doi.org/10.1093/annonc/mdw301.

[7] Hallberg B, Palmer RH. Mechanistic insight into ALK receptor tyrosine kinase in human cancer biology. Nat Rev Cancer 2013;13:685–700. https://doi.org/10.1038/nrc3580.

[8] Morris SW, Kirstein MN, Valentine MB, Dittmer KG, Shapiro DN, Saltman DL, Look AT. Fusion of a kinase gene, ALK, to a nucleolar protein gene, NPM, in non-Hodgkin's lymphoma. Science 1994;263:1281–4. https://doi.org/10.1126/science.8122112.

[9] Shiota M, Fujimoto J, Semba T, Satoh H, Yamamoto T, Mori S. Hyperphosphorylation of a novel 80 kDa protein-tyrosine kinase similar to Ltk in a human Ki-1 lymphoma cell line, AMS3. Oncogene 1994;9:1567–74.

[10] Morris SW, Naeve C, Mathew P, James PL, Kirstein MN, Cui X, Witte DP. ALK, the chromosome 2 gene locus altered by the t(2;5) in non-Hodgkin's lymphoma, encodes a novel neural receptor tyrosine kinase that is highly related to leukocyte tyrosine kinase (LTK). Oncogene 1997;14:2175–88. https://doi.org/10.1038/sj.onc.1201062.

[11] Iwahara T, Fujimoto J, Wen D, Cupples R, Bucay N, Arakawa T, Mori S, Ratzkin B, Yamamoto T. Molecular characterization of ALK, a receptor tyrosine kinase expressed specifically in the nervous system. Oncogene 1997;14:439–49. https://doi.org/10.1038/sj.onc.1200849.

[12] Stoica GE, Kuo A, Aigner A, Sunitha I, Souttou B, Malerczyk C, Caughey DJ, Wen D, Karavanov A, Riegel AT, Wellstein A. Identification of anaplastic lymphoma kinase as a receptor for the growth factor pleiotrophin. J Biol Chem 2001;276:16772–9. https://doi.org/10.1074/jbc.M010660200.

[13] Palmer RH, Vernersson E, Grabbe C, Hallberg B. Anaplastic lymphoma kinase: signalling in development and disease. Biochem J 2009;420:345–61. https://doi.org/10.1042/BJ20090387.

[14] Lorén CE, Englund C, Grabbe C, Hallberg B, Hunter T, Palmer RH. A crucial role for the anaplastic lymphoma kinase receptor tyrosine kinase in gut development in *Drosophila melanogaster*. EMBO Rep 2003;4:781–6. https://doi.org/10.1038/sj.embor.embor897.

[15] Donella-Deana A, Marin O, Cesaro L, Gunby RH, Ferrarese A, Coluccia AML, Tartari CJ, Mologni L, Scapozza L, Gambacorti-Passerini C, Pinna LA. Unique substrate specificity of anaplastic lymphoma kinase (ALK): development of phosphoacceptor peptides for the assay of ALK activity. Biochemistry 2005;44:8533–42. https://doi.org/10.1021/bi0472954.

[16] Tartari CJ, Gunby RH, Coluccia AML, Sottocornola R, Cimbro B, Scapozza L, Donella-Deana A, Pinna LA, Gambacorti-Passerini C. Characterization of some molecular mechanisms governing autoactivation of the catalytic domain of the anaplastic lymphoma kinase. J Biol Chem 2008;283:3743–50. https://doi.org/10.1074/jbc.M706067200.

[17] Chiarle R, Voena C, Ambrogio C, Piva R, Inghirami G. The anaplastic lymphoma kinase in the pathogenesis of cancer. Nat Rev Cancer 2008;8:11–23. https://doi.org/10.1038/nrc2291.

[18] Ambrogio C, Voena C, Manazza AD, Martinengo C, Costa C, Kirchhausen T, Hirsch E, Inghirami G, Chiarle R. The anaplastic lymphoma kinase controls cell shape and growth of anaplastic large cell lymphoma through Cdc42 activation. Cancer Res 2008;68:8899–907. https://doi.org/10.1158/0008-5472.CAN-08-2568.

[19] Ambrogio C, Voena C, Manazza AD, Piva R, Riera L, Barberis L, Costa C, Tarone G, Defilippi P, Hirsch E, Boeri Erba E, Mohammed S, Jensen ON, Palestro G, Inghirami G, Chiarle R. p130Cas mediates the transforming properties of the anaplastic lymphoma kinase. Blood 2005;106:3907–16. https://doi.org/10.1182/blood-2005-03-1204.

[20] Barreca A, Lasorsa E, Riera L, Machiorlatti R, Piva R, Ponzoni M, Kwee I, Bertoni F, Piccaluga PP, Pileri SA, Inghirami G. European T-cell lymphoma study group, anaplastic lymphoma kinase in human cancer. J Mol Endocrinol 2011;47:R11–23. https://doi.org/10.1530/JME-11-0004.

[21] Guan J, Umapathy G, Yamazaki Y, Wolfstetter G, Mendoza P, Pfeifer K, Mohammed A, Hugosson F, Zhang H, Hsu AW, Halenbeck R, Hallberg B, Palmer RH. FAM150A and FAM150B are activating ligands for anaplastic lymphoma kinase. Elife 2015;4:e09811. https://doi.org/10.7554/eLife.09811.

[22] Reshetnyak AV, Murray PB, Shi X, Mo ES, Mohanty J, Tome F, Bai H, Gunel M, Lax I, Schlessinger J. Augmentor α and β (FAM150) are ligands of the receptor tyrosine kinases ALK and LTK: hierarchy and specificity of ligand-receptor interactions. Proc Natl Acad Sci U S A 2015;112:15862–7. https://doi.org/10.1073/pnas.1520099112.

[23] Zhang H, Pao LI, Zhou A, Brace AD, Halenbeck R, Hsu AW, Bray TL, Hestir K, Bosch E, Lee E, Wang G, Liu H, Wong BR, Kavanaugh WM, Williams LT. Deorphanization of the human leukocyte tyrosine kinase (LTK) receptor by a signaling screen of the extracellular proteome. PNAS 2014;111:15741–5. https://doi.org/10.1073/pnas.1412009111.

[24] Englund C, Lorén CE, Grabbe C, Varshney GK, Deleuil F, Hallberg B, Palmer RH. Jeb signals through the Alk receptor tyrosine kinase to drive visceral muscle fusion. Nature 2003;425:512–6. https://doi.org/10.1038/nature01950.

[25] Lee H-H, Norris A, Weiss JB, Frasch M. Jelly belly protein activates the receptor tyrosine kinase Alk to specify visceral muscle pioneers. Nature 2003;425:507–12. https://doi.org/10.1038/nature01916.

[26] Weiss JB, Suyama KL, Lee HH, Scott MP. Jelly belly: a Drosophila LDL receptor repeat-containing signal required for mesoderm migration and differentiation. Cell 2001;107:387–98. https://doi.org/10.1016/s0092-8674(01)00540-2.

[27] Stute C, Schimmelpfeng K, Renkawitz-Pohl R, Palmer RH, Holz A. Myoblast determination in the somatic and visceral mesoderm depends on notch signalling as well as on milliways(miliAlk) as receptor for Jeb signalling. Development 2004;131:743–54. https://doi.org/10.1242/dev.00972.

[28] Ishihara T, Iino Y, Mohri A, Mori I, Gengyo-Ando K, Mitani S, Katsura I. HEN-1, a secretory protein with an LDL receptor motif, regulates sensory integration and learning in Caenorhabditis elegans. Cell 2002;109:639–49. https://doi.org/10.1016/s0092-8674(02)00748-1.

[29] Reiner DJ, Ailion M, Thomas JH, Meyer BJ. C. elegans anaplastic lymphoma kinase ortholog SCD-2 controls dauer formation by modulating TGF-beta signaling. Curr Biol 2008;18:1101–9. https://doi.org/10.1016/j.cub.2008.06.060.

[30] Zhang H, Pao LI, Zhou A, Brace AD, Halenbeck R, Hsu AW, Bray TL, Hestir K, Bosch E, Lee E, Wang G, Liu H, Wong BR, Kavanaugh WM, Williams LT. Deorphanization of the human leukocyte tyrosine kinase (LTK) receptor by a signaling screen of the extracellular proteome. Proc Natl Acad Sci U S A 2014;111:15741–5. https://doi.org/10.1073/pnas.1412009111.

[31] Fadeev A, Mendoza-Garcia P, Irion U, Guan J, Pfeifer K, Wiessner S, Serluca F, Singh AP, Nüsslein-Volhard C, Palmer RH. ALKALs are in vivo ligands for ALK family receptor tyrosine kinases in the neural crest and derived cells. PNAS 2018;115:E630–8. https://doi.org/10.1073/pnas.1719137115.

[32] Murray PB, Lax I, Reshetnyak A, Ligon GF, Lillquist JS, Natoli EJ, Shi X, Folta-Stogniew E, Gunel M, Alvarado D, Schlessinger J. Heparin is an activating ligand of the orphan receptor tyrosine kinase ALK. Sci Signal 2015;8:ra6. https://doi.org/10.1126/scisignal.2005916.

[33] Gouzi JY, Moressis A, Walker JA, Apostolopoulou AA, Palmer RH, Bernards A, Skoulakis EMC. The receptor tyrosine kinase Alk controls neurofibromin functions in Drosophila growth and learning. PLoS Genet 2011;7. https://doi.org/10.1371/journal.pgen.1002281.

[34] Weiss JB, Weber SJ, Torres ERS, Marzulla T, Raber J. Genetic inhibition of anaplastic lymphoma kinase rescues cognitive impairments in neurofibromatosis 1 mutant mice. Behav Brain Res 2017;321:148–56. https://doi.org/10.1016/j.bbr.2017.01.003.

[35] Bazigou E, Apitz H, Johansson J, Lorén CE, Hirst EMA, Chen P-L, Palmer RH, Salecker I. Anterograde Jelly belly and Alk receptor tyrosine kinase signaling mediates retinal axon targeting in Drosophila. Cell 2007;128:961–75. https://doi.org/10.1016/j.cell.2007.02.024.

[36] Pecot MY, Chen Y, Akin O, Chen Z, Tsui CYK, Zipursky SL. Sequential axon-derived signals couple target survival and layer specificity in the Drosophila visual system. Neuron 2014;82:320–33. https://doi.org/10.1016/j.neuron.2014.02.045.

[37] Rohrbough J, Broadie K. Anterograde Jelly belly ligand to Alk receptor signaling at developing synapses is regulated by mind the gap. Development 2010;137:3523–33. https://doi.org/10.1242/dev.047878.

[38] Cheng LY, Bailey AP, Leevers SJ, Ragan TJ, Driscoll PC, Gould AP. Anaplastic lymphoma kinase spares organ growth during nutrient restriction in Drosophila. Cell 2011;146:435–47. https://doi.org/10.1016/j.cell.2011.06.040.

[39] Woodling NS, Aleyakpo B, Dyson MC, Minkley LJ, Rajasingam A, Dobson AJ, Leung KHC, Pomposova S, Fuentealba M, Alic N, Partridge L. The neuronal receptor tyrosine kinase Alk is a target for longevity. Aging Cell 2020;e13137. https://doi.org/10.1111/acel.13137.

[40] Gouzi JY, Bouraimi M, Roussou IG, Moressis A, Skoulakis EMC. The Drosophila receptor tyrosine kinase Alk constrains long-term memory formation. J Neurosci 2018;38:7701–12. https://doi.org/10.1523/JNEUROSCI.0784-18.2018.

[41] Shirinian M, Varshney G, Lorén CE, Grabbe C, Palmer RH. Drosophila anaplastic lymphoma kinase regulates Dpp signalling in the developing embryonic gut. Differentiation 2007;75:418–26. https://doi.org/10.1111/j.1432-0436.2006.00148.x.

[42] Kitazono T, Hara-Kuge S, Matsuda O, Inoue A, Fujiwara M, Ishihara T. Multiple signaling pathways coordinately regulate forgetting of olfactory adaptation through control of sensory responses in Caenorhabditis elegans. J Neurosci 2017;37:10240–51. https://doi.org/10.1523/JNEUROSCI.0031-17.2017.

[43] Lopes SS, Yang X, Müller J, Carney TJ, McAdow AR, Rauch G-J, Jacoby AS, Hurst LD, Delfino-Machín M, Haffter P, Geisler R, Johnson SL, Ward A, Kelsh RN. Leukocyte tyrosine kinase functions in pigment cell development. PLoS Genet 2008;4. https://doi.org/10.1371/journal.pgen.1000026.

[44] Kelsh RN, Brand M, Jiang YJ, Heisenberg CP, Lin S, Haffter P, Odenthal J, Mullins MC, van Eeden FJ, Furutani-Seiki M, Granato M, Hammerschmidt M, Kane DA, Warga RM, Beuchle D, Vogelsang L, Nüsslein-Volhard C. Zebrafish pigmentation mutations and the processes of neural crest development. Development 1996;123:369–89.

[45] Fadeev A, Krauss J, Singh AP, Nüsslein-Volhard C. Zebrafish leucocyte tyrosine kinase controls iridophore establishment, proliferation and survival. Pigment Cell Melanoma Res 2016;29:284–96. https://doi.org/10.1111/pcmr.12454.

[46] Vernersson E, Khoo NKS, Henriksson ML, Roos G, Palmer RH, Hallberg B. Characterization of the expression of the ALK receptor tyrosine kinase in mice. Gene Expr Patterns 2006;6:448–61. https://doi.org/10.1016/j.modgep.2005.11.006.

[47] Hurley SP, Clary DO, Copie V, Lefcort F. Anaplastic lymphoma kinase is dynamically expressed on subsets of motor neurons and in the peripheral nervous system. J Comp Neurol 2006;495:202–12. https://doi.org/10.1002/cne.20887.

[48] Degoutin J, Brunet-de Carvalho N, Cifuentes-Diaz C, Vigny M. ALK (anaplastic lymphoma kinase) expression in DRG neurons and its involvement in neuron-Schwann cells interaction. Eur J Neurosci 2009;29:275–86. https://doi.org/10.1111/j.1460-9568.2008.06593.x.

[49] Pulford K, Morris SW, Turturro F. Anaplastic lymphoma kinase proteins in growth control and cancer. J Cell Physiol 2004;199:330–58. https://doi.org/10.1002/jcp.10472.

[50] Bilsland JG, Wheeldon A, Mead A, Znamenskiy P, Almond S, Waters KA, Thakur M, Beaumont V, Bonnert TP, Heavens R, Whiting P, McAllister G, Munoz-Sanjuan I. Behavioral and neurochemical alterations in mice deficient in anaplastic lymphoma kinase suggest therapeutic potential for psychiatric indications. Neuropsychopharmacology 2008;33:685–700. https://doi.org/10.1038/sj.npp.1301446.

[51] Chiarle R, Gong JZ, Guasparri I, Pesci A, Cai J, Liu J, Simmons WJ, Dhall G, Howes J, Piva R, Inghirami G. NPM-ALK transgenic mice spontaneously develop T-cell lymphomas and plasma cell tumors. Blood 2003;101:1919–27. https://doi.org/10.1182/blood-2002-05-1343.

[52] Chiarle R, Simmons WJ, Cai H, Dhall G, Zamo A, Raz R, Karras JG, Levy DE, Inghirami G. Stat3 is required for ALK-mediated lymphomagenesis and provides a possible therapeutic target. Nat Med 2005;11:623–9. https://doi.org/10.1038/nm1249.

[53] Piva R, Chiarle R, Manazza AD, Taulli R, Simmons W, Ambrogio C, D'Escamard V, Pellegrino E, Ponzetto C, Palestro G, Inghirami G. Ablation of oncogenic ALK is a viable therapeutic approach for anaplastic large-cell lymphomas. Blood 2006;107:689–97. https://doi.org/10.1182/blood-2005-05-2125.

[54] Carén H, Abel F, Kogner P, Martinsson T. High incidence of DNA mutations and gene amplifications of the ALK gene in advanced sporadic neuroblastoma tumours. Biochem J 2008;416:153–9. https://doi.org/10.1042/bj20081834.

[55] Chen Y, Takita J, Choi YL, Kato M, Ohira M, Sanada M, Wang L, Soda M, Kikuchi A, Igarashi T, Nakagawara A, Hayashi Y, Mano H, Ogawa S. Oncogenic mutations of ALK kinase in neuroblastoma. Nature 2008;455:971–4. https://doi.org/10.1038/nature07399.

[56] George RE, Sanda T, Hanna M, Fröhling S, Luther W, Zhang J, Ahn Y, Zhou W, London WB, McGrady P, Xue L, Zozulya S, Gregor VE, Webb TR, Gray NS, Gilliland DG, Diller L, Greulich H, Morris SW, Meyerson M, Look AT. Activating mutations in ALK provide a therapeutic target in neuroblastoma. Nature 2008;455:975–8. https://doi.org/10.1038/nature07397.

[57] Janoueix-Lerosey I, Lequin D, Brugières L, Ribeiro A, de Pontual L, Combaret V, Raynal V, Puisieux A, Schleiermacher G, Pierron G, Valteau-Couanet D, Frebourg T, Michon J, Lyonnet S, Amiel J, Delattre O. Somatic and germline activating mutations of the ALK kinase receptor in neuroblastoma. Nature 2008;455:967–70. https://doi.org/10.1038/nature07398.

[58] Mossé YP, Laudenslager M, Longo L, Cole KA, Wood A, Attiyeh EF, Laquaglia MJ, Sennett R, Lynch JE, Perri P, Laureys G, Speleman F, Kim C, Hou C, Hakonarson H, Torkamani A, Schork NJ, Brodeur GM, Tonini GP, Rappaport E, Devoto M, Maris JM. Identification of ALK as a major familial neuroblastoma predisposition gene. Nature 2008;455:930–5. https://doi.org/10.1038/nature07261.

[59] Zhu S, Lee J-S, Guo F, Shin J, Perez-Atayde AR, Kutok JL, Rodig SJ, Neuberg DS, Helman D, Feng H, Stewart RA, Wang W, George RE, Kanki JP, Look AT. Activated ALK collaborates with MYCN in neuroblastoma pathogenesis. Cancer Cell 2012;21:362–73. https://doi.org/10.1016/j.ccr.2012.02.010.

[60] Berry T, Luther W, Bhatnagar N, Jamin Y, Poon E, Sanda T, Pei D, Sharma B, Vetharoy WR, Hallsworth A, Ahmad Z, Barker K, Moreau L, Webber H, Wang W, Liu Q, Perez-Atayde A, Rodig S, Cheung N-K, Raynaud F, Hallberg B, Robinson SP, Gray NS, Pearson ADJ, Eccles SA, Chesler L, George RE. The ALK(F1174L) mutation potentiates the oncogenic activity of MYCN in neuroblastoma. Cancer Cell 2012;22:117–30. https://doi.org/10.1016/j.ccr.2012.06.001.

[61] Heukamp LC, Thor T, Schramm A, De Preter K, Kumps C, De Wilde B, Odersky A, Peifer M, Lindner S, Spruessel A, Pattyn F, Mestdagh P, Menten B, Kuhfittig-Kulle S, Künkele A, König K, Meder L, Chatterjee S, Ullrich RT, Schulte S, Vandesompele J, Speleman F, Büttner R, Eggert A, Schulte JH. Targeted expression of mutated ALK induces neuroblastoma in transgenic mice. Sci Transl Med 2012;4:141ra91. https://doi.org/10.1126/scitranslmed.3003967.

[62] Cazes A, Lopez-Delisle L, Tsarovina K, Pierre-Eugène C, De Preter K, Peuchmaur M, Nicolas A, Provost C, Louis-Brennetot C, Daveau R, Kumps C, Cascone I, Schleiermacher G, Prignon A, Speleman F, Rohrer H, Delattre O, Janoueix-Lerosey I. Activated Alk triggers prolonged neurogenesis and Ret upregulation providing a therapeutic target in ALK-mutated neuroblastoma. Oncotarget 2014;5:2688–702. https://doi.org/10.18632/oncotarget.1883.

[63] Ono S, Saito T, Terui K, Yoshida H, Enomoto H. Generation of conditional ALK F1174L mutant mouse models for the study of neuroblastoma pathogenesis. Genesis 2019;57:e23323. https://doi.org/10.1002/dvg.23323.

[64] Soda M, Choi YL, Enomoto M, Takada S, Yamashita Y, Ishikawa S, Fujiwara S, Watanabe H, Kurashina K, Hatanaka H, Bando M, Ohno S, Ishikawa Y, Aburatani H, Niki T, Sohara Y, Sugiyama Y, Mano H. Identification of the transforming EML4-ALK fusion gene in non-small-cell lung cancer. Nature 2007;448:561–6. https://doi.org/10.1038/nature05945.

[65] Soda M, Takada S, Takeuchi K, Choi YL, Enomoto M, Ueno T, Haruta H, Hamada T, Yamashita Y, Ishikawa Y, Sugiyama Y, Mano H. A mouse model for EML4-ALK-positive lung cancer. Proc Natl Acad Sci U S A 2008;105:19893–7. https://doi.org/10.1073/pnas.0805381105.

[66] Pyo KH, Lim SM, Kim HR, Sung YH, Yun MR, Kim S-M, Kim H, Kang HN, Lee JM, Kim SG, Park CW, Chang H, Shim HS, Lee H-W, Cho BC. Establishment of a conditional transgenic mouse model recapitulating EML4-ALK-positive human non-small cell lung cancer. J Thorac Oncol 2017;12:491–500. https://doi.org/10.1016/j.jtho.2016.10.022.

[67] Voena C, Menotti M, Mastini C, Di Giacomo F, Longo DL, Castella B, Merlo MEB, Ambrogio C, Wang Q, Minero VG, Poggio T, Martinengo C, D'Amico L, Panizza E, Mologni L, Cavallo F, Altruda F, Butaney M, Capelletti M, Inghirami G, Jänne PA, Chiarle R. Efficacy of a cancer vaccine against ALK-rearranged lung tumors. Cancer Immunol Res 2015;3:1333–43. https://doi.org/10.1158/2326-6066.CIR-15-0089.

[68] Ou S-HI, Zhu VW, Nagasaka M. Catalog of 5′ fusion partners in ALK-positive NSCLC circa 2020. JTO Clin Res Rep 2020;1:100015. https://doi.org/10.1016/j.jtocrr.2020.100015.

[69] Maddalo D, Manchado E, Concepcion CP, Bonetti C, Vidigal JA, Han Y-C, Ogrodowski P, Crippa A, Rekhtman N, de Stanchina E, Lowe SW, Ventura A. In vivo engineering of oncogenic chromosomal rearrangements with the CRISPR/Cas9 system. Nature 2014;516:423–7. https://doi.org/10.1038/nature13902.

[70] Blasco RB, Karaca E, Ambrogio C, Cheong T-C, Karayol E, Minero VG, Voena C, Chiarle R. Simple and rapid in vivo generation of chromosomal rearrangements using CRISPR/Cas9 technology. Cell Rep 2014;9:1219–27. https://doi.org/10.1016/j.celrep.2014.10.051.

[71] Wiesner T, Lee W, Obenauf AC, Ran L, Murali R, Zhang QF, Wong EWP, Hu W, Scott SN, Shah RH, Landa I, Button J, Lailler N, Sboner A, Gao D, Murphy DA, Cao Z, Shukla S, Hollmann TJ, Wang L, Borsu L, Merghoub T, Schwartz GK, Postow MA, Ariyan CE, Fagin JA, Zheng D, Ladanyi M, Busam KJ, Berger MF, Chen Y, Chi P. Alternative transcription initiation leads to expression of a novel ALK isoform in cancer. Nature 2015;526:453–7. https://doi.org/10.1038/nature15258.

[72] Dirks WG, Fähnrich S, Lis Y, Becker E, MacLeod RAF, Drexler HG. Expression and functional analysis of the anaplastic lymphoma kinase (ALK) gene in tumor cell lines. Int J Cancer 2002;100:49–56. https://doi.org/10.1002/ijc.10435.

[73] Passoni L, Longo L, Collini P, Coluccia AML, Bozzi F, Podda M, Gregorio A, Gambini C, Garaventa A, Pistoia V, Del Grosso F, Tonini GP, Cheng M, Gambacorti-Passerini C, Anichini A, Fossati-Bellani F, Di Nicola M, Luksch R. Mutation-independent anaplastic lymphoma kinase overexpression in poor prognosis neuroblastoma patients. Cancer Res 2009;69:7338–46. https://doi.org/10.1158/0008-5472.CAN-08-4419.

[74] Salido M, Pijuan L, Martínez-Avilés L, Galván AB, Cañadas I, Rovira A, Zanui M, Martínez A, Longarón R, Sole F, Serrano S, Bellosillo B, Wynes MW, Albanell J, Hirsch FR, Arriola E. Increased ALK gene copy number and amplification are frequent in non-small cell lung cancer. J Thorac Oncol 2011;6:21–7. https://doi.org/10.1097/JTO.0b013e3181fb7cd6.

[75] Chiarle R, Martinengo C, Mastini C, Ambrogio C, D'Escamard V, Forni G, Inghirami G. The anaplastic lymphoma kinase is an effective oncoantigen for lymphoma vaccination. Nat Med 2008;14:676–80. https://doi.org/10.1038/nm1769.

[76] Cessna MH, Zhou H, Sanger WG, Perkins SL, Tripp S, Pickering D, Daines C, Coffin CM. Expression of ALK1 and p80 in inflammatory myofibroblastic tumor and its mesenchymal mimics: a study of 135 cases. Mod Pathol 2002;15:931–8. https://doi.org/10.1097/01.MP.0000026615.04130.1F.

[77] Perez-Pinera P, Chang Y, Astudillo A, Mortimer J, Deuel TF. Anaplastic lymphoma kinase is expressed in different subtypes of human breast cancer. Biochem Biophys Res Commun 2007;358:399–403. https://doi.org/10.1016/j.bbrc.2007.04.137.

[78] De Brouwer S, De Preter K, Kumps C, Zabrocki P, Porcu M, Westerhout EM, Lakeman A, Vandesompele J, Hoebeeck J, Van Maerken T, De Paepe A, Laureys G, Schulte JH, Schramm A, Van Den Broecke C, Vermeulen J, Van Roy N, Beiske K, Renard M, Noguera R, Delattre O, Janoueix-Lerosey I, Kogner P, Martinsson T, Nakagawara A, Ohira M, Caron H, Eggert A, Cools J, Versteeg R, Speleman F. Meta-analysis of neuroblastomas reveals a skewed ALK mutation spectrum in tumors with MYCN amplification. Clin Cancer Res 2010;16:4353–62. https://doi.org/10.1158/1078-0432.CCR-09-2660.

[79] Brodeur GM. Neuroblastoma: biological insights into a clinical enigma. Nat Rev Cancer 2003;3:203–16. https://doi.org/10.1038/nrc1014.

[80] Lee CC, Jia Y, Li N, Sun X, Ng K, Ambing E, Gao M-Y, Hua S, Chen C, Kim S, Michellys P-Y, Lesley SA, Harris JL, Spraggon G. Crystal structure of the ALK (anaplastic lymphoma kinase) catalytic domain. Biochem J 2010;430:425–37. https://doi.org/10.1042/BJ20100609.

[81] Bresler SC, Weiser DA, Huwe PJ, Park JH, Krytska K, Ryles H, Laudenslager M, Rappaport EF, Wood AC, McGrady PW, Hogarty MD, London WB, Radhakrishnan R, Lemmon MA, Mossé YP. ALK mutations confer differential oncogenic activation and sensitivity to ALK inhibition therapy in neuroblastoma. Cancer Cell 2014;26:682–94. https://doi.org/10.1016/j.ccell.2014.09.019.

[82] Guan J, Wolfstetter G, Siaw J, Chand D, Hugosson F, Palmer RH, Hallberg B. Anaplastic lymphoma kinase L1198F and G1201E mutations identified in anaplastic thyroid cancer patients are not ligand-independent. Oncotarget 2017;8:11566–78. https://doi.org/10.18632/oncotarget.14141.

[83] Takita J. The role of anaplastic lymphoma kinase in pediatric cancers. Cancer Sci 2017;108:1913–20. https://doi.org/10.1111/cas.13333.

[84] Gualandi M, Iorio M, Engeler O, Serra-Roma A, Gasparre G, Schulte JH, Hohl D, Shakhova O. Oncogenic ALK F1174L drives tumorigenesis in cutaneous squamous cell carcinoma. Life Sci Alliance 2020;3. https://doi.org/10.26508/lsa.201900601.

[85] Okubo J, Takita J, Chen Y, Oki K, Nishimura R, Kato M, Sanada M, Hiwatari M, Hayashi Y, Igarashi T, Ogawa S. Aberrant activation of ALK kinase by a novel truncated form ALK protein in neuroblastoma. Oncogene 2012;31:4667–76. https://doi.org/10.1038/onc.2011.616.

[86] Cazes A, Louis-Brennetot C, Mazot P, Dingli F, Lombard B, Boeva V, Daveau R, Cappo J, Combaret V, Schleiermacher G, Jouannet S, Ferrand S, Pierron G, Barillot E, Loew D, Vigny M, Delattre O, Janoueix-Lerosey I. Characterization of rearrangements involving the ALK gene reveals a novel truncated form associated with tumor aggressiveness in neuroblastoma. Cancer Res 2013;73:195–204. https://doi.org/10.1158/0008-5472.CAN-12-1242.

[87] Fleuren EDG, Vlenterie M, van der Graaf WTA, Hillebrandt-Roeffen MHS, Blackburn J, Ma X, Chan H, Magias MC, van Erp A, van Houdt L, Cebeci SAS, van de Ven A, Flucke UE, Heyer EE, Thomas DM, Lord CJ, Marini KD, Vaghjiani V, Mercer TR, Cain JE, Wu J, Versleijen-Jonkers YMH, Daly RJ. Phosphoproteomic profiling reveals ALK and MET as novel actionable targets across synovial sarcoma subtypes. Cancer Res 2017;77:4279–92. https://doi.org/10.1158/0008-5472.CAN-16-2550.

[88] Fukuhara S, Nomoto J, Kim S-W, Taniguchi H, Maeshima AM, Tobinai K, Kobayashi Y. Partial deletion of the ALK gene in ALK-positive anaplastic large cell lymphoma. Hematol Oncol 2018;36:150–8. https://doi.org/10.1002/hon.2455.

[89] Gainor JF, Dardaei L, Yoda S, Friboulet L, Leshchiner I, Katayama R, Dagogo-Jack I, Gadgeel S, Schultz K, Singh M, Chin E, Parks M, Lee D, DiCecca RH, Lockerman E, Huynh T, Logan J, Ritterhouse LL, Le LP, Muniappan A, Digumarthy S, Channick C, Keyes C, Getz G, Dias-Santagata D, Heist RS, Lennerz J, Sequist LV, Benes CH, Iafrate AJ, Mino-Kenudson M, Engelman JA, Shaw AT. Molecular mechanisms of resistance to first- and second-generation ALK inhibitors in ALK-rearranged lung cancer. Cancer Discov 2016;6:1118–33. https://doi.org/10.1158/2159-8290.CD-16-0596.

[90] Yoda S, Lin JJ, Lawrence MS, Burke BJ, Friboulet L, Langenbucher A, Dardaei L, Prutisto-Chang K, Dagogo-Jack I, Timofeevski S, Hubbeling H, Gainor JF, Ferris LA, Riley AK, Kattermann KE, Timonina D, Heist RS, Iafrate AJ, Benes CH, Lennerz JK, Mino-Kenudson M, Engelman JA, Johnson TW, Hata AN, Shaw AT. Sequential ALK inhibitors can select for lorlatinib-resistant compound ALK mutations in ALK-positive lung cancer. Cancer Discov 2018;8:714–29. https://doi.org/10.1158/2159-8290.CD-17-1256.

[91] Tabbò F, Pizzi M, Kyriakides PW, Ruggeri B, Inghirami G. Oncogenic kinase fusions: an evolving arena with innovative clinical opportunities. Oncotarget 2016;7:25064–86. https://doi.org/10.18632/oncotarget.7853.

[92] Schram AM, Chang MT, Jonsson P, Drilon A. Fusions in solid tumours: diagnostic strategies, targeted therapy, and acquired resistance. Nat Rev Clin Oncol 2017;14:735–48. https://doi.org/10.1038/nrclinonc.2017.127.

[93] Siegel RL, Miller KD, Jemal A. Cancer statistics, 2020. CA Cancer J Clin 2020;70:7–30. https://doi.org/10.3322/caac.21590.

[94] Medves S, Demoulin J-B. Tyrosine kinase gene fusions in cancer: translating mechanisms into targeted therapies. J Cell Mol Med 2012;16:237–48. https://doi.org/10.1111/j.1582-4934.2011.01415.x.

[95] Bayliss R, Choi J, Fennell DA, Fry AM, Richards MW. Molecular mechanisms that underpin EML4-ALK driven cancers and their response to targeted drugs. Cell Mol Life Sci 2016;73:1209–24. https://doi.org/10.1007/s00018-015-2117-6.

[96] Maes B, Vanhentenrijk V, Wlodarska I, Cools J, Peeters B, Marynen P, de Wolf-Peeters C. The NPM-ALK and the ATIC-ALK fusion genes can be detected in non-neoplastic cells. Am J Pathol 2001;158:2185–93. https://doi.org/10.1016/S0002-9440(10)64690-1.

[97] Lee JJ-K, Park S, Park H, Kim S, Lee J, Lee J, Youk J, Yi K, An Y, Park IK, Kang CH, Chung DH, Kim TM, Jeon YK, Hong D, Park PJ, Ju YS, Kim YT. Tracing oncogene rearrangements in the mutational history of lung adenocarcinoma. Cell 2019;177:1842–57. e21 https://doi.org/10.1016/j.cell.2019.05.013.

[98] Jankovic GM, Pavlovic M, Vukomanovic DJ, Colovic MD, Lazarevic V. The fundamental prevalence of chronic myeloid leukemia-generating clonogenic cells in the light of the neutrality theory of evolution. Blood Cells Mol Dis 2001;27:913–7. https://doi.org/10.1006/bcmd.2001.0462.

[99] Abate F, Todaro M, van der Krogt J-A, Boi M, Landra I, Machiorlatti R, Tabbò F, Messana K, Abele C, Barreca A, Novero D, Gaudiano M, Aliberti S, Di Giacomo F, Tousseyn T, Lasorsa E, Crescenzo R, Bessone L, Ficarra E, Acquaviva A, Rinaldi A, Ponzoni M, Longo DL, Aime S, Cheng M, Ruggeri B, Piccaluga PP, Pileri S, Tiacci E, Falini B, Pera-Gresely B, Cerchietti L, Iqbal J, Chan WC, Shultz LD, Kwee I, Piva R, Wlodarska I, Rabadan R, Bertoni F, Inghirami G. European T-cell lymphoma study group, a novel patient-derived tumorgraft model with TRAF1-ALK anaplastic large-cell lymphoma translocation. Leukemia 2015;29:1390–401. https://doi.org/10.1038/leu.2014.347.

[100] Box JK, Paquet N, Adams MN, Boucher D, Bolderson E, O'Byrne KJ, Richard DJ. Nucleophosmin: from structure and function to disease development. BMC Mol Biol 2016;17:19. https://doi.org/10.1186/s12867-016-0073-9.

[101] Cui Y-X, Kerby A, McDuff FKE, Ye H, Turner SD. NPM-ALK inhibits the p53 tumor suppressor pathway in an MDM2 and JNK-dependent manner. Blood 2009;113:5217–27. https://doi.org/10.1182/blood-2008-06-160168.

[102] Wu F, Wang P, Young LC, Lai R, Li L. Proteome-wide identification of novel binding partners to the oncogenic fusion gene protein, NPM-ALK, using tandem affinity purification and mass spectrometry. Am J Pathol 2009;174:361–70. https://doi.org/10.2353/ajpath.2009.080521.

[103] Young LC, Bone KM, Wang P, Wu F, Adam BA, Hegazy S, Gelebart P, Holovati J, Li L, Andrew SE, Lai R. Fusion tyrosine kinase NPM-ALK deregulates MSH2 and suppresses DNA mismatch repair function novel insights into a potent oncoprotein. Am J Pathol 2011;179:411–21. https://doi.org/10.1016/j.ajpath.2011.03.045.

[104] Boi M, Rinaldi A, Kwee I, Bonetti P, Todaro M, Tabbò F, Piva R, Rancoita PMV, Matolcsy A, Timar B, Tousseyn T, Rodríguez-Pinilla SM, Piris MA, Beà S, Campo E, Bhagat G, Swerdlow SH, Rosenwald A, Ponzoni M, Young KH, Piccaluga PP, Dummer R, Pileri S, Zucca E, Inghirami G, Bertoni F. PRDM1/BLIMP1 is commonly inactivated in anaplastic large T-cell lymphoma. Blood 2013;122:2683–93. https://doi.org/10.1182/blood-2013-04-497933.

[105] Lobello C, Bikos V, Janikova A, Pospisilova S. The role of oncogenic tyrosine kinase NPM-ALK in genomic instability. Cancers (Basel) 2018;10. https://doi.org/10.3390/cancers10030064.

[106] Zhang Q, Wang HY, Woetmann A, Raghunath PN, Odum N, Wasik MA. STAT3 induces transcription of the DNA methyltransferase 1 gene (DNMT1) in malignant T lymphocytes. Blood 2006;108:1058–64. https://doi.org/10.1182/blood-2005-08-007377.

[107] Hassler MR, Pulverer W, Lakshminarasimhan R, Redl E, Hacker J, Garland GD, Merkel O, Schiefer A-I, Simonitsch-Klupp I, Kenner L, Weisenberger DJ, Weinhaeusel A, Turner SD, Egger G. Insights into the pathogenesis of anaplastic large-cell lymphoma through genome-wide DNA methylation profiling. Cell Rep 2016;17:596–608. https://doi.org/10.1016/j.celrep.2016.09.018.

[108] Zhang Q, Wang HY, Liu X, Wasik MA. STAT5A is epigenetically silenced by the tyrosine kinase NPM1-ALK and acts as a tumor suppressor by reciprocally inhibiting NPM1-ALK expression. Nat Med 2007;13:1341–8. https://doi.org/10.1038/nm1659.

[109] Ambrogio C, Martinengo C, Voena C, Tondat F, Riera L, di Celle PF, Inghirami G, Chiarle R. NPM-ALK oncogenic tyrosine kinase controls T-cell identity by transcriptional regulation and epigenetic silencing in lymphoma cells. Cancer Res 2009;69:8611–9. https://doi.org/10.1158/0008-5472.CAN-09-2655.

[110] Merkel O, Hamacher F, Laimer D, Sifft E, Trajanoski Z, Scheideler M, Egger G, Hassler MR, Thallinger C, Schmatz A, Turner SD, Greil R, Kenner L. Identification of differential and functionally active miRNAs in both anaplastic lymphoma kinase (ALK)+ and ALK- anaplastic large-cell lymphoma. Proc Natl Acad Sci U S A 2010;107:16228–33. https://doi.org/10.1073/pnas.1009719107.

[111] Turner SD, Yeung D, Hadfield K, Cook SJ, Alexander DR. The NPM-ALK tyrosine kinase mimics TCR signalling pathways, inducing NFAT and AP-1 by RAS-dependent mechanisms. Cell Signal 2007;19:740–7. https://doi.org/10.1016/j.cellsig.2006.09.007.

[112] Leventaki V, Drakos E, Medeiros LJ, Lim MS, Elenitoba-Johnson KS, Claret FX, Rassidakis GZ. NPM-ALK oncogenic kinase promotes cell-cycle progression through activation of JNK/cJun signaling in anaplastic large-cell lymphoma. Blood 2007;110:1621–30. https://doi.org/10.1182/blood-2006-11-059451.

[113] Armstrong F, Lamant L, Hieblot C, Delsol G, Touriol C. TPM3-ALK expression induces changes in cytoskeleton organisation and confers higher metastatic capacities than other ALK fusion proteins. Eur J Cancer 2007;43:640–6. https://doi.org/10.1016/j.ejca.2006.12.005.

[114] Richards MW, Law EWP, Rennalls LP, Busacca S, O'Regan L, Fry AM, Fennell DA, Bayliss R. Crystal structure of EML1 reveals the basis for Hsp90 dependence of oncogenic EML4-ALK by disruption of an atypical β-propeller domain. Proc Natl Acad Sci U S A 2014;111:5195–200. https://doi.org/10.1073/pnas.1322892111.

[115] Richards MW, O'Regan L, Roth D, Montgomery JM, Straube A, Fry AM, Bayliss R. Microtubule association of EML proteins and the EML4-ALK variant 3 oncoprotein require an N-terminal trimerization domain. Biochem J 2015;467:529–36. https://doi.org/10.1042/BJ20150039.

[116] Woo CG, Seo S, Kim SW, Jang SJ, Park KS, Song JY, Lee B, Richards MW, Bayliss R, Lee DH, Choi J. Differential protein stability and clinical responses of EML4-ALK fusion variants to various ALK inhibitors in advanced ALK-rearranged non-small cell lung cancer. Ann Oncol 2017;28:791–7. https://doi.org/10.1093/annonc/mdw693.

[117] Childress MA, Himmelberg SM, Chen H, Deng W, Davies MA, Lovly CM. ALK fusion partners impact response to ALK inhibition: differential effects on sensitivity, cellular phenotypes, and biochemical properties. Mol Cancer Res 2018;16:1724–36. https://doi.org/10.1158/1541-7786.MCR-18-0171.

[118] Lin JJ, Zhu VW, Yoda S, Yeap BY, Schrock AB, Dagogo-Jack I, Jessop NA, Jiang GY, Le LP, Gowen K, Stephens PJ, Ross JS, Ali SM, Miller VA, Johnson ML, Lovly CM, Hata AN, Gainor JF, Iafrate AJ, Shaw AT, Ou S-HI. Impact of EML4-ALK variant on resistance mechanisms and clinical outcomes in ALK-positive lung cancer. J Clin Oncol 2018;36:1199–206. https://doi.org/10.1200/JCO.2017.76.2294.

[119] Camidge DR, Dziadziuszko R, Peters S, Mok T, Noe J, Nowicka M, Gadgeel SM, Cheema P, Pavlakis N, de Marinis F, Cho BC, Zhang L, Moro-Sibilot D, Liu T, Bordogna W, Balas B, Müller B, Shaw AT. Updated efficacy and safety data and impact of the EML4-ALK fusion variant on the efficacy of Alectinib in untreated ALK-positive advanced non-small cell lung cancer in the global phase III ALEX study. J Thorac Oncol 2019;14:1233–43. https://doi.org/10.1016/j.jtho.2019.03.007.

[120] Cairo MS, Pinkerton R. Childhood, adolescent and young adult non-Hodgkin lymphoma: state of the science. Br J Haematol 2016;173:507–30. https://doi.org/10.1111/bjh.14035.

[121] Sabattini E, Bacci F, Sagramoso C, Pileri SA. WHO classification of tumours of haematopoietic and lymphoid tissues in 2008: an overview. Pathologica 2010;102:83–7.

[122] Swerdlow SH, Campo E, Pileri SA, Harris NL, Stein H, Siebert R, Advani R, Ghielmini M, Salles GA, Zelenetz AD, Jaffe ES. The 2016 revision of the World Health Organization classification of lymphoid neoplasms. Blood 2016;127:2375–90. https://doi.org/10.1182/blood-2016-01-643569.

[123] Adam P, Katzenberger T, Seeberger H, Gattenlöhner S, Wolf J, Steinlein C, Schmid M, Müller-Hermelink H-K, Ott G. A case of a diffuse large B-cell lymphoma of plasmablastic type associated with the t(2;5)(p23;q35) chromosome translocation. Am J Surg Pathol 2003;27:1473–6. https://doi.org/10.1097/00000478-200311000-00012.

[124] Cools J, Wlodarska I, Somers R, Mentens N, Pedeutour F, Maes B, De Wolf-Peeters C, Pauwels P, Hagemeijer A, Marynen P. Identification of novel fusion partners of ALK, the anaplastic lymphoma kinase, in anaplastic large-cell lymphoma and inflammatory myofibroblastic tumor. Genes Chromosomes Cancer 2002;34:354–62. https://doi.org/10.1002/gcc.10033.

[125] Hernández L, Pinyol M, Hernández S, Beà S, Pulford K, Rosenwald A, Lamant L, Falini B, Ott G, Mason DY, Delsol G, Campo E. TRK-fused gene (TFG) is a new partner of ALK in anaplastic large cell lymphoma producing two structurally different TFG-ALK translocations. Blood 1999;94:3265–8.

[126] Tort F, Pinyol M, Pulford K, Roncador G, Hernandez L, Nayach I, Kluin-Nelemans HC, Kluin P, Touriol C, Delsol G, Mason D, Campo E. Molecular characterization of a new ALK translocation involving moesin (MSN-ALK) in anaplastic large cell lymphoma. Lab Invest 2001;81:419–26. https://doi.org/10.1038/labinvest.3780249.

[127] Lamant L, Dastugue N, Pulford K, Delsol G, Mariamé B. A new fusion gene TPM3-ALK in anaplastic large cell lymphoma created by a (1;2)(q25;p23) translocation. Blood 1999;93:3088–95.

[128] Meech SJ, McGavran L, Odom LF, Liang X, Meltesen L, Gump J, Wei Q, Carlsen S, Hunger SP. Unusual childhood extramedullary hematologic malignancy with natural killer cell properties that contains tropomyosin 4 – anaplastic lymphoma kinase gene fusion. Blood 2001;98:1209–16. https://doi.org/10.1182/blood.v98.4.1209.

[129] Ma Z, Cools J, Marynen P, Cui X, Siebert R, Gesk S, Schlegelberger B, Peeters B, De Wolf-Peeters C, Wlodarska I, Morris SW. Inv(2)(p23q35) in anaplastic large-cell lymphoma induces constitutive anaplastic lymphoma kinase (ALK) tyrosine kinase activation by fusion to ATIC, an enzyme involved in purine nucleotide biosynthesis. Blood 2000;95:2144–9.

[130] Lamant L, Gascoyne RD, Duplantier MM, Armstrong F, Raghab A, Chhanabhai M, Rajcan-Separovic E, Raghab J, Delsol G, Espinos E. Non-muscle myosin heavy chain (MYH9): a new partner fused to ALK in anaplastic large cell lymphoma. Genes Chromosomes Cancer 2003;37:427–32. https://doi.org/10.1002/gcc.10232.

[131] Feldman AL, Vasmatzis G, Asmann YW, Davila J, Middha S, Eckloff BW, Johnson SH, Porcher JC, Ansell SM, Caride A. Novel TRAF1-ALK fusion identified by deep RNA sequencing of anaplastic large cell lymphoma. Genes Chromosomes Cancer 2013;52:1097–102. https://doi.org/10.1002/gcc.22104.

[132] Touriol C, Greenland C, Lamant L, Pulford K, Bernard F, Rousset T, Mason DY, Delsol G. Further demonstration of the diversity of chromosomal changes involving 2p23 in ALK-positive lymphoma: 2 cases expressing ALK kinase fused to CLTCL (clathrin chain polypeptide-like). Blood 2000;95:3204–7.

[133] De Paepe P, Baens M, van Krieken H, Verhasselt B, Stul M, Simons A, Poppe B, Laureys G, Brons P, Vandenberghe P, Speleman F, Praet M, De Wolf-Peeters C, Marynen P, Wlodarska I. ALK activation by the CLTC-ALK fusion is a recurrent event in large B-cell lymphoma. Blood 2003;102:2638–41. https://doi.org/10.1182/blood-2003-04-1050.

[134] Takeuchi K, Soda M, Togashi Y, Ota Y, Sekiguchi Y, Hatano S, Asaka R, Noguchi M, Mano H. Identification of a novel fusion, SQSTM1-ALK, in ALK-positive large B-cell lymphoma. Haematologica 2011;96:464–7. https://doi.org/10.3324/haematol.2010.033514.

[135] Bedwell C, Rowe D, Moulton D, Jones G, Bown N, Bacon CM. Cytogenetically complex SEC31A-ALK fusions are recurrent in ALK-positive large B-cell lymphomas. Haematologica 2011;96:343–6. https://doi.org/10.3324/haematol.2010.031484.

[136] Maesako Y, Izumi K, Okamori S, Takeoka K, Kishimori C, Okumura A, Honjo G, Akasaka T, Ohno H. inv(2)(p23q13)/RAN-binding protein 2 (RANBP2)-ALK fusion gene in myeloid leukemia that developed in an elderly woman. Int J Hematol 2014;99:202–7. https://doi.org/10.1007/s12185-013-1482-x.

[137] Lee SE, Kang SY, Takeuchi K, Ko YH. Identification of RANBP2-ALK fusion in ALK positive diffuse large B-cell lymphoma. Hematol Oncol 2014;32:221–4. https://doi.org/10.1002/hon.2125.

[138] Palacios G, Shaw TI, Li Y, Singh RK, Valentine M, Sandlund JT, Lim MS, Mulligan CG, Leventaki V. Novel ALK fusion in anaplastic large cell lymphoma involving EEF1G, a subunit of the eukaryotic elongation factor-1 complex. Leukemia 2017;31:743–7. https://doi.org/10.1038/leu.2016.331.

[139] Ise M, Kageyama H, Araki A, Itami M. Identification of a novel GORASP2-ALK fusion in an ALK-positive large B-cell lymphoma. Leuk Lymphoma 2019;60:493–7. https://doi.org/10.1080/10428194.2018.1493731.

[140] Sakamoto K, Nakasone H, Togashi Y, Sakata S, Tsuyama N, Baba S, Dobashi A, Asaka R, Tsai C-C, Chuang S-S, Izutsu K, Kanda Y, Takeuchi K. ALK-positive large B-cell lymphoma: identification of EML4-ALK and a review of the literature focusing on the ALK immunohistochemical staining pattern. Int J Hematol 2016;103:399–408. https://doi.org/10.1007/s12185-016-1934-1.

[141] Bischof D, Pulford K, Mason DY, Morris SW. Role of the nucleophosmin (NPM) portion of the non-Hodgkin's lymphoma-associated NPM-anaplastic lymphoma kinase fusion protein in oncogenesis. Mol Cell Biol 1997;17:2312–25. https://doi.org/10.1128/mcb.17.4.2312.

[142] Salaverria I, Beà S, Lopez-Guillermo A, Lespinet V, Pinyol M, Burkhardt B, Lamant L, Zettl A, Horsman D, Gascoyne R, Ott G, Siebert R, Delsol G, Campo E. Genomic profiling reveals different genetic aberrations in systemic ALK-positive and ALK-negative anaplastic large cell lymphomas. Br J Haematol 2008;140:516–26. https://doi.org/10.1111/j.1365-2141.2007.06924.x.

[143] Shaw AT, Hsu PP, Awad MM, Engelman JA. Tyrosine kinase gene rearrangements in epithelial malignancies. Nat Rev Cancer 2013;13:772–87. https://doi.org/10.1038/nrc3612.

[144] Tanaka M, Kohashi K, Kushitani K, Yoshida M, Kurihara S, Kawashima M, Ueda Y, Souzaki R, Kinoshita Y, Oda Y, Takeshima Y, Hiyama E, Taguchi T, Tanaka Y. Inflammatory myofibroblastic tumors of the lung

carrying a chimeric A2M-ALK gene: report of 2 infantile cases and review of the differential diagnosis of infantile pulmonary lesions. Hum Pathol 2017;66:177–82. https://doi.org/10.1016/j.humpath.2017.06.013.

[145] Debiec-Rychter M, Marynen P, Hagemeijer A, Pauwels P. ALK-ATIC fusion in urinary bladder inflammatory myofibroblastic tumor. Genes Chromosomes Cancer 2003;38:187–90. https://doi.org/10.1002/gcc.10267.

[146] Bridge JA, Kanamori M, Ma Z, Pickering D, Hill DA, Lydiatt W, Lui MY, Colleoni GW, Antonescu CR, Ladanyi M, Morris SW. Fusion of the ALK gene to the clathrin heavy chain gene, CLTC, in inflammatory myofibroblastic tumor. Am J Pathol 2001;159:411–5. https://doi.org/10.1016/S0002-9440(10)61711-7.

[147] Zarei S, Abdul-Karim FW, Chase DM, Astbury C, Policarpio-Nicolas MLC. Uterine inflammatory myofibroblastic tumor showing an atypical ALK signal pattern by FISH and DES-ALK fusion by RNA sequencing: a case report. Int J Gynecol Pathol 2020;39:152–6. https://doi.org/10.1097/PGP.0000000000000588.

[148] Inamura K, Kobayashi M, Nagano H, Sugiura Y, Ogawa M, Masuda H, Yonese J, Ishikawa Y. A novel fusion of HNRNPA1-ALK in inflammatory myofibroblastic tumor of urinary bladder. Hum Pathol 2017;69:96–100. https://doi.org/10.1016/j.humpath.2017.04.022.

[149] Haimes JD, Stewart CJR, Kudlow BA, Culver BP, Meng B, Koay E, Whitehouse A, Cope N, Lee J-C, Ng T, McCluggage WG, Lee C-H. Uterine inflammatory myofibroblastic tumors frequently harbor ALK fusions with IGFBP5 and THBS1. Am J Surg Pathol 2017;41:773–80. https://doi.org/10.1097/PAS.0000000000000801.

[150] Maruggi M, Malicki DM, Levy ML, Crawford JR. A novel KIF5B-ALK fusion in a child with an atypical central nervous system inflammatory myofibroblastic tumour. BMJ Case Rep 2018;2018. https://doi.org/10.1136/bcr-2018-226431.

[151] Rao N, Iwenofu H, Tang B, Woyach J, Liebner DA. Inflammatory myofibroblastic tumor driven by novel NUMA1-ALK fusion responds to ALK inhibition. J Natl Compr Canc Netw 2018;16:115–21. https://doi.org/10.6004/jnccn.2017.7031.

[152] Takeuchi K, Soda M, Togashi Y, Sugawara E, Hatano S, Asaka R, Okumura S, Nakagawa K, Mano H, Ishikawa Y. Pulmonary inflammatory myofibroblastic tumor expressing a novel fusion, PPFIBP1-ALK: reappraisal of anti-ALK immunohistochemistry as a tool for novel ALK fusion identification. Clin Cancer Res 2011;17:3341–8. https://doi.org/10.1158/1078-0432.CCR-11-0063.

[153] Ma Z, Hill DA, Collins MH, Morris SW, Sumegi J, Zhou M, Zuppan C, Bridge JA. Fusion of ALK to the Ran-binding protein 2 (RANBP2) gene in inflammatory myofibroblastic tumor. Genes Chromosomes Cancer 2003;37:98–105. https://doi.org/10.1002/gcc.10177.

[154] Panagopoulos I, Nilsson T, Domanski HA, Isaksson M, Lindblom P, Mertens F, Mandahl N. Fusion of the SEC31L1 and ALK genes in an inflammatory myofibroblastic tumor. Int J Cancer 2006;118:1181–6. https://doi.org/10.1002/ijc.21490.

[155] Honda K, Kadowaki S, Kato K, Hanai N, Hasegawa Y, Yatabe Y, Muro K. Durable response to the ALK inhibitor alectinib in inflammatory myofibroblastic tumor of the head and neck with a novel SQSTM1-ALK fusion: a case report. Invest New Drugs 2019;37:791–5. https://doi.org/10.1007/s10637-019-00742-2.

[156] Lawrence B, Perez-Atayde A, Hibbard MK, Rubin BP, Dal Cin P, Pinkus JL, Pinkus GS, Xiao S, Yi ES, Fletcher CD, Fletcher JA. TPM3-ALK and TPM4-ALK oncogenes in inflammatory myofibroblastic tumors. Am J Pathol 2000;157:377–84. https://doi.org/10.1016/S0002-9440(10)64550-6.

[157] Tian Q, Deng W-J, Li Z-W. Identification of a novel crizotinib-sensitive BCL11A-ALK gene fusion in a nonsmall cell lung cancer patient. Eur Respir J 2017;49. https://doi.org/10.1183/13993003.02149-2016.

[158] Tan DS-W, Kim D-W, Thomas M, Pantano S, Wang Y, Szpakowski SL, Yovine AJ, Mehra R, Chow LQ, Sharma S, Solomon BJ, Felip E, Camidge DR, Vansteenkiste JF, Bitter H, Petruzzelli LM, Dugan MH, Shaw AT. Genetic landscape of ALK+ non-small cell lung cancer (NSCLC) patients (pts) and response to ceritinib in ASCEND-1. JCO 2016;34:9064. https://doi.org/10.1200/JCO.2016.34.15_suppl.9064.

[159] Du X, Shao Y, Gao H, Zhang X, Zhang H, Ban Y, Qin H, Tai Y. CMTR1-ALK: an ALK fusion in a patient with no response to ALK inhibitor crizotinib. Cancer Biol Ther 2018;19:962–6. https://doi.org/10.1080/15384047.2018.1480282.

[160] Zhang M, Wang Q, Ding Y, Wang G, Chu Y, He X, Wu X, Shao YW, Lu K. CUX1-ALK, a novel ALK rearrangement that responds to crizotinib in non-small cell lung cancer. J Thorac Oncol 2018;13:1792–7. https://doi.org/10.1016/j.jtho.2018.07.008.

[161] Iyevleva AG, Raskin GA, Tiurin VI, Sokolenko AP, Mitiushkina NV, Aleksakhina SN, Garifullina AR, Strelkova TN, Merkulov VO, Ivantsov AO, Kuligina ES, Pozharisski KM, Togo AV, Imyanitov EN. Novel ALK fusion partners in lung cancer. Cancer Lett 2015;362:116–21. https://doi.org/10.1016/j.canlet.2015.03.028.

[162] Jiang J, Wu X, Tong X, Wei W, Chen A, Wang X, Shao YW, Huang J. GCC2-ALK as a targetable fusion in lung adenocarcinoma and its enduring clinical responses to ALK inhibitors. Lung Cancer 2018;115:5–11. https://doi.org/10.1016/j.lungcan.2017.10.011.

[163] Hong M, Kim RN, Song J-Y, Choi S-J, Oh E, Lira ME, Mao M, Takeuchi K, Han J, Kim J, Choi Y-L. HIP1-ALK, a novel fusion protein identified in lung adenocarcinoma. J Thorac Oncol 2014;9:419–22. https://doi.org/10.1097/JTO.0000000000000061.

[164] Takeuchi K, Choi YL, Togashi Y, Soda M, Hatano S, Inamura K, Takada S, Ueno T, Yamashita Y, Satoh Y, Okumura S, Nakagawa K, Ishikawa Y, Mano H. KIF5B-ALK, a novel fusion oncokinase identified by an immunohistochemistry-based diagnostic system for ALK-positive lung cancer. Clin Cancer Res 2009;15:3143–9. https://doi.org/10.1158/1078-0432.CCR-08-3248.

[165] Togashi Y, Soda M, Sakata S, Sugawara E, Hatano S, Asaka R, Nakajima T, Mano H, Takeuchi K. KLC1-ALK: a novel fusion in lung cancer identified using a formalin-fixed paraffin-embedded tissue only. PLoS ONE 2012;7:e31323. https://doi.org/10.1371/journal.pone.0031323.

[166] Bu K, Lu Y, Liu X, Cheng C, Li B. Lysocardiolipin acyltransferase 1-anaplastic lymphoma receptor tyrosine kinase: a novel crizotinib-sensitive fusion gene in lung adenocarcinoma. J Thorac Oncol 2020;15:e55–7. https://doi.org/10.1016/j.jtho.2019.11.016.

[167] Wang L, Yao S, Teng L, Zhang W, Chen L. Melanoregulin-anaplastic lymphoma kinase (ALK), a novel ALK rearrangement that responds to crizotinib in lung adenocarcinoma. J Thorac Oncol 2020;15:e44–6. https://doi.org/10.1016/j.jtho.2019.11.019.

[168] Wu X, Wang W, Zou B, Li Y, Yang X, Liu N, Ma Q, Zhang X, Wang Y, Li D. Novel NLRC4-ALK and EML4-ALK double fusion mutations in a lung adenocarcinoma patient: a case report. Thorac Cancer 2020. https://doi.org/10.1111/1759-7714.13389.

[169] Jung Y, Kim P, Jung Y, Keum J, Kim S-N, Choi YS, Do I-G, Lee J, Choi S-J, Kim S, Lee J-E, Kim J, Lee S, Kim J. Discovery of ALK-PTPN3 gene fusion from human non-small cell lung carcinoma cell line using next generation RNA sequencing. Genes Chromosomes Cancer 2012;51:590–7. https://doi.org/10.1002/gcc.21945.

[170] Kim RN, Choi Y-L, Lee M-S, Lira ME, Mao M, Mann D, Stahl J, Licon A, Choi SJ, Van Vrancken M, Han J, Wlodarska I, Kim J. SEC31A-ALK fusion gene in lung adenocarcinoma. Cancer Res Treat 2016;48:398–402. https://doi.org/10.4143/crt.2014.254.

[171] Chen H-F, Wang W-X, Xu C-W, Huang L-C, Li X-F, Lan G, Zhai Z-Q, Zhu Y-C, Du K-Q, Lei L, Fang M-Y. A novel SOS1-ALK fusion variant in a patient with metastatic lung adenocarcinoma and a remarkable response to crizotinib. Lung Cancer 2020;142:59–62. https://doi.org/10.1016/j.lungcan.2020.02.012.

[172] Majewski IJ, Mittempergher L, Davidson NM, Bosma A, Willems SM, Horlings HM, de Rink I, Greger L, Hooijer GKJ, Peters D, Nederlof PM, Hofland I, de Jong J, Wesseling J, Kluin RJC, Brugman W, Kerkhoven R, Nieboer F, Roepman P, Broeks A, Muley TR, Jassem J, Niklinski J, van Zandwijk N, Brazma A, Oshlack A, van den Heuvel M, Bernards R. Identification of recurrent FGFR3 fusion genes in lung cancer through kinome-centred RNA sequencing. J Pathol 2013;230:270–6. https://doi.org/10.1002/path.4209.

[173] Rikova K, Guo A, Zeng Q, Possemato A, Yu J, Haack H, Nardone J, Lee K, Reeves C, Li Y, Hu Y, Tan Z, Stokes M, Sullivan L, Mitchell J, Wetzel R, Macneill J, Ren JM, Yuan J, Bakalarski CE, Villen J, Kornhauser JM, Smith B, Li D, Zhou X, Gygi SP, Gu T-L, Polakiewicz RD, Rush J, Comb MJ. Global survey of phosphotyrosine signaling identifies oncogenic kinases in lung cancer. Cell 2007;131:1190–203. https://doi.org/10.1016/j.cell.2007.11.025.

[174] Feng T, Chen Z, Gu J, Wang Y, Zhang J, Min L. The clinical responses of TNIP2-ALK fusion variants to crizotinib in ALK-rearranged lung adenocarcinoma. Lung Cancer 2019;137:19–22. https://doi.org/10.1016/j.lungcan.2019.08.032.

[175] Choi Y-L, Lira ME, Hong M, Kim RN, Choi S-J, Song J-Y, Pandy K, Mann DL, Stahl JA, Peckham HE, Zheng Z, Han J, Mao M, Kim J. A novel fusion of TPR and ALK in lung adenocarcinoma. J Thorac Oncol 2014;9:563–6. https://doi.org/10.1097/JTO.0000000000000093.

[176] Hu S, Li Q, Peng W, Feng C, Zhang S, Li C. VIT-ALK, a novel alectinib-sensitive fusion gene in lung adenocarcinoma. J Thorac Oncol 2018;13:e72–4. https://doi.org/10.1016/j.jtho.2017.11.134.

[177] He Z, Wu X, Ma S, Zhang C, Zhang Z, Wang S, Yu S, Wang Q. Next-generation sequencing identified a novel WDPCP-ALK fusion sensitive to crizotinib in lung adenocarcinoma. Clin Lung Cancer 2019;20:e548–51. https://doi.org/10.1016/j.cllc.2019.06.001.

[178] Pinsolle J, Mondet J, Duruisseaux M, d'Alnoncourt S, Magnat N, de Fraipont F, Moro-Sibilot D, Toffart A-C, Brambilla E, McLeer-Florin A. A rare fusion of CLIP1 and ALK in a case of non-small-cell lung cancer with neuroendocrine features. Clin Lung Cancer 2019;20:e535–40. https://doi.org/10.1016/j.cllc.2019.05.001.

[179] Lipson D, Capelletti M, Yelensky R, Otto G, Parker A, Jarosz M, Curran JA, Balasubramanian S, Bloom T, Brennan KW, Donahue A, Downing SR, Frampton GM, Garcia L, Juhn F, Mitchell KC, White E, White J, Zwirko Z, Peretz T, Nechushtan H, Soussan-Gutman L, Kim J, Sasaki H, Kim HR, Park S, Ercan D, Sheehan CE, Ross JS, Cronin MT, Jänne PA, Stephens PJ. Identification of new ALK and RET gene fusions from colorectal and lung cancer biopsies. Nat Med 2012;18:382–4. https://doi.org/10.1038/nm.2673.

[180] Lee J, Kim HC, Hong JY, Wang K, Kim SY, Jang J, Kim ST, Park JO, Lim HY, Kang WK, Park YS, Lee J, Lee WY, Park YA, Huh JW, Yun SH, Do I-G, Kim SH, Balasubramanian S, Stephens PJ, Ross JS, Li GG, Hornby Z, Ali SM, Miller VA, Kim K-M, Ou S-HI. Detection of novel and potentially actionable anaplastic lymphoma kinase (ALK) rearrangement in colorectal adenocarcinoma by immunohistochemistry screening. Oncotarget 2015;6:24320–32. https://doi.org/10.18632/oncotarget.4462.

[181] Lin E, Li L, Guan Y, Soriano R, Rivers CS, Mohan S, Pandita A, Tang J, Modrusan Z. Exon array profiling detects EML4-ALK fusion in breast, colorectal, and non-small cell lung cancers. Mol Cancer Res 2009;7:1466–76. https://doi.org/10.1158/1541-7786.MCR-08-0522.

[182] Stransky N, Cerami E, Schalm S, Kim JL, Lengauer C. The landscape of kinase fusions in cancer. Nat Commun 2014;5:4846. https://doi.org/10.1038/ncomms5846.

[183] Sukov WR, Hodge JC, Lohse CM, Akre MK, Leibovich BC, Thompson RH, Cheville JC. ALK alterations in adult renal cell carcinoma: frequency, clinicopathologic features and outcome in a large series of consecutively treated patients. Mod Pathol 2012;25:1516–25. https://doi.org/10.1038/modpathol.2012.107.

[184] Cajaiba MM, Jennings LJ, George D, Perlman EJ. Expanding the spectrum of ALK-rearranged renal cell carcinomas in children: identification of a novel HOOK1-ALK fusion transcript. Genes Chromosomes Cancer 2016;55:814–7. https://doi.org/10.1002/gcc.22382.

[185] Kusano H, Togashi Y, Akiba J, Moriya F, Baba K, Matsuzaki N, Yuba Y, Shiraishi Y, Kanamaru H, Kuroda N, Sakata S, Takeuchi K, Yano H. Two cases of renal cell carcinoma harboring a novel STRN-ALK fusion gene. Am J Surg Pathol 2016;40:761–9. https://doi.org/10.1097/PAS.0000000000000610.

[186] Sugawara E, Togashi Y, Kuroda N, Sakata S, Hatano S, Asaka R, Yuasa T, Yonese J, Kitagawa M, Mano H, Ishikawa Y, Takeuchi K. Identification of anaplastic lymphoma kinase fusions in renal cancer: large-scale immunohistochemical screening by the intercalated antibody-enhanced polymer method. Cancer 2012;118:4427–36. https://doi.org/10.1002/cncr.27391.

[187] Debelenko LV, Raimondi SC, Daw N, Shivakumar BR, Huang D, Nelson M, Bridge JA. Renal cell carcinoma with novel VCL-ALK fusion: new representative of ALK-associated tumor spectrum. Mod Pathol 2011;24:430–42. https://doi.org/10.1038/modpathol.2010.213.

[188] Davis LE, Nusser KD, Przybyl J, Pittsenbarger J, Hofmann NE, Varma S, Vennam S, Debiec-Rychter M, van de Rijn M, Davare MA. Discovery and characterization of recurrent, targetable ALK fusions in leiomyosarcoma. Mol Cancer Res 2019;17:676–85. https://doi.org/10.1158/1541-7786.MCR-18-1075.

[189] Lee J, Singh A, Ali SM, Lin DI, Klempner SJ. TNS1-ALK fusion in a recurrent, metastatic uterine mesenchymal tumor originally diagnosed as leiomyosarcoma. Acta Med Acad 2019;48:116–20. https://doi.org/10.5644/ama2006-124.248.

[190] Perron E, Pissaloux D, Charon Barra C, Karanian M, Lamant L, Parfait S, Alberti L, de la Fouchardière A. Melanocytic myxoid spindle cell tumor with ALK rearrangement (MMySTAR): report of 4 cases of a nevus variant with potential diagnostic challenge. Am J Surg Pathol 2018;42:595–603. https://doi.org/10.1097/PAS.0000000000000973.

[191] Kazakov DV, Kyrpychova L, Martinek P, Grossmann P, Steiner P, Vanecek T, Pavlovsky M, Bencik V, Michal M, Michal M. ALK gene fusions in epithelioid fibrous histiocytoma: a study of 14 cases, with new histopathological findings. Am J Dermatopathol 2018;40:805–14. https://doi.org/10.1097/DAD.0000000000001085.

[192] Craig JW, Quade BJ, Muto MG, MacConaill LE. Endometrial cancer with an EML4-ALK rearrangement. Cold Spring Harb Mol Case Stud 2018;4. https://doi.org/10.1101/mcs.a003020.

[193] Du X-L, Hu H, Lin D-C, Xia S-H, Shen X-M, Zhang Y, Luo M-L, Feng Y-B, Cai Y, Xu X, Han Y-L, Zhan Q-M, Wang M-R. Proteomic profiling of proteins dysregulted in Chinese esophageal squamous cell carcinoma. J Mol Med 2007;85:863–75. https://doi.org/10.1007/s00109-007-0159-4.

[194] Onoda T, Kanno M, Sato H, Takahashi N, Izumino H, Ohta H, Emura T, Katoh H, Ohizumi H, Ohtake H, Asao H, Dehner LP, Hill AD, Hayasaka K, Mitsui T. Identification of novel ALK rearrangement A2M-ALK in a neonate with fetal lung interstitial tumor. Genes Chromosomes Cancer 2014;53:865–74. https://doi.org/10.1002/gcc.22199.

[195] Panagopoulos I, Gorunova L, Lund-Iversen M, Lobmaier I, Bjerkehagen B, Heim S. Recurrent fusion of the genes FN1 and ALK in gastrointestinal leiomyomas. Mod Pathol 2016;29:1415–23. https://doi.org/10.1038/modpathol.2016.129.

[196] Aghajan Y, Levy ML, Malicki DM, Crawford JR. Novel PPP1CB-ALK fusion protein in a high-grade glioma of infancy. BMJ Case Rep 2016;2016. https://doi.org/10.1136/bcr-2016-217189.

[197] Ren H, Tan Z-P, Zhu X, Crosby K, Haack H, Ren J-M, Beausoleil S, Moritz A, Innocenti G, Rush J, Zhang Y, Zhou X-M, Gu T-L, Yang Y-F, Comb MJ. Identification of anaplastic lymphoma kinase as a potential therapeutic target in ovarian cancer. Cancer Res 2012;72:3312–23. https://doi.org/10.1158/0008-5472.CAN-11-3931.

[198] Hung YP, Dong F, Watkins JC, Nardi V, Bueno R, Dal Cin P, Godleski JJ, Crum CP, Chirieac LR. Identification of ALK rearrangements in malignant peritoneal mesothelioma. JAMA Oncol 2018;4:235–8. https://doi.org/10.1001/jamaoncol.2017.2918.

[199] Fujimoto M, Togashi Y, Matsuzaki I, Baba S, Takeuchi K, Inaba Y, Jinnin M, Murata S-I. A case report of atypical Spitz tumor harboring a novel MLPH-ALK gene fusion with discordant ALK immunohistochemistry results. Hum Pathol 2018;80:99–103. https://doi.org/10.1016/j.humpath.2018.02.021.

[200] Yeh I, de la Fouchardiere A, Pissaloux D, Mully TW, Garrido MC, Vemula SS, Busam KJ, LeBoit PE, McCalmont TH, Bastian BC. Clinical, histopathologic, and genomic features of Spitz tumors with ALK fusions. Am J Surg Pathol 2015;39:581–91. https://doi.org/10.1097/PAS.0000000000000387.

[201] Pérot G, Soubeyran I, Ribeiro A, Bonhomme B, Savagner F, Boutet-Bouzamondo N, Hostein I, Bonichon F, Godbert Y, Chibon F. Identification of a recurrent STRN/ALK fusion in thyroid carcinomas. PLoS ONE 2014;9:e87170. https://doi.org/10.1371/journal.pone.0087170.

[202] Coffin CM, Patel A, Perkins S, Elenitoba-Johnson KS, Perlman E, Griffin CA. ALK1 and p80 expression and chromosomal rearrangements involving 2p23 in inflammatory myofibroblastic tumor. Mod Pathol 2001;14:569–76. https://doi.org/10.1038/modpathol.3880352.

[203] Lovly CM, Gupta A, Lipson D, Otto G, Brennan T, Chung CT, Borinstein SC, Ross JS, Stephens PJ, Miller VA, Coffin CM. Inflammatory myofibroblastic tumors harbor multiple potentially actionable kinase fusions. Cancer Discov 2014;4:889–95. https://doi.org/10.1158/2159-8290.CD-14-0377.

[204] Shaw AT, Yeap BY, Mino-Kenudson M, Digumarthy SR, Costa DB, Heist RS, Solomon B, Stubbs H, Admane S, McDermott U, Settleman J, Kobayashi S, Mark EJ, Rodig SJ, Chirieac LR, Kwak EL, Lynch TJ, Iafrate AJ. Clinical features and outcome of patients with non-small-cell lung cancer who harbor EML4-ALK. J Clin Oncol 2009;27:4247–53. https://doi.org/10.1200/JCO.2009.22.6993.

[205] Lin JJ, Riely GJ, Shaw AT, Targeting ALK. Precision medicine takes on drug resistance. Cancer Discov 2017;7:137–55. https://doi.org/10.1158/2159-8290.CD-16-1123.

[206] Remon J, Tabbò F, Jimenez B, Collazo A, de Castro J, Novello S. Sequential blinded treatment decisions in ALK-positive non-small cell lung cancers in the era of precision medicine. Clin Transl Oncol 2020;22(9):1425–9. https://doi.org/10.1007/s12094-020-02290-1.

[207] Rosenbaum JN, Bloom R, Forys JT, Hiken J, Armstrong JR, Branson J, McNulty S, Velu PD, Pepin K, Abel H, Cottrell CE, Pfeifer JD, Kulkarni S, Govindan R, Konnick EQ, Lockwood CM, Duncavage EJ. Genomic heterogeneity of ALK fusion breakpoints in non-small-cell lung cancer. Mod Pathol 2018;31:791–808. https://doi.org/10.1038/modpathol.2017.181.

[208] Zhang Y, Zeng L, Zhou C, Li Y, Wu L, Xia C, Jiang W, Hu Y, Liao D, Xiao L, Liu L, Yang H, Xiong Y, Guan R, Lizaso A, Mansfield AS, Yang N. Detection of nonreciprocal/reciprocal ALK translocation as poor predictive marker in patients with first-line crizotinib-treated ALK-rearranged NSCLC. J Thorac Oncol 2020. https://doi.org/10.1016/j.jtho.2020.02.007.

[209] Mossé YP, Lim MS, Voss SD, Wilner K, Ruffner K, Laliberte J, Rolland D, Balis FM, Maris JM, Weigel BJ, Ingle AM, Ahern C, Adamson PC, Blaney SM. Safety and activity of crizotinib for paediatric patients with refractory solid tumours or anaplastic large-cell lymphoma: a Children's oncology group phase 1 consortium study. Lancet Oncol 2013;14:472–80. https://doi.org/10.1016/S1470-2045(13)70095-0.

[210] Sano R, Krytska K, Larmour CE, Raman P, Martinez D, Ligon GF, Lillquist JS, Cucchi U, Orsini P, Rizzi S, Pawel BR, Alvarado D, Mossé YP. An antibody-drug conjugate directed to the ALK receptor demonstrates efficacy in preclinical models of neuroblastoma. Sci Transl Med 2019;11. https://doi.org/10.1126/scitranslmed.aau9732.

ALK rearranged lung cancer: TKI treatment and outcome

Ana Collazo-Lorduy[a,b], Beatriz Jiménez[a],
María Castro-Henriques[c], and Jordi Remon[c]

[a]Department of Medical Oncology, Centro Integral Oncología Clara Campal Madrid, HM-Sanchinarro, Medical Oncology Department, Madrid, Spain [b]Department of Medical Oncology, Hospital Universitario Puerta de Hierro, Madrid, Spain [c]Department of Medical Oncology, Centro Integral Oncológico Clara Campal (HM-CIOCC), Hospital HM Delfos, HM Hospitales, Barcelona, Spain

Abstract

Introduction of tyrosine kinase inhibitors (TKIs) have changed the treatment paradigm and improved the prognosis of oncogenic addicted non-small cell lung cancer patients (NSCLC). Nowadays, for *ALK*-positive lung cancers three ALK TKI have been approved: crizotinib, alectinib and brigatinib. Both, alectinib and crizotinib are preferred first-line treatment options based on significant improvement in progression free survival and higher intracranial activity compared with crizotinib. However, the new ALK TKIs in first-line setting are being explored such as brigatinib and lorlatinib, which may shift again the treatment paradigm of *ALK*-positive NSCLC. The mechanisms of acquired resistance to ALK TKI and the most suitable approach upfront as well as at progression, the role of *ALK* fusion partners for treatment decisions making and how liquid biopsy may improve the knowledge for this disease are current challenges. In this chapter, we summarize the current evidence for the efficacy of the different ALK TKIs in first-line setting and discuss current and future challenges in this disease.

Abbreviations

AEs	adverse events
BBB	blood-brain barrier
BM	brain metastases
CI	confidence interval
ctDNA	circulating tumor DNA
HR	hazard ratio
Ic	intra-cranial

ICI	immune checkpoint inhibitors
MTD	maximum tolerated dose
NGS	next generation sequencing
NR	not reached
NSCLC	nonsmall cell lung cancer
OS	overall survival
PFS	progression free survival
QoL	quality of life
RR	response rate
SoC	standard of care
tBM	treated brain metastses
TKI	tyrosine kinase inhibitors
TTD	time to deterioration

Conflict of interest

ACL, BJ, MCH: None related to current manuscript or outside.

JR: None related to current manuscript, outside of current manuscript: advisory: Boehringer-Ingelheim, MSD, Pfizer, BMS, Astra-Zeneca. Travel: OSE immunotherapeutics, BMS, Astra-Zeneca.

Introduction

ALK gene rearrangements occur in 5% of patients with advanced non-small cell lung cancer (NSCLC). First-generation ALK tyrosine kinase inhibitor (TKI), crizotinib, and, more recently next generation ALK TKIs (second generation: ceritinib, ensartinib, alectinib, brigatinib and third generation: lorlatinib) have enlarged the therapeutic arsenal in this population. In Table 1 we report the ALK activity of each of these compounds. Similar to other oncogenic addicted lung adenocarcinomas, personalized treatment with ALK TKIs in the first-line setting is the standard of care in daily clinical practice [1,2], and initiating personalized treatment in a timely manner in *ALK*-positive tumors may have a positive impact on survival outcomes [3]. Table 2 summarizes the outcome of ALK TKI in first-line setting in *ALK*-positive NSCLC patients.

TABLE 1   ALK tyrosine kinase inhibitors, target and cellular IC50 for inhibiting ALK.

Drug	Target	IC50 EML4-ALK variant 1 WT (nM)
Crizotinib	c-MET, ALK, ROS1	23
Ceritinib	ALK, ROS1	12.5
Ensartinib	ALK, ROS, MET, NTRK, AXL	1.28
Alectinib	ALK	1.377
Brigatinib	ALK, ROS1, EGFR	0.959
Lorlatinib	ALK, ROS1	0.267

Based on Horn L, Whisenant JG, Wakelee H et al. Monitoring therapeutic response and resistance: analysis of circulating tumor DNA in patients with ALK+ lung cancer. J Thorac Oncol 2019;14(11):1901–11.

TABLE 2 Efficacy of ALK tyrosine kinase inhibitors in first-line setting in ALK-positive non-small cell lung cancer patients.

Trial	Treatment	N	RR (%)	PFS (months)	OS (months)
PROFILE 1014 [4,5]	Crizotinib vs. Chemotherapy	343	74 vs. 45 (P < .001)	10.9 vs. 7.0[a] (HR 0.45; P < .001)	NR vs. 47.5 (HR 0.76; P = .098)
PROFILE 1029 [6]	Crizotinib vs. Chemotherapy	207	87.5 vs. 45.6 (P < .001)	11.1 vs. 6.2[a] (HR 0,402; P < .001)	28.7 vs. 27.7 (HR 0.897; P = .33)
ASCEND-4 [46]	Ceritinib vs. Chemotherapy	376	73.0 vs. 26.3	16.6 vs. 8.1[a] (HR 0.55; P < .00001)	NR vs. 26.2 (HR 0.73; P = .056)
ALEX [7,8]	Alectinib$_{600}$ vs. Crizotinib	303	83% vs. 76 (P = .09)	34.8 vs. 10.9 (HR 0.43; P < .001)	Not estimable (HR 0.69; 0.47–1.02)
J-ALEX [9,10]	Alectinib$_{300}$ vs. Crizotinib	207	92% vs. 79%	34.1 vs. 10.2[a] (HR 0.37; 0.26–0.52)	NR vs. 43.7 (HR 0.80; P = .386)
ALESSIA [11]	Alectinib$_{600}$ vs. Crizotinib	187	92% vs. 79% (P = .0095)	NR vs. 11.1[a] (HR 0.22; P < .0001)	Not estimable (HR 0.28; P = .0027)
ALTA-1L [12,13]	Brigatinib vs. Crizotinib	275	74% vs. 63% (P = .0342)	24.0 vs. 11.0[a] (HR 0.49; P < .001)	Not estimable (HR 0.92; 0.57–1.47)
CROWN	Lorlatinib vs. Crizotinib	280	Pending	Pending	Pending
EXALT-3	Ensartinib vs. Crizotinib	290	Pending	Pending	Pending

[a] PFS assessed by independent review. NR, not reached.

ALK TKI in first line setting

Crizotinib

Crizotinib is an ATP-competitive small-molecule inhibitor of the receptor tyrosine kinases c-MET, ALK and ROS1. Although initially developed as a potent small molecule inhibitor of c-MET, in ALK-translocated cell lines, crizotinib inhibited downstream effector functions and induced apoptosis [14–16], supporting its activity as a potent inhibitor of ALK. The mean apparent plasma terminal elimination half-life of crizotinib in patients with ALK-positive NSCLC was 42h following a single dose of crizotinib 250mg [17]. A phase I clinical trial was designed to include a dose-escalation phase, followed by a dose-expansion phase at the maximum tolerated dose (MTD) in patients with MET-amplification or ALK-rearrangement. Of note, this trial was already enrolling patients in the dose-escalation phase when ALK-rearrangement in NSCLC was first reported in August 2007. Two patients with NSCLC harboring ALK-rearrangement were treated with crizotinib during dose escalation and showed dramatic improvement in their symptoms. This observation led to large-scale prospective screening of NSCLC patients and recruitment of *ALK*-positive NSCLC patients into an expanded molecular cohort at the MTD of 250 mg twice daily. The trial reported a meaningful clinical activity, with a response rate (RR) of 61% and median progression free survival (PFS) of 9.7 months [18,19], launching the phase II PROFILE 1005 trial [20] and the phase III PROFILE 1007 registration trial [21] in previously treated advanced *ALK*-positive NSCLC patients. In the PROFILE 1007 trial, crizotinib was associated with a longer PFS (7.7 months versus 3.0 months, hazard ratio [HR], 0.49, 95% confidence interval [CI], 0.37–0.64, $P < .001$), higher RR (65% versus 20% for chemotherapy; $P < .001$), and significantly greater improvement in patient-reported symptoms, global quality of life (QOL), and delayed time to deterioration (TTD) of prespecified lung cancer symptoms compared with standard second-line chemotherapy [21,22].

Based on these results, crizotinib was approved by FDA in 2011 and EMA in 2012 as standard treatment for lung cancers harboring *ALK* fusions, identified through a companion diagnostic assay. The efficacy of crizotinib in first-line setting in advanced *ALK*-positive NSCLC patients has been assessed in two randomized phase III clinical trials, the PROFILE 1014 [4,5] and the PROFILE 1029 [6] only performed in Asian patients.

The phase III PROFILE 1014 trial randomized 343 treatment-naïve *ALK*-positive NSCLC patients to crizotinib or platinum-pemetrexed chemotherapy up to six cycles. Crossover to crizotinib treatment after disease progression was permitted for patients receiving chemotherapy. Crizotinib achieved the PFS primary endpoint assessed by independent radiologic review (10.9 months with crizotinib versus 7.0 months with chemotherapy, HR 0.40; 95% CI: 0.35–0.60, $P < .001$). The HR for PFS favored crizotinib across most subgroups defined according to stratification factors (ECOG performance status, race, presence or absence of brain metastases) and other baseline characteristics such as age or gender. Likewise, crizotinib reported higher RR compared with chemotherapy (74% versus 45%, $P < .001$). The most common adverse events (AEs) with crizotinib were vision disorders (71%), diarrhea (61%), and edema (49%), however, most AEs in the two treatment groups were grade 1 or 2 in severity. Grade 3 or 4 elevations of aminotransferase levels occurred in 24 patients in the crizotinib group (14%), but these elevations were managed primarily with dose interruptions or dose reductions. Permanent discontinuation due to AEs occurred in 14% and 12% of patients in the crizotinib and chemotherapy arm, respectively. As compared with chemotherapy, crizotinib was associated with greater reduction in lung cancer symptoms and greater improvement in quality of life [4]. These data led to crizotinib being the first *ALK* TKI to achieve worldwide approval for *ALK*-positive NSCLC patients in first-line setting.

Overall survival (OS) data were not mature at the time of the PFS analysis for the PROFILE 1014 trial with only 26% of deaths reported and a median follow-up duration of 17 months. At that time OS did not differ significantly between the two arms (HR 0.83; 95% CI: 0.54–1.26, $P = .36$) [4]. Likewise, after a median follow-up of 46 months in both arms, median OS was not reached (NR) in crizotinib arm compared with 47.5 months with chemotherapy (HR 0.760; 95% CI: 0.548–1.053; $P = .0978$). Survival probability at 4 years was 56.6% with crizotinib and 49.1% with chemotherapy [5].

The absence of OS benefit with crizotinib in the PROFILE 1014 trial could be explained as 84.2% of patients in the chemotherapy arm received crizotinib at the time of progression. When OS was adjusted for crossover, HR favored crizotinib with a median OS of 59.8 months compared with 19.2 months with chemotherapy (HR 0.346; 95% CI: 0.081 to 0.718) [5].

Likewise, in the crizotinib arm, 24% received a subsequent ALK TKI and 32.0% of patients received treatment other than an ALK TKI (60% platinum-based chemotherapy) as a first subsequent treatment. An exploratory analyses according to the subsequent treatment arm reported that the longest OS was observed in crizotinib-treated patients who received a subsequent ALK TKI ($N = 57$, median OS not reached), whereas patients in the crizotinib arm who received treatment that did not include an ALK TKI in any line of subsequent treatment ($N = 37$) had a median OS of 20.8 months. In the chemotherapy arm, patients who received at least one ALK TKI in any line of subsequent treatment ($N = 145$) had a median OS of 49.5 months. Lastly, the small subset ($N = 3$) of patients who received treatment that did not include an ALK TKI in any line of subsequent treatment had a median OS of 12.1 months [5]. These data suggests that treatment with ALK TKI such as crizotinib may have a positive impact in the patients' outcome. There is further indirect evidence that crizotinib prolongs OS in patients harboring an *ALK*-rearrangement. In a retrospective comparison of 82 patients with *ALK*-rearrangement receiving crizotinib, 36 patients with *ALK*-rearrangement not receiving crizotinib, 67 patients with an activating *EGFR*-mutation and 253 patients with wild-type *EGFR* and *ALK* survival were compared. *ALK*-positive patients treated with second- or third-line with crizotinib had a 1-year survival of 70% compared with 1-year survival of 44% for those *ALK*-positive patients treated with any other second- or third-line therapy (HR 0.36; 95% CI 0.17–0.75; $P = .004$) [23].

One subgroup of patients with special interest in oncogenic addicted tumors is those with brain metastasis (BM). In more recent series with central nervous system (CNS) screening, baseline incidence of BM in *ALK*-positive NSCLC patients ranges from 14% to 42% [7,9,11,12]. However, although in the PROFILE 1014 trial baseline CNS screening was not mandatory 79 out of 343 (23%) patients in had stable and treated BM (tBM) at baseline [4]. Intracranial disease control rate at 12-weeks (85% versus 45%, $P < .001$) and at 24-weeks (56% versus 25%, $P = .006$, respectively) was higher with crizotinib than with chemotherapy, respectively. Indeed, PFS was significantly longer with crizotinib versus chemotherapy in tBM present (9.0 months versus 3.0 months, HR 0.40, $P < .001$) and BM absent (11.1 months versus 7.2 months, HR 0.51, $P < .001$). However, percentages of patients with brain as the sole site of progression were higher with crizotinib than with chemotherapy in the intention to treat population (24% versus 10%), in tBM present (38% versus 23%) and in BM absent subgroup (19% versus 6%) [24]. Likewise, the OS with crizotinib only favored patients without BM at baseline (HR 0.672; 95% CI: 0.457–0.987; $P = .0413$) but for the subgroup of patients with tBM at baseline, the difference in OS did not reach statistical significance (HR 1.285; 95% CI: 0.716 to 2.306; $P = .3991$) [5].

The PROFILE 1014 study strongly suggested that with optimal ALK TKI sequencing remarkable and clinically meaningful prolonged survival can be obtained [5], and these data were also confirmed by the French retrospective CLINALK study [25] and other cohorts [26] with a median OS up to 7.5 years from metastatic disease diagnosis for those patients who received different ALK TKI in the therapeutic strategy after upfront crizotinib. Therefore, it suggests that there is a subgroup of *ALK*-positive tumors with continued ALK-dependency at the time of crizotinib-acquired resistance (AR). Intra-tyrosine kinase *ALK*-mutations are the main mechanism AR to ALK TKI [27]. Approximately one quarter of crizotinib-refractory patients have *ALK*-mutations detected by plasma or tissue genotyping [28]. However, the efficacy of next generation ALK TKI in crizotinib-refractory setting is independent of the occurrence of acquired *ALK*-mutations [28–31], and the efficacy was comparable among patients with and without *ALK* mutations [28,29]. In crizotinib-refractory patients the lack of AR *ALK*-mutations could be explained due to the inability of crizotinib for blocking the *ALK*-rearranged tumors that may be overcome with a more potent and selective next generation ALK TKI. Similarly in crizotinib-refractory patients with AR *ALK*-mutations, the efficacy of next-generation ALK TKI could be explained by the fact that majority of these mutations impart sensitivity to most of next-generation ALK TKI [27]. Therefore, majority of crizotinib-resistant tumors are still driven by ALK [28], and that effective post-crizotinib sequential strategies with next generation ALK TKI could be proposed also without knowing the specific mechanism of AR [32].

Finally, treatment beyond progression is a daily clinical strategy that aims to extend the benefit of personalized treatment before switching to other therapeutic strategy. In the PROFILE 1014, among patients randomly assigned to crizotinib, 65 out of 89 patients with progressive disease (73%) continued to receive crizotinib beyond disease progression for a median of 3.1 months [4]. This potential clinical benefit of crizotinib beyond progression was also reported in a cohort of 120 *ALK*-positive NSCLC patients (51% with brain progression). Treatment beyond progression with crizotinib was correlated with longer median OS compared with those patients who received other chemotherapy (16.4 vs. 5.4 months) [33], although this benefit could also be related to local therapies and more indolent disease in the crizotinib arm. For patients with oligoprogressive disease, even with isolated CNS progression, treatment with local therapies, either surgery or radiotherapy, while continuing crizotinib could be viewed as an acceptable option [34]. In the PROFILE 1014 study, among 25 patients with intracranial progression on crizotinib, 19 received radiotherapy for their BM while continued crizotinib. Median treatment time beyond intracranial progression in these 22 patients was 20.4 weeks, which was numerically longer than the 11.7 weeks median treatment time beyond extracranial progression [24]. Similarly, in a retrospective single-institution study, local therapy (either surgery or radiotherapy) for BM in *EGFR*-mutant (17 treated with erlotinib) or *ALK*-rearranged (38 treated with crizotinib) NSCLC patients and CNS progression allowed continuation of therapy for an additional 7.1 months [35]. These data reinforce local therapies as potential strategies in oncogenic addicted NSCLC patients, including *ALK*-positive tumors.

Although the PROFILE 1014 reported the significant superiority of crizotinib in untreated *ALK*-positive NSCLC patients in the first-line setting [4,5], this benefit was only consistent in smaller subset of East Asian patients. The randomized phase III PROFILE 1029 trial, prospectively evaluated crizotinib versus platinum-pemetrexed chemotherapy in 207 *ALK*-positive East Asian NSCLC patients. Crizotinib achieved the PFS primary endpoint assessed by independent review, (11.1 months versus 6.2 months, HR 0,402; 95% CI: 0.286–0.565; $P < .001$), and also improved the RR (87.5% versus 45.6%, $p < 0.001$) compared with chemotherapy. There were no differences in the OS (28.7 months versus 27.7 months, HR 0.897; 95% CI: 0.556–1.445, $P = .33$), most likely because of the low percentage of deaths across both treatment arms (35.1%) and the high rate of patients in the chemotherapy arm who subsequently received crizotinib treatment (81%). Safety profile in PROFILE 1029 [6] were consistent with the PROFILE 1014 [4], and crizotinib did also improve QoL compared with chemotherapy [6]. Of note, neither PROFILE 1014 [4,5] nor PROFILE 1029 [6] included maintenance treatment with pemetrexed because both studies were designed and initiated before maintenance pemetrexed was established as a standard of care. However, the pemetrexed maintenance would probably have a minor impact on the PFS endpoint of both trials.

According to current data, crizotinib is a potential therapeutic strategy in first-line setting for advanced *ALK*-positive NSCLC patients even for those patients with asymptomatic and treated baseline BM. However, most patients develop drug resistance after approximately 11 months of crizotinib treatment and a frequent site of treatment failure is the CNS. The limited intracranial activity with crizotinib is due to the fact that crizotinib concentration in cerebrospinal fluid is much lower than in blood plasma (cerebrospinal fluid to plasma ratio 0.0026 [36]), and this could be explained as crizotinib is a substrate to the p-glycoprotein transporter (ABCB1), a transmembrane protein delivering the drug to the extracellular space, hampering the brain accumulation of the drug [37]. Indeed, nearly half of patients have at least four BM at the time of intracranial progression with crizotinib [38]. Although patients with BM may achieve a prolonged OS (median OS after development of brain metastases of 49.5 months) [26,38], BM is the main cause of poor QoL among NSCLC patients [39]. Therefore, newer and more potent selective ALK TKI with higher intracranial activity have been developed with the aim to increase the disease control in the CNS and to improve patients' outcomes.

Ceritinib

Ceritinib is an oral ATP-competitive ALK and ROS1 TKI, but contrary to crizotinib does not have MET TKI activity, nevertheless may inhibit other receptors to a lesser extent such as the insulin-like growth factor 1 receptor (IGF-1) [40]. In vitro, ceritinib inhibits *ALK* with a 20-fold greater potency than crizotinib, crosses the blood-brain barrier (BBB) in vivo and shows nanomolar potency against patient-derived crizotinib-resistant tumor cell lines [29,41]. Ceritinib has a half-life of 40 h, and steady state is achieved after 15 days [29] with increased absorption with food, especially when including high fat content. Ceritinib is primarily metabolized by CYP3A4, therefore, the concomitant use either inhibitors or inducers of 3A4 substrates is not recommended [40].

The clinical development of ceritinib has been carried through several clinical trials (ASCEND-1 to ASCEND-8). In the phase I ASCEND-1 trial among 246 patients with *ALK*-rearranged NSCLC, including both ALK TKI-naïve ($N = 83$) and ALK TKI-pretreated patients ($N = 163$), ceritinib at the recommended dose (RD) of 750 mg/day reported a very promising RR (72.3% and 56.3%) and median PFS (18.4 months and 6.9 months, naïve an pre-treated, respectively), even among 94 patients with baseline BM with an intracranial RR (icRR) of 79% and 65%, respectively [29,42]. Of note, ceritinib activity was observed in patients with or without crizotinib resistance mutations [29]. Based on these data, ceritinib obtained an accelerated FDA approval for crizotinib-refractory patients in 2014 and by EMA in 2015. Later, both phase II trials, the ASCEND-2 [43] trial and the ASCEND 3 trial [44], confirmed the clinical benefit of ceritinib in previously treated and TKI-naïve population, respectively. Finally, the phase III ASCEND-5 trial confirmed the efficacy of ceritinib over standard second-line chemotherapy in crizotinib and platinum-based chemotherapy-refractory *ALK*-positive NSCLC patient [45]. Of note, the ASCEND-3 trial reported a median OS of 51.3 months, which confirms the prolonged and clinical meaningful survival benefit of *ALK*-positive TKI-naïve patients treated with personalized treatment [44], endorsing the survival data from previous studies or cohorts with upfront crizotinib [5,25,26].

The phase III ASCEND-4 trial [46] confirmed the efficacy of ceritinib in the first-line setting. In the trial, 376 *ALK*-positive (assessed by immunohistochemistry) and TKI-naïve NSCLC patients were randomly assigned to ceritinib ($n = 189$) or platinum-pemetrexed chemotherapy with pemetrexed maintenance ($n = 187$). In the chemotherapy arm, crossover to ceritinib in case of progression was allowed. Ceritinib achieved the primary endpoint, reporting a significant longer PFS by independent review compared with chemotherapy (16.6 months versus 8.1 months, HR 0.55, 95% CI: 0.42–0.73]; $P < .00001$), which reached 26.3 months compared with 6.8 months (HR 0.48; 95% CI: 0.33–0.69) for those patients without baseline BM. However, ceritinib did not improve PFS among 121 patients with baseline asymptomatic and stable BM with or without previous radiotherapy (10.7 months versus 6.7 months, HR 0.70; 95% CI: 0.44–1.12). Ceritinib also improved the RR (73% versus 27%). The median OS was not reached in the ceritinib arm and was 26.2 months in the chemotherapy arm (HR 0·73, 95%: CI 0.50–1.08; $P = .056$), with 2-year OS of 71% versus 58.2% respectively. Lack of survival benefit could be explained as 105 out of 145 (72%) patients received an ALK TKI after discontinuation of chemotherapy. Treatment beyond progression with ceritinib achieved an additional exposure of 9.6 weeks. Among 44 patients with measurable BM at baseline, the icRR with ceritinib was 73% compared with 27.3% with chemotherapy. The median duration of intracranial response with ceritinib was of 16.6 months. The most common AEs in the ceritinib arm were diarrhea (85%), nausea (69%), vomiting (66%), and an increase in alanine aminotransferase (60%), whereas in the chemotherapy arm were nausea (55%), vomiting (36%) and anemia (35%); with a grade 3–4 AEs in 65% and 40% of cases, respectively. Up to 80% of patients in ceritinib arm required dose adjustment or interruptions due to toxicity, mainly gastrointestinal toxicity (28%). However, only 5% of patients discontinued ceritinib due to AEs, and ceritinib significantly improved lung-cancer specific symptoms and prolonged TTD [46]. Based on these results, the FDA approved ceritinib as first-line treatment in *ALK*-positive NSCLC patients in May 2017.

Ceritinib at the dose of 750 mg/day fasted might be associated with a high frequency of gastrointestinal toxicity and raised liver enzymes, which may compromise the treatment compliance. In the phase I ASCEND-8 trial, 306 patients were randomized to ceritinib 450-mg fed ($N = 108$) or 600-mg fed ($N = 87$) or 750-mg fasted ($N = 111$), of which 304 patients were included in safety analysis and 198 treatment-naive *ALK*-positive NSCLC patients were included in the efficacy analysis (450-mg fed [$N = 73$], 600-mg fed [$N = 51$], and 750-mg fasted [$N = 74$]). Brain metastases at baseline were present in 32.9%, 29.4%, and 28.4% of patients in the 450-mg fed, 600-mg fed, and 750-mg fasted arms, respectively [47]. The RR by independent review was 78.1%, 72.5% and 75.7%, respectively, whereas the median duration of response was not estimable, 20.7 months and 15.7 months, respectively. Likewise, the median PFS was not estimable, 17.0 months and 12.2 months, respectively. The safety analysis in the whole population reported that the arm at 450-mg fed showed the highest median relative dose intensity (100% versus 78.5% versus 83.7%), lowest proportion of patients with dose reductions (24.0% versus 65.1% versus 60.9%) as well as dose interruptions (50.9% versus 74.4% versus 72.7%), and lowest proportion of patients with gastrointestinal toxicities (75.9% versus 82.6% versus 91.8%). From a pharmacokinetic perspective, the two doses (450 mg with food and 750 mg fasted) give nearly the same systemic exposure (area under the curve [0 to 24 h], geometric mean ratio = 1.04; maximum serum concentration, geometric mean ratio = 1.03); therefore, the penetration into the brain and intracranial antitumor activity are expected to be equivalent. The reason for similar systemic exposure with the lower dose is the higher absorption of ceritinib when administered with food [47]. ASCEND-8 study confirms previous results from ASCEND-1 trial, which reported that ceritinib at a dose of at least 400 mg or more daily was similarly effective than 750 mg daily in crizotinib-refractory setting (RR 56% versus 56%, respectively), as well as in crizotinib-naïve patients, as RR with at least 400 mg of ceritinib daily in this population was of 62% [29]. Therefore, ceritinib at a dose of 450 mg with food compared to 750-mg fasted showed consistent efficacy and less gastrointestinal toxicity and should be the preferred dose regimen in daily clinical practice in *ALK*-positive NSCLC patients even among those with brain metastases.

Based on the onset of BM in *ALK*-positive NSCLC patients and the intracranial activity of ceritinib [41,42], the phase II ASCEND-7 trial was designed to prospective evaluate the activity of ceritinib (750 mg daily fasted) in *ALK*-positive NSCLC patients with active BM either newly diagnosed or progressive, and at least one extracranial measurable lesion by RECIST v1.1 criteria. Patients may have been pretreated with chemotherapy and/or crizotinib, and were assigned to four different arms depending on prior treatment: arm 1 ($N = 42$) previously treated with brain radiotherapy and an ALK TKI; arm 2 ($N = 40$) no previous radiotherapy but previously treated with an ALK TKI; arm 3 ($N = 12$) previously treated with brain radiotherapy and ALK TKI, arm 4 ($N = 44$) treatment naïve without previous radiotherapy. The RR, DoR and PFS in the whole population was 36%, 10.8 months and 7.2 months in arm 1, 30%, 12.8 months and 5.6 months in arm 2, 50%, not estimable and not estimable in arm 3, and 59%, 9.2 months, and 7.9 months in arm 4, respectively. The icRR in patients with measurable BM was 39.3%, 27.6%, 28.6%, and 51.5% in arms 1 to 4 respectively. The ASCEND-7 mirrors the icRR reported in *ALK*-positive NSCLC patients with active BM and crizotinib refractory in the ASCEND-5 trial (~35%) [45], as well as partially in TKI-naïve patients in the ASCEND-4 trial (~73%) [46]. The ASCEND-7 trial had a fifth arm that enrolled 18 patients with leptomeningeal carcinomatosis with or without active metastases [48]. Out of 18 patients, 78% also had brain metastases and 89% had received prior crizotinib. The RR was 17% with an icRR of 13%. The median DoR and PFS was 5.5 and 5.2 months, respectively with a median OS of 7.2 months. These results suggest a meaningful clinically significant benefit with ceritinib in poor prognosis *ALK*-positive population.

Alectinib

Alectinib is a highly selective and potent TKI targeting *ALK* (IC_{50}: 1.9 nM) and *RET* (IC_{50}: 4.8 nM) [49,50]. Alectinib has shown systemic and intracranial efficacy in *ALK*-positive lung cancer tumors [50,51], and alectinib has reported a high brain-to-plasma ratios (0.63–0.94) on animal models. Unlike crizotinib and ceritinib, preclinical studies suggest that alectinib is not a substrate of glycoprotein-P, a key drug efflux pump typically expressed in the BBB, and that it has greater CNS activity than other ALK TKIs [52]. Indeed, alectinib achieved significant reduction in the size of tumors remaining after treatment with crizotinib in xenograft models, as well as tumors harboring secondarily acquired ALK resistance mutations [53]. According to the results of phase I/II studies, the dose of alectinib is 600 mg twice daily under fed conditions in non-Japanese patients [54], whereas the dose is 300 mg twice daily in Japanese *ALK*-positive NSCLC patients [55,56].

The efficacy of alectinib (600 mg twice daily) in crizotinib-refractory *ALK*-positive NSCLC patients was evaluated in two pivotal, single-arm, open-label phase II trials: NP28673 [57] and NP28761 [58]. Pooled analyses of both studies reported a RR of 51%, median PFS of 8.3 months [59], and median OS of 29.1 months [60], with an icRR of 64% [61]. The efficacy of alectinib in the crizotinib-refractory setting was later confirmed in the phase III ALUR study [62]. Results of both phase II trials launched alectinib approval in crizotinib-refractory population by FDA on December 2015. Two years later, results from ALEX trial [7] led alectinib approval in first-line setting by both, the EMA and FDA in 12th October 2017 and in 6th November 2017, respectively.

The phase III ALEX study randomized treatment-naïve *ALK*-positive NSCLC to receive either alectinib (600 mg twice daily, $N = 152$) or crizotinib (250 mg twice daily, $N = 151$). Patients with asymptomatic BM or leptomeningeal carcinomatosis were eligible, and baseline as well as subsequent brain imaging was performed every 8 weeks until disease progression. The primary endpoint was investigator-assessed PFS and crossover was not allowed per protocol. Alectinib significantly improved the PFS compared with crizotinib [7], with a final median PFS of 34.8 months versus 10.9 months, respectively (HR 0.43; 95% CI: 0.32–0.58; $P < .0001$), and this benefit occurred in patients with baseline BM ($N = 122$, 25.4 months versus 7.4 months, respectively, HR 0.37, 95% CI: 0.23–0.58) and in those without BM ($N = 181$, 38.6 months versus 14.8 months, respectively, HR 0.46, 95% CI: 0.31–0.68) [8]. The efficacy of alectinib was demonstrated in practically all patients' subgroups, regardless of the type of *EML4-ALK* fusion variant and the method used for its determination [63]. Although crossover in ALEX trial was not allowed, no differences in OS have been reported between alectinib versus crizotinib (HR 0.69, 95% CI: 0.47–1.02, with a similar HR in patients with BM at baseline, HR 0.60, 95% CI: 0.34–1.05 and in patients without BM at baseline, HR 0.77; 95% CI: 0.45–1.32). However, recently it has been reported that the 5-year OS rate with alectinib was 62.5% versus 45.5% with crizotinib; the OS data remain immature with 37% of events recorded (stratified HR, 0.67; 95% CI: 0.46–0.98), supporting alectinib as stronger first-line treatment option in ALK-positive NSCLC patients. Likewise, in the ALEX trial, alectinib compared with crizotinib improved the RR but the difference was not statistically significant (83% versus 76%, $P = .09$) [7,63], although the DoR was meaningful longer in alectinib arm (33.1 months versus 11.1 months, respectively). Despite the longer median treatment durations for alectinib compared with crizotinib (27.0 months versus 10.8 months), fewer patients had grade 3–5 AEs (48.7% versus 55%, respectively) [8], and the percentage of treatment discontinuations (13.2% versus 13.2%), dose reductions (16.4% versus 20.5%) and dose interruptions (22.4% and 25.4%) were slightly lower with alectinib than with crizotinib [63], and AEs of any grade favored alectinib compared with crizotinib, except for myalgia and anemia [7]. Patients treated with alectinib reported clinically meaningful improvement in QoL and multiple lung cancer symptoms for a longer duration than those treated with crizotinib, but these differences were not statistically significant [64].

As extracranial efficacy of alectinib is similar to crizotinib, the longer PFS benefit with alectinib seems mainly driven by the higher intracranial activity of alectinib. Out of 122 patients with baseline BM, 43 had measurable lesions (alectinib, $N = 21$; crizotinib, $N = 22$). The icRR was 85.7% with alectinib versus 71.4% with crizotinib in patients who received prior radiotherapy and 78.6% versus 40.0%, respectively, in those who had not. Indeed, the 12-months cumulative incidence rate of CNS progression was higher in crizotinib arm than in alectinib arm in both, patients with BM at baseline (59% versus 16%) and among those without BM at baseline (32% versus 4.6%, respectively). Time to CNS progression was significantly longer with alectinib versus crizotinib and comparable between patients with and without baseline BM ($P < .0001$) [65]. CNS results from ALEX study confirmed alectinib as a new SoC for untreated *ALK*-positive NSCLC patients regardless of presence or not of baseline BM.

The efficacy of alectinib compared with crizotinib in Asian population has been tested in two randomized phase III clinical trials, the J-ALEX with alectinib at 300 mg twice daily [9,10], and ALESSIA trial with alectinib at 600 mg twice daily [11]. These studies allowed the approval of alectinib in first-line setting of *ALK*-positive NSCLC patients in China on August 2018 and in May 2018 in Japan.

In the J-ALEX study, 207 Japanese *ALK*-positive treatment naïve NSCLC patients were enrolled. Although baseline characteristics were generally balanced, there was a greater proportion of patients with baseline BM in the crizotinib arm (29/104 patients [27.9%]) than in the alectinib arm (14/103 patients [13.6%]). The primary endpoint was PFS assessed by independent review, and a sustained improvement in PFS with alectinib relative to crizotinib was shown (34.1 months vs. 10.2 months; HR 0.37; 95% CI: 0.26–0.52) and the imbalance in baseline BM did not impact on the primary-endpoint results [9,10]. The independent review-assessed cumulative incidence rates of CNS and non-CNS progression were lower with alectinib than with crizotinib at all time points in patients both with and without baseline CNS metastases [66]. After a median follow-up of 42 months, median OS was not reached with alectinib and it was 43.7 months with crizotinib (HR 0.80; 99.87% CI: 0.35–1.82; $P = .3860$). Treatment crossover after study withdrawal was allowed and 41% of patients in the alectinib arm and 89% of patients in the crizotinib arm received at least one post-progression therapy, mainly an ALK TKI, and in the crizotinib arm, 84% received alectinib as a post-progression treatment. Compared with alectinib, crizotinib-treated patients had a higher incidence of related serious AEs (25.0% vs. 13.6%, respectively) and grade ≥ 3 AEs (60.6% vs. 36.9%, respectively [9,10].

ALESIA was a randomized, open-label, phase III trial performed in 21 investigational sites in China, South Korea, and Thailand, and enrolled 187 patients who were randomized (2:1) to receive either alectinib ($N = 125$) or crizotinib ($N = 62$). The study reached its primary endpoint of PFS assessed by independent review and was statistically significantly longer with alectinib than crizotinib (NR versus 11.1 month, HR 0.22, 95% CI 0.13–0.38; $P < .0001$). The RR also favored alectinib compared with crizotinib (91% versus 77%, $P = .0095$), as well as the duration of response (HR 0.22, 95% CI 0.12–0.40; $P < .0001$). The time to CNS progression (CNS-HR 0.14) and icRR were also better with alectinib than crizotinib (73% versus. 22%, respectively). Again, OS data were immature at the primary analysis (HR 0.28; 95%CI: 0.12–0.68; $P = .0027$; median OS not estimable in both arms), with an event rate of 6.4% in the alectinib arm and 21.0% in the crizotinib arm. In ALESIA trial, crossover was not allowed and only 24% of patients in the control arm received a next-generation ALK TKI at crizotinib-progression. Toxicity was as expected and similar to other studies [11].

Based on the results of these phase III clinical trials, alectinib appears as a new potential standard treatment in first-line setting based on prolonged PFS compared with crizotinib, and the capability to avert the CNS progression in *ALK*-positive NSCLC patients with or without BM at baseline. The mature OS with alectinib remains unknown, and currant data does not show survival differences compared with crizotinib. Of note, crossover is always desirable in settings where a drug has already proven of benefit in a subsequent line of therapy and exists an attempt to be advanced to an earlier line [22]. Therefore, under current standards, the control arm from ALEX and ALESIA trials should be considered suboptimal and under-treated as crossover was not allowed. Finally, indirect cross-trial comparisons suggest that efficacy of alectinib regarding to RR and PFS is similar regardless of the dose, and whether higher doses would lead to longer PFS or higher tumor penetration has been raised.

Brigatinib

Brigatinib (AP26113) is another next-generation ALK TKI (also active against *ROS1*, *EGFR-T790M*, *IGFR* and *FLT3* mutations). Brigatinib potently inhibits ALK and ROS1, with a high degree of selectivity over more than 250 kinases. Across a panel of *ALK*-positive cell lines, brigatinib inhibited native *ALK* (IC_{50}, 10 nmol/L) with 12-fold greater potency than crizotinib [67], and also reported a broader spectrum of preclinical activity than ceritinib and alectinib against known crizotinib-resistant ALK-mutants [68,69]. Superior efficacy of brigatinib was also observed in mice with *ALK*-positive tumors implanted subcutaneously or intracranial [67].

In the phase I/II trial, brigatinib was assessed at doses of 30–300 mg in different patient cohorts. In *ALK*-rearranged NSCLC patients cohort, brigatinib showed promising clinical activity and an acceptable safety profile in crizotinib-treated ($N=71$, RR 63% and median PFS and OS of 13.2 months and 30.1 months, respectively) and crizotinib-naive patients ($N=8$, RR 100%, and median PFS and OS of 34.2 months and not reached, respectively) [70,71]. According to the efficacy, the dose of 180 mg once daily was the recommended phase II dose. However, in light of accumulating early pulmonary events with this dose, which occurred within 7 days of treatment initiation, two alternative regimens were investigated in the phase II trial [70]. Based on the results of this phase I/II trial, brigatinib was granted break-through therapy designation by the FDA in October 2014.

In the ALTA phase II trial, 222 crizotinib-refractory *ALK*-positive patients were randomized to either oral brigatinib 90 mg once daily (arm A) or 180 mg once daily with a 7-day lead-in at 90 mg (arm B). Patients were stratified by CNS metastases and best response to crizotinib. At baseline, 71% and 67% had BM in arm A and arm B, respectively. The RR and median PFS were 46% and 9.2 months in arm A and 56% and 16.7 months in arm B, with a median OS of 29.5 months and 34.1 months, respectively. Brigatinib showed a high icRR reaching 50% and 67% respectively; and icPFS was 12.8 versus 18.4 months [72]. For instance, brigatinib is the ALK TKI with the longest PFS and icPFS in the post-crizotinib setting. Globally, these promising results launched to assess the efficacy of brigatinib in treatment-naïve population.

The ALTA-1 L phase III trial randomized 235 *ALK*-positive TKI-naïve NSCLC patients to receive brigatinib at a dose of 180 mg once daily (with a 7-day lead-in period at 90 mg) or crizotinib at a dose of 250 mg twice daily. Unlike previous phase III clinical trials previous chemotherapy was allowed and 26% of patients had previously received chemotherapy in advanced disease. Likewise, one-third of patients had baseline asymptomatic BM. Brigatinib achieved the PFS primary endpoint assessed by independent review, and the most recent updated data in PFS reported a median PFS of 24.0 months compared with 11.0 months with crizotinib (HR 0.49, 95% CI 0.35–0.68, $P < .001$), and this benefice occurred regardless of previous chemotherapy (HR 0.44, 95% CI 0.23–0.83 for 73 patients previously treated with chemotherapy; and HR 0.52, 95% CI: 0.35–0.70 for 202 patients without previous chemotherapy) or the occurrence of baseline BM (HR 0.25, 95% CI: 0.14–0.46, $P < .0001$ for 81 patients with BM; and HR 0.65, 95% CI: 0.44–0.97, $P = .0298$ among 194 patients without BM at baseline). Confirmed RR was of 74% and 62% for brigatinib and crizotinib ($P = .0342$), respectively, and median DoR was not reached for brigatinib and was of 13.8 months with crizotinib. OS data is still immature (HR 0.92; 95% 0.57–1.47, $P = .771$), but crossover was allowed in ALTA-1L trial and 44% of patients in crizotinib arm crossed over to brigatinib. For these crossing patients, the RR and median PFS with second-line brigatinib were 54.1% and 15.6 months respectively [12,13]. In the ALTA-1L trial, brigatinib compared with crizotinib improved the icRR (in the group of patients with any BM at baseline [$N = 96$]: 66% versus 16%, $P = .0001$, respectively; as well as in the measurable BM group [$N = 41$]: 78% and 26%, $P < .0001$, respectively) and icPFS (24 months versus 5.6 months, HR 0.31, 95% CI: 0.17–0.56, $P < .0001$) [12]. The most common AEs with brigatinib were an increased creatine kinase level (39%), cough (25%), hypertension (23%) and an increased lipase level (19%). Grade ≥ 3 AEs were reported in 61% of patients in the brigatinib and 55% of patients in the crizotinib arm [13]. Early onset (within 14 days after treatment initiation) of interstitial lung disease or pneumonitis occurred in 5% and 2%, respectively. Finally, the dose reductions rate was of 38% and 25% in brigatinib arm and crizotinib arm respectively, whereas discontinuation rate due to AEs was of 13% and 9%, respectively [12]. Despite these differences, brigatinib was associated with improved health related QoL compared with crizotinib [12,73]. Based on these results, on May 22, 2020, U.S. FDA has approved brigatinib as first-line treatment in ALK-positive advanced NSCLC patients. Simultaneously, the EMA Committee for Medicinal Products for Human Use (CHMP) adopted a positive opinion recommending the approval of brigatinib as upfront treatment in the same population.

Ensartinib

Ensartinib (X-396) is a novel, aminopyridazine-based small molecule that potently inhibits *ALK*. Ensartinib is 10-fold more potent than crizotinib for inhibiting the growth of *ALK*-positive lung cancer cell lines, and reported activity in a broad spectrum of ALK-mutations. Additionally, ensartinib have reported some degree of activity against *ROS1, cMET, AXL* and *TRK* A, B, C [74]. Ensartinib has been tested in both, crizotinib refractory [75,76] and ALK TKI-naïve patients [75].

In the phase I/II eXalt-2 trial, 37 patients were enrolled in dose escalation, and 60 patients were enrolled in dose expansion. The most common treatment-related AEs were rash (56%), nausea (36%), pruritus (28%), vomiting (26%), and fatigue (22%); and 23% of patients experienced a grade 3–4 AEs, mainly rash and pruritus. The maximum tolerated dose was not reached, but the recommended phase 2 dose (RP2D) chosen was 225 mg daily. Among the *ALK*-positive efficacy evaluable patients treated at ≥ 200 mg, the RR was 60% and median

PFS was 9.2 months. Of note, in ALK TKI-naïve patients the RR was 80% and the median PFS was 26.2 months. Ensartinib also reported intracranial activity with an icRR of 64% [75].

Plasma from 76 patients (17 TKI-naïve) enrolled in the phase I/II eXalt-2 trial was collected for circulating tumor DNA (ctDNA) analyses. Disease-associated genetic alterations were detected in 74% (56 of 76) of patients, the most common being *EML4-ALK*. Of note, the concordance rate of *ALK* fusion between plasma and tissue was 91% (20 of 22 blood and tissue samples). The efficacy of ensartinib was evaluated according to different *EML4-ALK* variants detected in the ctDNA analyses. Those patients with a detectable EML4-ALK variant 1 (V1) fusion compared with variant 3 (V3) had improved RR (53% and 14%) and longer PFS (8.2 months and 1.9 months) to ensartinib [77]. Although, the study did not report the efficacy in TKI-naïve patients according to the *EML4-ALK* V1 and V3, the data reported suggests that *ALK*-positive lung cancer is a more complex and heterogeneous disease and other factors should bear in mind for making treatment decisions in the first-line setting and beyond.

These promising results in the phase I/II trial launched the ongoing phase III eXalt-3 clinical trial (NCT02767804, Fig. 1) assessing the PFS benefit of ensartinib compared with crizotinib in ALK TKI-naïve NSCLC patients in first-line setting. Of note, patients with asymptomatic BM and previous chemotherapy are allowed.

Lorlatinib

Lorlatinib is an oral and selective third-generation ALK and ROS1 TKI specifically developed to penetrate the BBB through reduction of P-glycoprotein-1 mediated efflux with broad *ALK* mutational coverage [78]. Lorlatinib has also shown antitumor activity in *ALK*-positive intracranial tumor models [79], and in the clinic lorlatinib has reported a mean ratio of cerebrospinal fluid (CSF)/plasma (unbound) of 0.75, and a half-life ranging from 19 to 28 h [80].

In a phase 1, dose-escalation study, lorlatinib showed both systemic and intracranial activity in 54 patients with advanced *ALK*- or *ROS1*-positive NSCLC, most of whom had CNS metastases and had previously had two or more TKI treatments fail. Among 41 *ALK*-positive

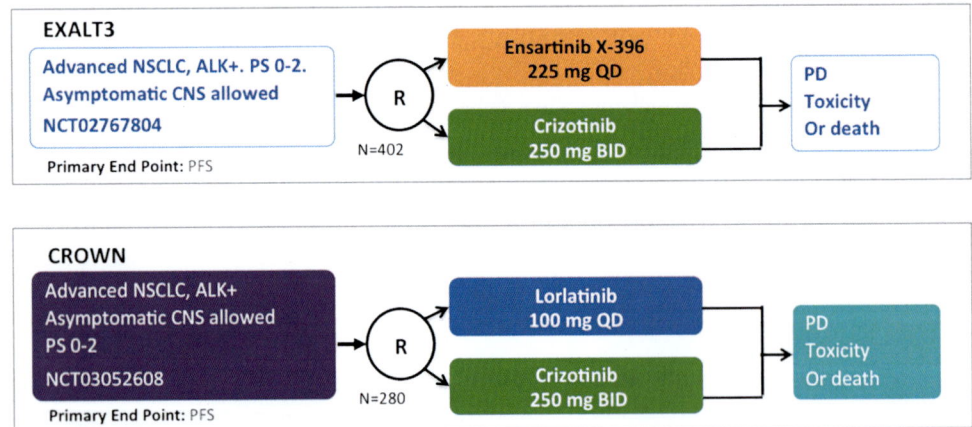

FIG. 1 Ongoing phase III clinical trials with second- and third-generation ALK TKI compared with crizotinib in first-line setting in ALK-positive NSCLC patients.

NSCLC patients, the RR was 46% with a median DoR of 12.4 months and in patients with measurable BM, the icRR was 46%. The estimated median PFS was 9.6 months. The most common AEs in the safety population were hypercholesterolaemia (72%), hypertriglyceridaemia (39%), peripheral neuropathy (39%), and peripheral edema (39%). Cognitive, speech, and mood effects (~39%) were generally transient and reversible. Almost one-third of patients temporarily discontinued lorlatinib at any point and 24% of patients required at least one dose reduction because of treatment-related AEs. The RP2D selected was 100 mg once daily [80]. As some of these AEs are very drug-specific, a clinical guideline for the management of AEs associated with lorlatinib has been recently published [81].

Based on these data, the FDA granted lorlatinib accelerated approval status in November 2018 for *ALK*-positive NSCLC patients who had progressed on crizotinib and at least one additional ALK TKI or who had disease progression on alectinib or ceritinib as the first ALK TKI received. In May 2019, the EMA also approved lorlatinib for use in these patient populations.

In the phase II trial, 276 *ALK/ROS1*-positive NSCLC patients were enrolled into six different expansion cohorts (EXP1–6) on the basis of *ALK* and *ROS1* status and previous therapy [82,83]. In the cohort EXP1 of TKI-naïve *ALK*-positive NSCLC patients ($N = 30$), lorlatinib reported a RR and icRR of 90% and 67%, respectively, with a median PFS and icDoR not reached [82], supporting its efficacy as a potential strategy in first-line setting. The ongoing phase III CROWN trial (NCT03052608, Fig. 1) assesses the efficacy in terms of PFS of lorlatinib compared with crizotinib in first-line setting in TKI-naïve *ALK*-positive NSCLC patients (Fig. 1). Patients with asymptomatic BM are also allowed. The trial has completed the recruitment and results are awaited, although data coming from the phase II trial with lorlatinib in this population preclude that lorlatinib will achieve the primary endpoint.

Others ALK TKI

ASP3026, is a second-generation anaplastic lymphoma kinase (ALK) inhibitor that has potent in vitro activity against crizotinib-resistant *ALK*-positive tumors [84].

Repotrectinib (TPX-0005) is a next-generation ROS1, pan-TRK, and ALK TKI, overcomes resistance due to acquired solvent-front mutations involving *ROS1*, *NTRK1-3*, and *ALK*. Repotrectinib may represent an effective therapeutic option for patients with *ROS1*-, *NTRK1-3*-, or *ALK*-rearranged malignancies who have progressed on earlier-generation TKIs [85]. However, major development of this drug is among *ROS1* and *NTRK*-positive tumors.

WX0593 is a potent ALK and ROS TKI, which has been tested in a phase I trial enrolling 46 *ALK*- and 10 *ROS1*-positive NSCLC patients (39% previously treated with crizotinib). In *ALK*-positive population WX0593 reported a RR of 66% (81% in 21 crizotinib-naïve patients; 44% in 18 crizotinib-refractory patients; and 80% among 5 patients previously treated with crizotinib and a other ALK TKI), whereas the RR reached 30% in the whole *ROS1*-positive NSCLC patients. The most common AEs were hypercholesterolemia, nausea, hypertension, and increased liver enzymes. A phase II is ongoing and evaluates the RR at 120 and 180 mg, respectively [86].

TSR-011 (Belizatinib) is a dual ALK and TRK inhibitor, active against ALK inhibitor resistant tumors in preclinical studies. In a phase I clinical trial reported a partial response of 43% (6/14) among TKI-naive *ALK*-positive NSCLC patients. Based on the limited clinical activity was observed and the competitive ALK inhibitor landscape and benefit/risk considerations, further TSR-011 development was discontinued [87].

Future challenges in first-line setting

Optimal ALK TKI in first-line setting

ALK-positive NSCLC patients may have a prolonged OS, with a 4-year OS ranging from 52% to 65% [8]. Although the prolonged survival in *ALK*-positive NSCLC seems more related to the sequencing strategies at progression rather than the ALK TKI subtype in first line setting [5], data coming from ALEX trial and ALTA-1L support second-generation ALK TKIs as the new SoC in first line setting based on prolonged PFS and higher intracranial activity compared with crizotinib. Whether one second-generation ALK TKI is more suitable than other in first-line setting remains unknown, as there is no head-to-head clinical trial. Taking into account all the limitations and bias, cross trial comparison between brigatinib and alectinib has reported similar outcomes in terms of RR (74% versus 72%), PFS by investigator (29.4 months and HR 0.43 versus 34.8 months and HR 0.43), and icRR in measurable BM (78% versus 81%) [8,12,65]. Unlike to alectinib, brigatinib improved the health-related QoL [12,64], and brigatinib has also shown the longest median icPFS (24 months and 2-year icPFS rate of 48% [12]) reported with ALK TKI in the first-line setting. Therefore, it is possible that soon, not alectinib, but brigatinib will be the preferred ALK TKI to be used for the first-line therapy of *ALK*-positive NSCLC patients. Similarly, in the coming future, results of the ongoing phase III CROWN and EXALT trial (Fig. 1) may shift the treatment paradigm once again, towards upfront lorlatinib, the third generation ALK TKI. However, this new treatment approach with upfront next-generation ALK TKI has also challenged the current sequential strategy with ALK TKI at the time of progression.

Therefore, the resistance mechanisms to these next generation ALK TKIs need to be elucidated, especially if third-generation ALK TKI become the new standard of care (SoC), in order to establish subsequent personalized treatment strategies, when appropriate. Sequential personalized treatment strategies affect the management of patients with oncogene-addicted tumors, and although sequential strategy of second- followed by third-generation ALK TKI is optimal, there is no clinical data about the most suitable sequential approach after upfront third-generation ALK TKI [32].

In this new treatment paradigm, further improving treatment efficacy in the first-line setting is an additional challenge. *ALK* is an heterogeneous disease with several fusion-partners specially after the widespread use of next generation sequencing (NGS) [88], and *ALK*-rearranged NSCLC harbors a spectrum of concurrent genomic alterations at baseline, such as *TP53*-mutation, that may have prognostic and predictive significance [77,89,90]. It remains unknown whether the efficacy of ALK TKIs may vary according to the fusion partner or whether *ALK*-positive tumors with co-mutations may obtain higher benefit with combination strategies or with upfront third-generation ALK TKIs. Although not universal, different studies have shown differences in sensitivity to ALK TKIs according to the *EML4-ALK* fusion type [63,77,91–93]. Therefore, defining the phenotypes within *ALK*-rearranged tumors, either with tissue or liquid biopsies, which are associated with the greatest benefit from the different ALK

TKIs or associated with increased risk to onset specific *ALK* mutations such as *G1202R* [93] is also a challenge, with the aim to better personalize the first-line treatment approach and beyond. Highlighting the importance of potential combination strategies in first-line setting in *ALK*-positive NSCLC patients with either angiogenic agents (NCT03779191, NCT02521051) or chemotherapy (instead of ALK TKI monotherapy) in specific subgroups of patients is another real challenge.

Finally, at current fixed dose of ALK TKI, interpatient variability in exposure may reach 40%, and drug exposure is the main mechanism of primary acquired resistance with TKIs [94]. Although the degree of BBB penetration and CNS retention of specific TKIs depend on its molecular structure, both, extracranial and intracranial activities may also be related to drug exposure. Recently, a retrospective pharmacokinetic study enrolling 100 *ALK*-positive NSCLC patients (48 treated with crizotinib and 52 treated with alectinib), reported that 48% and 37% of patients treated with crizotinib and alectinib had plasma concentrations below the minimum threshold, and PFS on crizotinib or alectinib was shorter for those patients with not enough drug exposure (crizotinib: 5.7 months versus 17.4 months, $P = .08$; alectinib: 12.8 months versus not-reached, $P = .04$) [95]. Therefore, based on these results and the fact that almost 20% of patients enrolled in phase III clinical trials reported dose-reductions, interruptions or dose discontinuations, another future challenge is the therapeutic drug monitoring with the aim to individualize treatment and improve treatment outcomes.

Role of liquid biopsies in first-line setting

In the coming future baseline genomic profile according to liquid biopsy results may become a new standard as almost one-third of advanced NSCLC patients do not have adequate tumor tissue for genomic profiling [96]. Liquid biopsy can be used for knowing the broad genomic profile at baseline or at the time of progression in *ALK*-positive NSCLC [28,77,97]. Prospective validation of the efficacy of TKI in druggable oncogenic NSCLC population based on liquid biopsy results is also a challenge. The ongoing multicohort phase II/III BFAST trial assessed efficacy (RR) of targeted therapies or immune checkpoint inhibitors according to the results of liquid biopsy assessed by NGS. Out of 2188 patients with blood-based NGS results, 119 patients (5.4%) had *ALK*-positive disease and 87 patients received alectinib (43% *TP53* co-mutation). The RR and 12-months PFS assessed by investigator were 87.4% and 78.4% [98], which were very similar to data reported in ALEX trial [7]. Data from BFAST trial validates the clinical utility of blood-based NGS as an additional method for making clinical decision in *ALK*-positive NSCLC patients.

Brain metastases, upfront ALK TKI or radiotherapy

The optimal treatment management of patients with BM and the optimal place of radiotherapy is also a challenge. Almost one-third of *ALK*-positive NSCLC patients may have brain metastases at baseline, which may be associated with reduced quality of life. In oncogenic addicted NSCLC patients with BM, a multidisciplinary approach is the optimal strategy. However, based on the available clinical data and long OS in patients with asymptomatic synchronous BM at diagnosis and the high intracranial efficacy of next-generation ALK TKI, treatment with next-generation ALK TKIs alone should be considered as the initial treatment even in some patients with symptomatic BM. However, close CNS surveillance is strongly recommended for early intervention in patients with an inadequate CNS response. This strategy may defer CNS radiotherapy and avoid long-term neurologic side effects associated with local therapies, mainly radionecrosis. In other cases, sequential treatment initiated with local therapy followed by a TKI is appropriate. For patients who experience CNS progression with controlled extracranial disease while on TKI treatment, local therapy (preferably stereotactic radiosurgery, SRS) followed by the same TKI is an option in patients with a limited number of lesions or who are asymptomatic. In cases of multiple CNS progression, a switch to another ALK TKI with higher CNS-penetration activity may be appropriate [99]. However, prospective evidence of these strategies is required.

Conclusion

Nowadays, based on the improved PFS and better intracranial activity, alectinib and brigatinib are the most suitable first-line treatment options in *ALK*-positive patients. Indeed, alectinib has reported to improve the 5-year OS compared with crizotinib, endorsing the role of upfront second-generation ALK TKIs. In the future, new ALK TKIs may change the treatment sequential strategy and it is relevant to define mechanisms of resistance to establish personalized treatments at progression and improving the survival of *ALK* positive patients.

References

[1] Planchard D, Popat S, Kerr K, et al. Metastatic non-small cell lung cancer: ESMO Clinical Practice Guidelines for diagnosis, treatment and follow-up. Ann Oncol Off J Eur Soc Med Oncol 2018;29(Suppl 4):iv192–237.
[2] Hoang T, Myung S-K, Pham TT, Park B. Efficacy of crizotinib, ceritinib, and alectinib in ALK-positive non-small cell lung cancer treatment: a meta-analysis of clinical trials. Cancer 2020;12(3):526. https://doi.org/10.3390/cancers12030526.
[3] Sheinson D, Wong WB, Wu N, Mansfield AS. Impact of delaying initiation of anaplastic lymphoma kinase inhibitor treatment on survival in patients with advanced non-small-cell lung cancer. Lung Cancer Amst Neth 2020;143:86–92.
[4] Solomon BJ, Mok T, Kim D-W, et al. First-line crizotinib versus chemotherapy in ALK-positive lung cancer. N Engl J Med 2014;371(23):2167–77.
[5] Solomon BJ, Kim D-W, Wu Y-L, et al. Final overall survival analysis from a study comparing first-line crizotinib versus chemotherapy in ALK-mutation-positive non-small-cell lung cancer. J Clin Oncol Off J Am Soc Clin Oncol 2018;36(22):2251–8.
[6] Wu Y-L, Lu S, Lu Y, et al. Results of PROFILE 1029, a phase III comparison of first-line crizotinib versus chemotherapy in east Asian patients with ALK-positive advanced non-small cell lung cancer. J Thorac Oncol 2018;13(10):1539–48.
[7] Peters S, Camidge DR, Shaw AT, et al. Alectinib versus crizotinib in untreated ALK-positive non-small-cell lung cancer. N Engl J Med 2017;377(9):829–38.
[8] Mok TSK, Shaw AT, Camidge RD, et al. Final PFS, updated OS and safety data from the randomised, phase III ALEX study of alectinib (ALC) versus crizotinib (CRZ) in untreated advanced ALK+ NSCLC. Ann Oncol 2019;30:v607.
[9] Nakagawa K, Hida T, Nokihara H, et al. Final progression-free survival results from the J-ALEX study of alectinib versus crizotinib in ALK-positive non-small-cell lung cancer. Lung Cancer Amst Neth 2020;139:195–9.
[10] Hida T, Nokihara H, Kondo M, et al. Alectinib versus crizotinib in patients with ALK-positive non-small-cell lung cancer (J-ALEX): an open-label, randomised phase 3 trial. Lancet Lond Engl 2017;390(10089):29–39.

[11] Zhou C, Kim S-W, Reungwetwattana T, et al. Alectinib versus crizotinib in untreated Asian patients with ana-plastic lymphoma kinase-positive non-small-cell lung cancer (ALESIA): a randomised phase 3 study. Lancet Respir Med 2019;7(5):437–46.

[12] Camidge R, Kim HR, Ahn M-J, et al. Brigatinib vs crizotinib in patients with ALK inhibitor-naive advanced ALK+ NSCLC: updated results from the phase III ALTA-1L trial. Ann Oncol 2019. https://doi.org/10.1093/annonc/mdz446.

[13] Camidge DR, Kim HR, Ahn M-J, et al. Brigatinib versus crizotinib in ALK-positive non-small-cell lung cancer. N Engl J Med 2018;379(21):2027–39.

[14] Christensen JG, Zou HY, Arango ME, et al. Cytoreductive antitumor activity of PF-2341066, a novel inhibitor of anaplastic lymphoma kinase and c-Met, in experimental models of anaplastic large-cell lymphoma. Mol Cancer Ther 2007;6(12 Pt 1):3314–22.

[15] McDermott U, Iafrate AJ, Gray NS, et al. Genomic alterations of anaplastic lymphoma kinase may sensitize tu-mors to anaplastic lymphoma kinase inhibitors. Cancer Res 2008;68(9):3389–95.

[16] Zou HY, Li Q, Lee JH, et al. An orally available small-molecule inhibitor of c-Met, PF-2341066, exhibits cytoreductive antitumor efficacy through antiproliferative and antiangiogenic mechanisms. Cancer Res 2007;67(9):4408–17.

[17] Curran MP. Crizotinib: in locally advanced or metastatic non-small cell lung cancer. Drugs 2012;72(1):99–107.

[18] Kwak EL, Bang Y-J, Camidge DR, et al. Anaplastic lymphoma kinase inhibition in non-small-cell lung cancer. N Engl J Med 2010;363(18):1693–703.

[19] Camidge DR, Bang Y-J, Kwak EL, et al. Activity and safety of crizotinib in patients with ALK-positive non-small-cell lung cancer: updated results from a phase 1 study. Lancet Oncol 2012;13(10):1011–9.

[20] Blackhall F, Ross Camidge D, Shaw AT, et al. Final results of the large-scale multinational trial PROFILE 1005: efficacy and safety of crizotinib in previously treated patients with advanced/metastatic ALK-positive non-small-cell lung cancer. ESMO Open 2017;2(3), e000219.

[21] Shaw AT, Kim D-W, Nakagawa K, et al. Crizotinib versus chemotherapy in advanced ALK-positive lung cancer. N Engl J Med 2013;368(25):2385–94.

[22] Blackhall F, Kim D-W, Besse B, et al. Patient-reported outcomes and quality of life in PROFILE 1007: a random-ized trial of crizotinib compared with chemotherapy in previously treated patients with ALK-positive advanced non-small-cell lung cancer. J Thorac Oncol 2014;9(11):1625–33.

[23] Shaw AT, Yeap BY, Solomon BJ, et al. Effect of crizotinib on overall survival in patients with advanced non-small-cell lung cancer harbouring ALK gene rearrangement: a retrospective analysis. Lancet Oncol 2011;12(11):1004–12.

[24] Solomon BJ, Cappuzzo F, Felip E, et al. Intracranial efficacy of crizotinib versus chemotherapy in patients with advanced ALK-positive non-small-cell lung cancer: results from PROFILE 1014. J Clin Oncol Off J Am Soc Clin Oncol 2016;34(24):2858–65.

[25] Duruisseaux M, Besse B, Cadranel J, et al. Overall survival with crizotinib and next-generation ALK inhibitors in ALK-positive non-small-cell lung cancer (IFCT-1302 CLINALK): a French nationwide cohort retrospective study. Oncotarget 2017;8(13):21903–17.

[26] Pacheco JM, Gao D, Smith D, et al. Natural history and factors associated with overall survival in stage IV ALK-rearranged non-small cell lung cancer. J Thorac Oncol 2019;14(4):691–700.

[27] Gainor JF, Dardaei L, Yoda S, et al. Molecular mechanisms of resistance to first- and second-generation ALK inhibitors in ALK-rearranged lung cancer. Cancer Discov 2016;6(10):1118–33.

[28] Shaw AT, Solomon BJ, Besse B, et al. ALK resistance mutations and efficacy of lorlatinib in advanced anaplastic lymphoma kinase-positive non-small-cell lung cancer. J Clin Oncol Off J Am Soc Clin Oncol 2019;37(16):1370–9.

[29] Shaw AT, Kim D-W, Mehra R, et al. Ceritinib in ALK-rearranged non-small-cell lung cancer. N Engl J Med 2014;370(13):1189–97.

[30] Gettinger SN, Zhang S, Hodgson JG, et al. Activity of brigatinib (BRG) in crizotinib (CRZ) resistant patients (pts) according to ALK mutation status. J Clin Oncol 2016;34(15 suppl):9060.

[31] Wolf J, Helland A, Oh IJ, Migliorno MR, Dziadziuszko R, de Castro J, et al. Phase 3 ALUR study of alectinib in pretreated ALK+ NSCLC: final efficacy, safety and targeted genomic sequencing analyses. J Thorac Oncol 2019;14:S210. https://doi.org/10.1016/j.jtho.2019.08.416.

[32] Remon J, Tabbò F, Jimenez B, et al. Sequential blinded treatment decisions in ALK-positive non-small cell lung cancers in the era of precision medicine. Clin Transl Oncol 2020;22(9):1425–9. https://doi.org/10.1007/s12094-020-02290-1.

[33] Ou S-HI, Jänne PA, Bartlett CH, et al. Clinical benefit of continuing ALK inhibition with crizotinib beyond initial disease progression in patients with advanced ALK-positive NSCLC. Ann Oncol 2014;25(2):415–22.

[34] Takeda M, Okamoto I, Nakagawa K. Clinical impact of continued crizotinib administration after isolated central nervous system progression in patients with lung cancer positive for ALK rearrangement. J Thorac Oncol 2013; 8(5):654–7.

[35] Weickhardt AJ, Scheier B, Burke JM, et al. Local ablative therapy of oligoprogressive disease prolongs disease control by tyrosine kinase inhibitors in oncogene-addicted non-small-cell lung cancer. J Thorac Oncol 2012; 7(12):1807–14.

[36] Costa DB, Kobayashi S, Pandya SS, et al. CSF concentration of the anaplastic lymphoma kinase inhibitor crizotinib. J Clin Oncol Off J Am Soc Clin Oncol 2011;29(15):e443–5.

[37] Tang SC, Nguyen LN, Sparidans RW, et al. Increased oral availability and brain accumulation of the ALK inhibitor crizotinib by coadministration of the P-glycoprotein (ABCB1) and breast cancer resistance protein (ABCG2) inhibitor elacridar. Int J Cancer 2014;134(6):1484–94.

[38] Johung KL, Yeh N, Desai NB, et al. Extended survival and prognostic factors for patients with ALK-rearranged non-small-cell lung cancer and brain metastasis. J Clin Oncol Off J Am Soc Clin Oncol 2016;34(2):123–9.

[39] Peters S, Bexelius C, Munk V, Leighl N. The impact of brain metastasis on quality of life, resource utilization and survival in patients with non-small-cell lung cancer. Cancer Treat Rev 2016;45:139–62.

[40] Marsilje TH, Pei W, Chen B, et al. Synthesis, structure-activity relationships, and in vivo efficacy of the novel potent and selective anaplastic lymphoma kinase (ALK) inhibitor 5-chloro-N2-(2-isopropoxy-5-methyl-4-(piperidin-4-yl)phenyl)-N4-(2-(isopropylsulfonyl)phenyl)pyrimidine-2,4-diamine (LDK378) currently in phase 1 and phase 2 clinical trials. J Med Chem 2013;56(14):5675–90.

[41] Friboulet L, Li N, Katayama R, et al. The ALK inhibitor ceritinib overcomes crizotinib resistance in non-small cell lung cancer. Cancer Discov 2014;4(6):662–73.

[42] Kim D-W, Mehra R, Tan DSW, et al. Activity and safety of ceritinib in patients with ALK-rearranged non-small-cell lung cancer (ASCEND-1): updated results from the multicentre, open-label, phase 1 trial. Lancet Oncol 2016;17(4):452–63.

[43] Crinò L, Ahn M-J, De Marinis F, et al. Multicenter phase II study of whole-body and intracranial activity with ceritinib in patients with ALK-rearranged non-small-cell lung cancer previously treated with chemotherapy and crizotinib: results from ASCEND-2. J Clin Oncol Off J Am Soc Clin Oncol 2016;34(24):2866–73.

[44] Nishio M, Felip E, Orlov S, et al. Final overall survival and other efficacy and safety results from ASCEND-3: phase II study of ceritinib in ALKi-naive patients with ALK-rearranged NSCLC. J Thorac Oncol 2020;15 (4):609–17.

[45] Shaw AT, Kim TM, Crinò L, et al. Ceritinib versus chemotherapy in patients with ALK-rearranged non-small-cell lung cancer previously given chemotherapy and crizotinib (ASCEND-5): a randomised, controlled, open-label, phase 3 trial. Lancet Oncol 2017;18(7):874–86.

[46] Soria J-C, Tan DSW, Chiari R, et al. First-line ceritinib versus platinum-based chemotherapy in advanced ALK-rearranged non-small-cell lung cancer (ASCEND-4): a randomised, open-label, phase 3 study. Lancet Lond Engl 2017;389(10072):917–29.

[47] Cho BC, Obermannova R, Bearz A, et al. Efficacy and safety of Ceritinib (450 mg/d or 600 mg/d) with food versus 750-mg/d fasted in patients with ALK receptor tyrosine kinase (ALK)-positive NSCLC: primary efficacy results from the ASCEND-8 study. J Thorac Oncol 2019;14(7):1255–65.

[48] Barlesi F, Kim D-W, Bertino EM, et al. Efficacy and safety of ceritinib in ALK-positive non-small cell lung cancer (NSCLC) patients with leptomeningeal metastases (LM): results from the phase II, ASCEND-7 study. Ann Oncol 2019;30:v143–4.

[49] Sakamoto H, Tsukaguchi T, Hiroshima S, et al. CH5424802, a selective ALK inhibitor capable of blocking the resistant gatekeeper mutant. Cancer Cell 2011;19(5):679–90.

[50] Kodama T, Tsukaguchi T, Satoh Y, et al. Alectinib shows potent antitumor activity against RET-rearranged non-small cell lung cancer. Mol Cancer Ther 2014;13(12):2910–8.

[51] Kodama T, Hasegawa M, Takanashi K, et al. Antitumor activity of the selective ALK inhibitor alectinib in models of intracranial metastases. Cancer Chemother Pharmacol 2014;74(5):1023–8.

[52] Gainor JF, Sherman CA, Willoughby K, et al. Alectinib salvages CNS relapses in ALK-positive lung cancer patients previously treated with crizotinib and ceritinib. J Thorac Oncol 2015;10(2):232–6.

[53] Kodama T, Tsukaguchi T, Yoshida M, et al. Selective ALK inhibitor alectinib with potent antitumor activity in models of crizotinib resistance. Cancer Lett 2014;351(2):215–21.

[54] Gadgeel SM, Gandhi L, Riely GJ, et al. Safety and activity of alectinib against systemic disease and brain metastases in patients with crizotinib-resistant ALK-rearranged non-small-cell lung cancer (AF-002JG): results from the dose-finding portion of a phase 1/2 study. Lancet Oncol 2014;15(10):1119–28.

[55] Seto T, Kiura K, Nishio M, et al. CH5424802 (RO5424802) for patients with ALK-rearranged advanced non-small-cell lung cancer (AF-001JP study): a single-arm, open-label, phase 1-2 study. Lancet Oncol 2013;14(7):590–8.

[56] Tamura T, Kiura K, Seto T, et al. Three-year follow-up of an alectinib phase I/II study in ALK-positive non-small-cell lung cancer: AF-001JP. J Clin Oncol Off J Am Soc Clin Oncol 2017;35(14):1515–21.

[57] Ou S-HI, Ahn JS, De Petris L, et al. Alectinib in Crizotinib-refractory ALK-rearranged non-small-cell lung cancer: a phase II global study. J Clin Oncol Off J Am Soc Clin Oncol 2016;34(7):661–8.

[58] Shaw AT, Gandhi L, Gadgeel S, et al. Alectinib in ALK-positive, crizotinib-resistant, non-small-cell lung cancer: a single-group, multicentre, phase 2 trial. Lancet Oncol 2016;17(2):234–42.

[59] Yang JC-H, Ou S-HI, De Petris L, et al. Pooled systemic efficacy and safety data from the pivotal phase II studies (NP28673 and NP28761) of alectinib in ALK-positive non-small cell lung cancer. J Thorac Oncol 2017;12(10):1552–60.

[60] Ou S-HI, Gadgeel SM, Barlesi F, et al. Pooled overall survival and safety data from the pivotal phase II studies (NP28673 and NP28761) of alectinib in ALK-positive non-small-cell lung cancer. Lung Cancer Amst Neth 2020;139:22–7.

[61] Gadgeel SM, Shaw AT, Govindan R, et al. Pooled analysis of CNS response to alectinib in two studies of pretreated patients with ALK-positive non-small-cell lung cancer. J Clin Oncol Off J Am Soc Clin Oncol 2016;34(34):4079–85.

[62] Novello S, Mazières J, Oh I-J, et al. Alectinib versus chemotherapy in crizotinib-pretreated anaplastic lymphoma kinase (ALK)-positive non-small-cell lung cancer: results from the phase III ALUR study. Ann Oncol Off J Eur Soc Med Oncol 2018;29(6):1409–16.

[63] Camidge DR, Dziadziuszko R, Peters S, et al. Updated efficacy and safety data and impact of the EML4-ALK fusion variant on the efficacy of Alectinib in untreated ALK-positive advanced non-small cell lung cancer in the global phase III ALEX study. J Thorac Oncol 2019;14(7):1233–43.

[64] Pérol M, Pavlakis N, Levchenko E, et al. Patient-reported outcomes from the randomized phase III ALEX study of alectinib versus crizotinib in patients with ALK-positive non-small-cell lung cancer. Lung Cancer Amst Neth 2019;138:79–87.

[65] Gadgeel S, Peters S, Mok T, et al. Alectinib versus crizotinib in treatment-naive anaplastic lymphoma kinase-positive (ALK+) non-small-cell lung cancer: CNS efficacy results from the ALEX study. Ann Oncol Off J Eur Soc Med Oncol 2018;29(11):2214–22.

[66] Nishio M, Nakagawa K, Mitsudomi T, et al. Analysis of central nervous system efficacy in the J-ALEX study of alectinib versus crizotinib in ALK-positive non-small-cell lung cancer. Lung Cancer Amst Neth 2018;121:37–40.

[67] Zhang S, Anjum R, Squillace R, et al. The potent ALK inhibitor brigatinib (AP26113) overcomes mechanisms of resistance to first- and second-generation ALK inhibitors in preclinical models. Clin Cancer Res 2016;22(22):5527–38.

[68] Zhang S, Anjum R, Squillace R, et al. The potent ALK inhibitor brigatinib (AP26113) overcomes mechanisms of resistance to first- and second-generation ALK inhibitors in preclinical models. Clin Cancer Res 2016;22(22):5527–38.

[69] Gainor JF, Dardaei L, Yoda S, et al. Molecular mechanisms of resistance to first- and second-generation ALK inhibitors in ALK-rearranged lung cancer. Cancer Discov 2016;6(10):1118–33.

[70] Gettinger SN, Bazhenova LA, Langer CJ, et al. Activity and safety of brigatinib in ALK-rearranged non-small-cell lung cancer and other malignancies: a single-arm, open-label, phase 1/2 trial. Lancet Oncol 2016;17(12):1683–96.

[71] Bazhenova L, Gettinger SN, Langer CJ, et al. Brigatinib (BRG) in patients (pts) with ALK+ non-small cell lung cancer (NSCLC): updates from a phase 1/2 trial. J Clin Oncol 2017;35(15 suppl):e20682.

[72] Huber RM, Hansen KH, Paz-Ares Rodríguez L, et al. Brigatinib in crizotinib-refractory ALK+ NSCLC: 2-year follow-up on systemic and intracranial outcomes in the phase 2 ALTA trial. J Thorac Oncol 2020;15(3):404–15.

[73] Campelo RG, Lin HM, Perol M, et al. Health-related quality of life (HRQoL) results from ALTA-1L: phase 3 study of brigatinib vs crizotinib as first-line (1L) ALK therapy in advanced ALK+ non-small cell lung cancer (NSCLC). J Clin Oncol 2019;37(15 suppl):9084.

[74] Lovly CM, Heuckmann JM, de Stanchina E, et al. Insights into ALK-driven cancers revealed through development of novel ALK tyrosine kinase inhibitors. Cancer Res 2011;71(14):4920–31.

[75] Horn L, Infante JR, Reckamp KL, et al. Ensartinib (X-396) in ALK-positive non-small cell lung cancer: results from a first-in-human phase I/II, multicenter study. Clin Cancer Res 2018;24(12):2771–9.

[76] Yang Y, Zhou J, Zhou J, et al. Efficacy, safety, and biomarker analysis of ensartinib in crizotinib-resistant, ALK-positive non-small-cell lung cancer: a multicentre, phase 2 trial. Lancet Respir Med 2020;8(1):45–53.

[77] Horn L, Whisenant JG, Wakelee H, et al. Monitoring therapeutic response and resistance: analysis of circulating tumor DNA in patients with ALK+ lung cancer. J Thorac Oncol 2019;14(11):1901–11.

[78] Johnson TW, Richardson PF, Bailey S, et al. Discovery of (10R)-7-amino-12-fluoro-2,10,16-trimethyl-15-oxo-10,15,16,17-tetrahydro-2H-8,4-(metheno)pyrazolo[4,3-h][2,5,11]-benzoxadiazacyclotetradecine-3-carbonitrile (PF-06463922), a macrocyclic inhibitor of anaplastic lymphoma kinase (ALK) and c-ROS oncogene 1 (ROS1) with preclinical brain exposure and broad-spectrum potency against ALK-resistant mutations. J Med Chem 2014; 57(11):4720–44.

[79] Zou HY, Friboulet L, Kodack DP, et al. PF-06463922, an ALK/ROS1 inhibitor, overcomes resistance to first and second generation ALK inhibitors in preclinical models. Cancer Cell 2015;28(1):70–81.

[80] Shaw AT, Felip E, Bauer TM, et al. Lorlatinib in non-small-cell lung cancer with ALK or ROS1 rearrangement: an international, multicentre, open-label, single-arm first-in-man phase 1 trial. Lancet Oncol 2017;18(12):1590–9.

[81] Bauer TM, Felip E, Solomon BJ, et al. Clinical management of adverse events associated with lorlatinib. Oncologist 2019;24(8):1103–10.

[82] Solomon BJ, Besse B, Bauer TM, et al. Lorlatinib in patients with ALK-positive non-small-cell lung cancer: results from a global phase 2 study. Lancet Oncol 2018;19(12):1654–67.

[83] Bauer TM, Shaw AT, Johnson ML, et al. Brain penetration of lorlatinib: cumulative incidences of CNS and non-CNS progression with lorlatinib in patients with previously treated ALK-positive non-small-cell lung cancer. Target Oncol 2020;15(1):55–65.

[84] Li T, LoRusso P, Maitland ML, et al. First-in-human, open-label dose-escalation and dose-expansion study of the safety, pharmacokinetics, and antitumor effects of an oral ALK inhibitor ASP3026 in patients with advanced solid tumors. J Hematol Oncol 2016;9:23.

[85] Drilon A, Ou S-HI, Cho BC, et al. Repotrectinib (TPX-0005) is a next-generation ROS1/TRK/ALK inhibitor that potently inhibits ROS1/TRK/ALK solvent- front mutations. Cancer Discov 2018;8(10):1227–36.

[86] Shi Y-K, Fang J, Zhang S, et al. Safety and efficacy of WX-0593 in ALK-positive or ROS1-positive non-small cell lung cancer. Ann Oncol 2019;30:v607–8.

[87] Lin C-C, Arkenau H-T, Lu S, et al. A phase 1, open-label, dose-escalation trial of oral TSR-011 in patients with advanced solid tumours and lymphomas. Br J Cancer 2019;121(2):131–8.

[88] Ou S-HI, Zhu VW, Nagasaka M. Catalog of 5′ fusion partners in ALK-positive NSCLC circa 2020. JTO Clin Res Rep 2020;1. https://doi.org/10.1016/j.jtocrr.2020.100015.

[89] Kron A, Alidousty C, Scheffler M, et al. Impact of TP53 mutation status on systemic treatment outcome in ALK-rearranged non-small-cell lung cancer. Ann Oncol Off J Eur Soc Med Oncol 2018;29(10):2068–75.

[90] Skoulidis F, Heymach JV. Co-occurring genomic alterations in non-small-cell lung cancer biology and therapy. Nat Rev Cancer 2019;19(9):495–509.

[91] Yoshida T, Oya Y, Tanaka K, et al. Differential crizotinib response duration among ALK fusion variants in ALK-positive non–small-cell lung cancer. J Clin Oncol 2016;34(28):3383–9.

[92] Li Y, Zhang T, Zhang J, et al. Response to crizotinib in advanced ALK-rearranged non-small cell lung cancers with different ALK-fusion variants. Lung Cancer Amst Neth 2018;118:128–33.

[93] Lin JJ, Zhu VW, Yoda S, et al. Impact of EML4-ALK variant on resistance mechanisms and clinical outcomes in ALK-positive lung cancer. J Clin Oncol Off J Am Soc Clin Oncol 2018;36(12):1199–206.

[94] Camidge DR, Pao W, Sequist LV. Acquired resistance to TKIs in solid tumours: learning from lung cancer. Nat Rev Clin Oncol 2014;11(8):473–81.

[95] Groenland SL, Geel DR, Janssen JM, et al. Exposure-response analyses of ALK-inhibitors crizotinib and alectinib in NSCLC patients. Ann Oncol 2019;30:v608.

[96] Chouaid C, Dujon C, Do P, et al. Feasibility and clinical impact of re-biopsy in advanced non small-cell lung cancer: a prospective multicenter study in a real-world setting (GFPC study 12-01). Lung Cancer Amst Neth 2014;86(2):170–3.

[97] Mezquita L, Swalduz A, Jovelet C, et al. Clinical relevance of an amplicon-based liquid biopsy for detecting ALK and ROS1 fusion and resistance mutations in patients with non–small-cell lung cancer. JCO Precis Oncol 2020;4:272–82.

[98] Gadgeel SM, Mok TSK, Peters S, et al. Phase II/III blood first assay screening trial (BFAST) in patients (pts) with treatment-naïve NSCLC: initial results from the ALK+ cohort. Ann Oncol 2019;30:v918.

[99] Remon J, Besse B. Brain metastases in oncogene-addicted non-small cell lung cancer patients: incidence and treatment. Front Oncol 2018;8:88.

Systemic treatments other than TKI: Reflections on chemotherapy, immunotherapy and antiangiogenic agents in *ALK*-driven NSCLC

Alessandro Leonetti[a,b] *and Marcello Tiseo*[a,b]

[a]Medical Oncology Unit, University Hospital of Parma, Parma, Italy [b]Department of Medicine and Surgery, University of Parma, Parma, Italy

Abstract

Anaplastic lymphoma kinase (ALK) tyrosine kinase inhibitors (TKIs) resistance constitutes a relevant issue in *ALK*-positive non-small cell lung cancer (NSCLC). Beyond ALK-TKIs, other treatment strategies are envisaged during the course of disease, especially once ALK-TKI options are exhausted. Chemotherapy represents a valuable therapeutic weapon in this subgroup of patients, who are particularly sensitive to pemetrexed-based regimens. In addition, immune checkpoint inhibitors and antiangiogenics are being intensively investigated for the treatment of *ALK*-positive NSCLC. From the perspective of combination approaches, atezolizumab plus bevacizumab and carboplatin-paclitaxel demonstrated a promising activity and received Food and Drug Administration approval after failure of previous targeted therapy in *ALK*-driven NSCLC. The aim of the present chapter is to review the current knowledge about systemic treatments other than *ALK*-TKIs, ranging from preclinical evidences to clinical implications. Moreover, ongoing research focused on novel therapeutic strategies will be discussed.

Abbreviations

ALK	anaplastic lymphoma kinase
NSCLC	non-small cell lung cancer
PD-1	programmed cell death protein 1
PD-L1	programmed cell death ligand 1
TS	thymidylate synthase
VEGF	vascular endothelial growth factor
VEGFR	vascular endothelial growth factor receptor

Conflict of interest

Marcello Tiseo declares: Advisory boards and speakers' fee for Astra-Zeneca, Pfizer, Eli-Lilly, BMS, Novartis, Roche, MSD, Boehringer Ingelheim, Otsuka, Takeda, Pierre Fabre; Research Grants from: Astra-Zeneca, Boehringer Ingelheim. Alessandro Leonetti declares speakers' fee for Astra-Zeneca.

Introduction

The development of targeted therapies with anaplastic lymphoma kinase (ALK) tyrosine kinase inhibitors (TKIs) dramatically changed the management and prognosis of *ALK*-positive non-small cell lung cancer (NSCLC) patients. ALK-TKIs, ranging from the first-in-class crizotinib to the last FDA-approved lorlatinib, significantly improved progression-free survival (PFS), overall response rate (ORR) and quality of life in this peculiar subgroup of patients. To date, alectinib—which is a second generation ALK-TKI—is the current preferred first-line therapy in advanced *ALK*-positive NSCLC, with a median PFS reaching 34.8 months, based on the updated analysis of the phase III ALEX trial [1]. Despite the remarkable activity of ALK-TKIs, both in terms of extracranial and intracranial responses, acquired ALK-TKIs resistance constitutes a major challenge, and the optimal next-line therapy still remains to be assessed. In this regard, the landscape of resistance mechanisms driven by either *ALK* mutations or activation of alternative signaling pathways has become increasingly complex and patients could inexorably experience the failure of ALK inhibition. As a consequence, different treatment strategies beyond ALK-TKIs are envisaged, especially in the absence of actionable *ALK* resistance mutations.

The aim of this chapter is to review the current evidence about systemic treatments other than ALK-TKIs in advanced *ALK*-positive NSCLC, with a focus on chemotherapy, immune checkpoint inhibitors (ICIs) and antiangiogenic agents, whose rationale is summarized in Fig. 1.

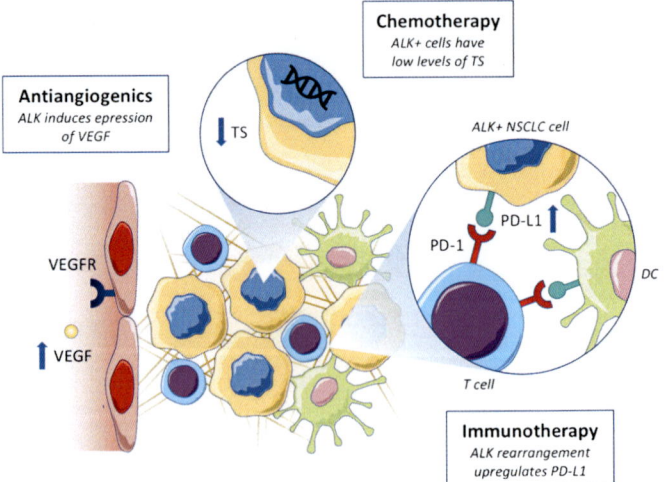

FIG. 1 Chemotherapy, immunotherapy, and antiangiogenic agents in *ALK*-driven NSCLC. *ALK*, anaplastic lymphoma kinase; *NSCLC*, non-small cell lung cancer; *PD-1*, programmed cell death protein 1; *PD-L1*, programmed cell death ligand 1; *TS*, thymidylate synthase; *VEGF*, vascular endothelial growth factor; *VEGFR*, vascular endothelial growth factor receptor.

The role of chemotherapy in **ALK**-positive NSCLC

Most advanced *ALK*-positive NSCLC patients will receive chemotherapy at some point in the course of the disease, mainly after failure of previous ALK-TKIs. Thus, it is important to enlighten the efficacy of chemotherapeutic agents in this molecularly distinct subgroup of patients.

When comparing ALK-TKIs to standard chemotherapy regimens for the treatment of metastatic *ALK*-positive NSCLC, both in first-line and in second-line, chemotherapy demonstrated inferior outcomes [2–7]. In this regard, available evidence comes from randomized phase III trials which are summarized in Table 1. In the second-line studies, chemotherapy consisted of pemetrexed $500\,mg/m^2$ or docetaxel $75\,mg/m^2$ every 3 weeks [2–4]. When ALK-TKIs' efficacy was assessed in first-line, chemotherapy comparator arms were based on pemetrexed $500\,mg/m^2$ with either cisplatin $75\,mg/m^2$ or carboplatin AUC 5–6, IV every 3 weeks for ≤6 cycles, or for 4 cycles followed by maintenance pemetrexed [5–7]. In these studies, chemotherapy granted a median PFS of 1.6–3.0 months and 6.8–8.1 months in second- and first-line, respectively [2–7]. ORR achieved by chemotherapy agents did not exceed 46% front-line, with worse results after failure of a previous treatment [2–7]. Furthermore, from the available data, chemotherapy demonstrated a limited efficacy in reducing existing brain metastases, as well as preventing the onset of new central nervous system (CNS) lesions [2–7]. However, it should be noted that none of the above-mentioned phase III trials identified statistically significant differences in OS between the ALK-TKI arm and chemotherapy arm, probably due to a cross-over in their relative chemotherapy arm and the immaturity of results [3,4,6,7,9–11]. Moreover, it is important to emphasize that

TABLE 1   Efficacy of chemotherapy in *ALK*-positive NSCLC: evidence from phase III randomized trials.

Study	Patients in CT arm	No of BM (%)	Experimental arm	CT arm	mPFS (months)	ORR	IRR	Reference
First line								
PROFILE 1014	171	47 (27)	Crizotinib 250 mg PO BID	Pemetrexed $500\,mg/m^2$ with either cisplatin $75\,mg/m^2$ or carboplatin AUC 5–6, IV q3w for ≤6 cycles	7.0	45%	Intracranial progression/ new lesions: 15%	[5]
Lu et al.	103 (Asian)	32 (31%)	Crizotinib 250 mg PO BID	Pemetrexed $500\,mg/m^2$ with either cisplatin $75\,mg/m^2$ or carboplatin AUC 5–6, IV q3w for ≤6 cycles	6.8	46%	NA	[8]

Continued

TABLE 1   Efficacy of chemotherapy in *ALK*-positive NSCLC: evidence from phase III randomized trials—cont'd

Study	Patients in CT arm	No of BM (%)	Experimental arm	CT arm	mPFS (months)	ORR	IRR	Reference
ASCEND-4	187	62 (33%)	Ceritinib 750 mg PO QD q3w	Pemetrexed 500 mg/m^2 with either cisplatin 75 mg/m^2 or carboplatin AUC 5–6 for 4 cycles followed by maintenance pemetrexed	8.1	20%	21.2%	[7]
Second line								
PROFILE 1007	174	60 (34%)	Crizotinib 250 mg PO BID	Pemetrexed 500 mg/m^2 or docetaxel 75 mg/m^2 q3w	3.0	20%	NA	[2]
ASCEND-5	116	69 (59%)	Ceritinib 750 mg PO QD q3w	Pemetrexed 500 mg/m^2 or docetaxel 75 mg/m^2 q3w	1.6	7%	NA	[3]
ALUR	35	26 (74%)	Alectinib 600 mg PO BID	Pemetrexed 500 mg/m^2 or docetaxel 75 mg/m^2 q3w	1.6	3%	0%	[4]

BID, bis in die; *BM*, brain metastases; *CT*, chemotherapy; *IRR*, intracranial response rate; *mPFS*, median progression-free survival; *NA*, not available; *No*, number; *ORR*, overall response rate; *PO*, per os; *QD*, quoque die.

the treatment outcomes, including ORR and PFS, following platinum-based combination chemotherapies in *ALK*-positive patients are similar to those of the historical controls in unselected NSCLC [12–14].

In a more recent multicenter retrospective analysis, Lin et al. explored the efficacy of platinum/pemetrexed chemotherapy in patients with advanced *ALK*-positive NSCLC refractory to at least one second-generation ALK-TKI. Among 58 patients enrolled, almost one third of patients achieved an objective response and median PFS was 4.3 months [15]. Although ORR was similar to what reported in prior phase III trials of treatment-naive *ALK*-positive patients [5–7], the relatively short median PFS suggested that *ALK*-positive tumors may be less sensitive to chemotherapy once they have become resistant to ALK-TKI [15]. In this clinical setting, lorlatinib may be preferable to chemotherapy based on the demonstrated efficacy after failure of second-generation ALK-TKI, especially on CNS, even if a direct comparison between lorlatinib and chemotherapy is lacking [16,17].

Regarding chemosensitivity of *ALK*-positive NSCLC, previous research showed that *ALK*-positive patients are sensitive to platinum-based chemotherapy in the same manner than *ALK*-negative ones. In a retrospective study by Shaw and collaborators including advanced NSCLC patients, median time to progression (TTP) for patients who received platinum-based chemotherapy was in the same range of 8 to 10 months, regardless of *ALK* status [18]. Similarly, when Lee and colleagues retrospectively investigated the efficacy of chemotherapy in 21 *ALK*-selected NSCLC patients, the PFS was similar with that of a matched cohort of *ALK*-negative patients [19]. In both studies, ORR of chemotherapy ranged from 25% to 35%, overlapping with unselected NSCLC populations [18,19]. Consistent with these findings, Takeda and collaborators documented a similar sensitivity to platinum doublet, both in terms of ORR and PFS, among 200 patients with advanced nonsquamous NSCLC who were grouped according to *ALK* and *EGFR* status [20].

Interestingly, different studies have suggested that *ALK*-positive patients could be particularly responsive to pemetrexed. In an *ALK* enriched population of 89 NSCLC patients, across different lines of therapy, *ALK* positivity predicted a favorable PFS to pemetrexed (monotherapy or platinum-based combination), and this finding was confirmed on a multivariate analysis adjusting for line of therapy, mono- versus platinum and non-platinum combination therapy, age, sex, histology, and smoking status. In this report, the median PFS was 9 months with pemetrexed regimens in 19 *ALK*-positive patients, exceeding that of 37 *ALK*/*EGFR*/*KRAS*-negative patients by 5 months [21]. Consistently, in a retrospective analysis, 15 *ALK*-positive patients who received second-line pemetrexed and beyond showed longer TTP than *EGFR*-mutated or *ALK*-negative patients (9.2 versus 1.4 versus 2.9 months, $P = .001$). The same positive trend was observed for ORR (47% versus 5% versus 16%, $P = .001$) [22]. Further larger cohort studies validated the notion that pemetrexed is an effective treatment strategy in *ALK*-positive NSCLC, both as a monotherapy and in combination with platinum [23–25].

Although it is unclear why *ALK*-positive patients derive a greater benefit from pemetrexed, a biological rational could be taken into account. Indeed, *ALK*-positive NSCLC expresses low levels of thymidylate synthase (TS), whose high expression is a well-known mechanism of resistance to pemetrexed [26,27]. In vitro data showed that *ALK*-positive cells had significantly lower TS mRNA levels with regard to control cells [22]. When assessing TS expression on tissue specimens from 257 NSCLC patients, a low expression level of TS mRNA was observed more frequently in *ALK*-positive specimens than *ALK*-negative controls ($P < .05$) [28]. Concerning this relevant association, the molecular mechanism underlying the link between *ALK* and TS remains to be explained.

The role of immunotherapy in *ALK*-positive NSCLC

The clinical implementation of ICIs targeting the PD-1/PD-L1 axis has certainly revolutionized the current treatment strategies for advanced NSCLC. However, ICIs' efficacy varies greatly according to different immune and molecular profiles of these tumors. In particular, in the case of NSCLC patients harboring actionable driver mutations (i.e., *EGFR*, *ALK*, *ROS1*), the role of ICIs is controversial. Large scale efficacy data in these molecular subgroups are not available, since oncogene-addicted NSCLC patients have been usually excluded from phase III immunotherapy trials. Overall, current evidence suggests that ICIs have rather a weak activity in oncogene-driven NSCLC, and this could be due to the low tumor mutational burden and low immunogenicity which is typical of these tumors [29,30]. As a consequence, to date, immunotherapy single agents should be usually proposed only after exhaustion of more validated treatments such as TKIs and chemotherapy [31].

Focusing on *ALK*-positive NSCLC, few studies with ICIs included these patients, thus precluding definitive conclusions (Table 2). From the molecular point of view, it has been demonstrated that *ALK* rearrangement upregulates PD-L1 expression in vitro. On the other hand, the inhibition of ALK by alectinib or by siRNA-mediated *ALK* knockdown decreased PD-L1 levels [44]. A positive correlation between *ALK* and PD-L1 expression has been also described in clinicopathological studies [44–47], but this observation has not been confirmed in further meta-analyses [48,49].

The large randomized phase III trials CheckMate 057, KEYNOTE 010 and OAK allowed *ALK*-positive patients to be enrolled [32,33,50]. In these studies, single-agent nivolumab

TABLE 2 *ALK*-positive NSCLC patients enrolled in clinical trials with ICIs.

Study	Phase	ALK+ patients (% of the total)	Clinical setting	Trial design	Reference
ICI monotherapy					
CheckMate 057	III	21 (4%)	2nd line	Nivolumab vs docetaxel	[32]
KEYNOTE 010	II/III	8 (<1%)	2nd line	Pembrolizumab vs docetaxel	[33]
OAK	III	2 (<1%)	2nd line	Atezolizumab vs docetaxel	[34]
ATLANTIC	II	15 (Cohort 1, 14%)	≥3rd line	Durvalumab	[35]
ICI combinations					
Antonia et al.	Ib	1 (1%)	≥1st line	Durvalumab + tremelimumab	[36]
CheckMate 370 (Cohort E)	I/II	13	1st line	Nivolumab + crizotinib	[37]
JAVELIN Lung 101	Ib	28	≥2nd line	Avelumab + lorlatinib	[38]
Felip et al.	Ib	36	≥1st line	Nivolumab + ceritinib	[39]
Kim et al.	Ib	21	1st line	Atezolizumab + alectinib	[40]
Chalmers et al.	I	3	1st line	Ipilimumab + crizotinib	[41]
IMpower150	III	35 (<1%)	1st line	Atezolizumab + CP vs bevacizumab + CP vs atezolizumab + bevacizumab + CP	[42]
IMpower130	III	47 (<1%)[a]	1st line	Atezolizumab + CnP vs CnP	[43]

[a] *This number includes patients with* EGFR *mutations.*
CnP, carboplatin + nab-paclitaxel; *CP*, carboplatin + paclitaxel; *CT*, chemotherapy.

(anti-PD-1), pembrolizumab (anti-PD-1) and atezolizumab (anti-PD-L1) were compared to docetaxel in second-line, respectively. However, the number of ALK-positive patients who were included was too low to speculate on the efficacy of ICI monotherapy in this subgroup of patients. Next, ATLANTIC trial assessed the effect of durvalumab (anti-PD-L1) treatment in three cohorts of patients with NSCLC, defined by EGFR/ALK status, who had been treated with at least two previous systemic regimens [35]. In particular, EGFR+/ALK+ subjects (cohort 1) achieved worse outcomes than EGFR-/ALK− patients (cohort 2 and 3), and all objective responses in cohort 1 occurred in EGFR+ patients, where high PD-L1 expression enriched for responses [35].

Moving to combination strategies, ICIs in addition to ALK-TKIs have been recently tested in early phase studies. Crizotinib plus nivolumab resulted in severe hepatic toxicities (38%) leading to treatment discontinuation and interruption of further enrollment [37]. The combination of crizotinib plus ipilimumab (anti-CTLA4), given at the dose of 3 mg/kg for 4 cycles concurrent with the ALK-TKI, induced hypophysitis and another grade 2 pneumonitis in two out of three patients. Against substantial toxicities, the study showed promising PFS and OS on a very small scale [41]. Crizotinib has also been investigated in combination with pembrolizumab, but the trial has been prematurely terminated due to low enrollment [51].

Other second-generation ALK-TKIs are being explored in combination with immunotherapy, showing the same concerns about tolerability. A not negligible incidence of grade 3–4 hepatic toxicities, as well as rash and lipase level increase, was documented when ceritinib was combined with nivolumab in both pretreated and treatment-naïve ALK-positive patients [39]. Again, despite relevant adverse events (AEs), the latter combination resulted in promising activity, particularly in patients with high PD-L1 expression [39]. Beyond ceritinib, Kim et al. recently published the data from a phase Ib trial of alectinib plus atezolizumab, in which alectinib was given for 7 days before combining with atezolizumab [40]. 21 previously-untreated ALK-positive patients, including those with asymptomatic brain metastases, were enrolled in this study. Taking into account the small number of patients, ORR achieved by alectinib plus atezolizumab was 81% and the median PFS and duration of response were 21.7 months (95% CI: 10.3–21.7) and 20.3 months (95% CI: 11.5–20.3), respectively. The incidence of grade 3 AEs with the combination therapy was 52.4%, with no new safety findings and grade 4–5 AEs reported [40].

Lorlatinib, a third-generation ALK-TKI, has been explored in combination with avelumab (anti-PD-L1) in the phase Ib JAVELIN Lung 101 study [38]. Among 28 heavily pretreated subjects with ALK-positive NSCLC, ORR was 46.7% [38], which was consistent with a prior study of lorlatinib alone [52]. In accordance with other ALK-TKIs-ICI combination studies, grade ≥3 AEs occurred in 53.6% of patients and they concerned mainly hepatic function and triglycerides levels [38].

Considering the remarkable results of chemo-immunotherapy combinations for the first-line treatment of EGFR/ALK-negative NSCLC, this strategy has also been extended in ALK-positive population in two large-scale randomized phase III trials [42,43].

IMpower130 study assessed the efficacy of the addition of atezolizumab to front-line carboplatin plus nab-paclitaxel (CnP) in advanced non-squamous NSCLC, including patients with EGFR/ALK genomic alterations after previous TKI ($n = 47$) [43]. Even though the study met its co-primary endpoints of PFS and OS in the intention-to-treat population, subgroup analyses showed no benefit of this strategy in EGFR/ALK-positive population over standard chemotherapy (atezolizumab + CnP mPFS: 7.0 versus 6.0 months; Hazard Ratio [HR], 0.75, 95% CI: 0.36–1.54; atezolizumab + CnP mOS: 14.4 versus 10.0 months; HR, 0.98, 95% CI: 0.41–2.31) [43,53].

In the IMpower150 study, treatment-naïve patients were randomized to receive atezolizumab plus carboplatin-paclitaxel (ACP), bevacizumab plus carboplatin-paclitaxel (BCP), or atezolizumab plus BCP (ABCP), followed by maintenance therapy with atezolizumab, bevacizumab, or both [42]. Oncogene-addicted NSCLC patients (both EGFR-mutated and ALK-positive) were enrolled in this study, after progression to at least one prior TKI, although most patients (87%) did not harbor these genetic variants. To date, only results obtained from comparison between ABCP and BCP have been published. The trial met its co-primary endpoints comparing ABCP versus BCP, which were PFS, both in ITT wild-type (WT) population (without EGFR mutations or ALK translocation) and among patients in the WT population who had high expression of an effector T-cell gene signature, as well as OS in ITT-WT population [42]. Remarkably, with the limitations of an exploratory analysis, mPFS was also longer with ABCP than with BCP among patients with EGFR mutations or ALK translocations (9.7 versus 6.1 months; HR, 0.59, 95% CI: 0.37–0.94). In addition, ABCP regimen prolonged mOS compared to BCP in EGFR/ALK-positive patients (Not Estimable versus 17.5 months; HR 0.54, 95% CI: 0.29–1.03) [54]. ACP arm was also superior to BCP arm in terms of mOS in this peculiar population (21.1 versus 17.5 months; HR 0.82, 95% CI 0.49–1.37), with a less extent than ABCP [54]. A trend towards a superior mOS in favor of ABCP versus BCP was confirmed in a specifically-addressed subgroup analysis of EGFR-mutated patients, but data regarding exclusively ALK-positive patients are not available yet [55]. As a result, for the first time, this trial led to the approval of ICI-CT combination strategy (ABCP) by Food and Drug Administration (FDA) for EGFR-mutated or ALK-positive patients, after failure to previous targeted therapies. Although IMpower150 included a relatively small number of patients, ABCP could be a promising therapeutic opportunity in pretreated ALK-positive patients, particularly at the point that ALK-TKI options have been exhausted. Considering that ALK-positive NSCLC is particularly sensitive to pemetrexed, other combination approaches including pemetrexed-based chemotherapy surely deserve further investigation. A study evaluating platinum-pemetrexed-atezolizumab triplet (+/− bevacizumab) for TKI-pretreated patients with oncogene-addicted NSCLC, including those with ALK rearrangement, is recruiting participants [56]. In the same clinical setting, the addition of pembrolizumab to platinum-pemetrexed doublet is currently under investigation in a phase II trial [57].

The role of antiangiogenics in ALK-positive NSCLC

Antiangiogenic drugs constitute an essential weapon in the landscape of NSCLC treatment. Since angiogenesis plays a fundamental role in NSCLC development and metastatic spread, targeted treatment against angiogenesis can lead to regression of neovessels resulting in tumor and metastases growth inhibition. Moreover, normalizing tumor vasculature with antiangiogenic therapy could result in increased tumor drug delivery [58]. These effects can be obtained by acting on the vascular endothelial growth factor (VEGF) pathway by inhibiting either its ligand or its receptor. Among antiangiogenics, the VEGF-A recombinant antibody bevacizumab is the only agent approved for the first-line treatment of nonsquamous NSCLC in combination with platinum-based chemotherapy. Ramucirumab, an antibody against VEGF receptor (VEGFR), and nintedanib, an antiangiogenic multi-TKI, are currently approved for the second-line treatment in combination with docetaxel [59].

To date, clinical trials of antiangiogenic agents specifically addressed to *ALK*-positive NSCLC are lacking. However, a biological rationale supporting a putative efficacy of these drugs in the presence of *ALK* alterations could be considered. In a study focused on *ALK*-positive anaplastic large-cell lymphomas (ALCL), it has been demonstrated that expression of ALK induced expression of VEGF through downregulation of miR-16 [60]. In addition, VEGF blood levels were higher in *ALK*-positive ALCL patients than healthy donors [60]. Still in ALCL, treatment with bevacizumab strongly impaired *ALK*-positive ALCL growth in mouse xenografts [61]. Regarding *ALK*-positive NSCLC, a crosstalk between ALK and VEGFR has been demonstrated in a preclinical model [62]. Taken together, these data support the involvement of angiogenesis in *ALK*-translocated malignancies, including NSCLC, which may enhance the efficacy of antiangiogenic agents.

The largest evidence of bevacizumab-chemotherapy combination in *ALK*-selected patients comes from the randomized phase III IMpower150 trial, which has been discussed above [42]. Beyond IMpower150, several clinical reports have highlighted the potential benefit of bevacizumab in NSCLC that harbor *ALK* rearrangements. Habib and colleagues firstly reported a prolonged response (>18 months) to bevacizumab in addition to weekly paclitaxel in an *ALK*-positive NSCLC patient who did not obtain objective responses to either of four previous regimens [63]. In another report, an exceptional complete response was documented in a 52 years-old *ALK*-positive female patient following the combination of bevacizumab and cisplatin-pemetrexed, including disappearance of brain lesions [64]. Liu et al. also described a case of *ALK*-positive NSCLC who achieved a clinical benefit of 7 months of PFS from the combined treatment of bevacizumab plus pemetrexed, after failure of two previous ALK-TKIs [65]. Next, bevacizumab plus pemetrexed granted long-term disease control in NSCLC patients with *ALK* rearrangements compared to wild-type population in a retrospective analysis [66].

Interestingly, previous research found that bevacizumab was also effective against perilesional edema typically found in brain radiation necrosis subsequent to stereotactic brain irradiation. This effect could be obtained through the inhibition of VEGF, which is typically produced by the reactive astrocytes in response to radiation damage [67]. In this sense, bevacizumab has been proven to reduce brain radiation necrosis in four *ALK*-positive NSCLC patients with irradiated brain metastases, in concomitance with ALK-TKI, and the combination was safe and well tolerated [68].

More recently, bevacizumab plus platinum-pemetrexed combination has been demonstrated to restore sensitivity to alectinib in a case report. Authors suggested that bevacizumab enhanced delivery of alectinib to the tumor. Indeed, they found that cell lines derived from the primary-resistant patient after failure of alectinib were still sensitive to the ALK-TKI ex vivo. Thus, they hypothesized that the putative mechanism of resistance to alectinib in this peculiar case was driven by the tumor microenvironment, which was modulated by subsequent bevacizumab [69]. The bigeminal inhibition of ALK and angiogenesis with alectinib plus apatinib (VEGFR-2 inhibitor) was able to reverse alectinib resistance in cell lines [62]. Promising results were also obtained in vivo with ramucirumab combined with alectinib, both in terms of reduced neoangiogenesis and enhanced antitumor activity [70].

Given this background, the combination of ALK-TKIs and antiangiogenics could be an appealing strategy in order to both enhance ALK-TKI efficacy and prevent resistance. Crizotinib plus bevacizumab is the only combination which has been prospectively investigated in *ALK*-positive NSCLC patients to date. Among 16 treatment-naïve subjects, this approach granted a mPFS of 13.0 months (95% CI: 9.8–6.2), with hepatic injury being the most frequent grade ≥ 3 AE reported (12.5%) [71]. Looking at the next future, novel combination approaches with alectinib plus bevacizumab and brigatinib plus bevacizumab in *ALK*-selected NSCLC populations are currently under investigation [72,73].

Conclusion

ALK-TKIs have definitely laid the foundation of the current treatment algorithm of advanced *ALK*-positive NSCLC. The rapidly-evolving landscape of resistance mechanisms and development of next-generation ALK-TKIs will ensure a customized manner to treat *ALK*-rearranged NSCLC by keeping the pressure on ALK. Nevertheless, since most of the patients will experience the onset of resistance, novel treatment strategies should be considered, especially once ALK-TKIs options are exhausted. In this regard, chemotherapy still remains a valuable opportunity, with a special mention to pemetrexed-based doublet. ICIs and antiangiogenics surely deserve more extensive investigation, due to the limited evidence available to date, in a view of combined approaches. Atezolizumab plus bevacizumab and carboplatin-paclitaxel is the only combination regimen that demonstrated a compelling activity and safety in *ALK*-driven NSCLC on a larger scale. Novel combinations, including those with ALK-TKIs, are being currently tested and might enrich the future weaponry to counteract this peculiar disease.

References

[1] Mok TSK, Shaw AT, Camidge RD, Gadgeel SM, Rosell R, Dziadziuszko R, Kim D-W, Perol M, Ou S-H, Bordogna W, Smoljanović V, Hilton M, Peters S. Final PFS, updated OS and safety data from the randomised, phase III ALEX study of alectinib (ALC) versus crizotinib (CRZ) in untreated advanced ALK+ NSCLC. Ann Oncol 2019;30:v607. https://doi.org/10.1093/annonc/mdz260.006.

[2] Shaw AT, Kim DW, Nakagawa K, Seto T, Crino L, Ahn MJ, De Pas T, Besse B, Solomon BJ, Blackhall F, Wu YL, Thomas M, O'Byrne KJ, Moro-Sibilot D, Camidge DR, Mok T, Hirsh V, Riely GJ, Iyer S, Tassell V, Polli A, Wilner KD, Janne PA. Crizotinib versus chemotherapy in advanced ALK-positive lung cancer. N Engl J Med 2013;368:2385–94. https://doi.org/10.1056/NEJMoa1214886.

[3] Shaw AT, Kim TM, Crinò L, Gridelli C, Kiura K, Liu G, Novello S, Bearz A, Gautschi O, Mok T, Nishio M, Scagliotti G, Spigel DR, Deudon S, Zheng C, Pantano S, Urban P, Massacesi C, Viraswami-Appanna K, Felip E. Ceritinib versus chemotherapy in patients with ALK-rearranged non-small-cell lung cancer previously given chemotherapy and crizotinib (ASCEND-5): a randomised, controlled, open-label, phase 3 trial. Lancet Oncol 2017;18:874–86. https://doi.org/10.1016/S1470-2045(17)30339-X.

[4] Novello S, Mazières J, Oh IJ, de Castro J, Migliorino MR, Helland A, Dziadziuszko R, Griesinger F, Kotb A, Zeaiter A, Cardona A, Balas B, Johannsdottir HK, Das-Gupta A, Wolf J. Alectinib versus chemotherapy in crizotinib-pretreated anaplastic lymphoma kinase (ALK)-positive non-small-cell lung cancer: results from the phase III ALUR study. Ann Oncol 2018;29:1409–16. https://doi.org/10.1093/annonc/mdy121.

[5] Solomon BJ, Mok T, Kim DW, Wu YL, Nakagawa K, Mekhail T, Felip E, Cappuzzo F, Paolini J, Usari T, Iyer S, Reisman A, Wilner KD, Tursi J, Blackhall F. First-line crizotinib versus chemotherapy in ALK-positive lung cancer. N Engl J Med 2014;371:2167–77. https://doi.org/10.1056/NEJMoa1408440.

[6] Lu S, Mok T, Lu Y, Zhou J, Shi Y, Sriuranpong V, Ho JCM, Ong CK, Tsai C-M, Chung C-H, Wilner KD, Tang Y, Masters E, Selaru P, Wu Y-L. Phase 3 study of first-line crizotinib versus pemetrexed – cisplatin/carboplatin (PCC) in East Asian patients (pts) with ALK + advanced non-squamous non-small cell lung cancer (NSCLC). J Clin Oncol 2016;34:9058. https://doi.org/10.1200/jco.2016.34.15_suppl.9058.

[7] Soria JC, Tan DSW, Chiari R, Wu YL, Paz-Ares L, Wolf J, Geater SL, Orlov S, Cortinovis D, Yu CJ, Hochmair M, Cortot AB, Tsai CM, Moro-Sibilot D, Campelo RG, McCulloch T, Sen P, Dugan M, Pantano S, Branle F, Massacesi C, de Castro G. First-line ceritinib versus platinum-based chemotherapy in advanced ALK-rearranged non-small-cell lung cancer (ASCEND-4): a randomised, open-label, phase 3 study. Lancet 2017;389:917–29. https://doi.org/10.1016/S0140-6736(17)30123-X.

[8] Lu S, Mok T, Lu Y, Zhou J, Shi Y, Sriuranpong V, Ho JCM, Ong CK, Tsai C-M, Chung C-H, Wilner KD, Tang Y, Masters E, Selaru P, Wu Y-L. Phase 3 study of first-line crizotinib vs pemetrexed – cisplatin/carboplatin (PCC) in east Asian patients (pts) with ALK + advanced non-squamous non-small cell lung cancer (NSCLC). J Clin Oncol 2016;34:9058. https://doi.org/10.1200/jco.2016.34.15_suppl.9058.

[9] Shaw AT, Janne PA, Besse B, Solomon BJ, Blackhall FH, Camidge DR, Mok T, Hirsh V, Scranton JR, Polli A, Tang Y, Wilner KD, Kim D-W. Crizotinib versus chemotherapy in ALK + advanced non-small cell lung cancer (NSCLC): final survival results from PROFILE 1007. J Clin Oncol 2016;34:9066. https://doi.org/10.1200/jco.2016.34.15_suppl.9066.

[10] Solomon BJ, Kim D-W, Wu Y-L, Nakagawa K, Mekhail T, Felip E, Cappuzzo F, Paolini J, Usari T, Tang Y, Wilner KD, Blackhall F, Mok TS. Final overall survival analysis from a study comparing first-line crizotinib versus chemotherapy in ALK-mutation-positive non–small-cell lung cancer. J Clin Oncol 2018;36:2251–8. https://doi.org/10.1200/JCO.2017.77.4794.

[11] Lee YC, Hsieh CC, Lee YL, Li CY. Which should be used first for alk-positive non-small-cell lung cancer: chemotherapy or targeted therapy? A meta-analysis of five randomized trials. Medicina (Kaunas) 2019;55:29. https://doi.org/10.3390/medicina55020029.

[12] Schiller JH, Harrington D, Belani CP, Langer C, Sandler A, Krook J, Zhu J, Johnson DH. Comparison of four chemotherapy regimens for advanced non–small-cell lung cancer. N Engl J Med 2002;346:92–8. https://doi.org/10.1056/NEJMoa011954.

[13] Scagliotti GV, De Marinis F, Rinaldi M, Crinò L, Gridelli C, Ricci S, Matano E, Boni C, Marangolo M, Failla G, Altavilla G, Adamo V, Ceribelli A, Clerici M, Di Costanzo F, Frontini L, Tonato M. Phase III randomized trial comparing three platinum-based doublets in advanced non-small-cell lung cancer. J Clin Oncol 2002;20:4285–91. https://doi.org/10.1200/JCO.2002.02.068.

[14] Ardizzoni A, Boni L, Tiseo M, Fossella FV, Schiller JH, Paesmans M, Radosavljevic D, Paccagnella A, Zatloukal P, Mazzanti P, Bisset D, Rosell R. Cisplatin- versus carboplatin-based chemotherapy in first-line treatment of advanced non-small-cell lung cancer: an individual patient data meta-analysis. J Natl Cancer Inst 2007. https://doi.org/10.1093/jnci/djk196.

[15] Lin JJ, Schoenfeld AJ, Zhu VW, Yeap BY, Chin E, Rooney M, Plodkowski AJ, Digumarthy SR, Dagogo-Jack I, Gainor JF, Ou SHI, Riely GJ, Shaw AT. Efficacy of platinum/pemetrexed combination chemotherapy in ALK-positive NSCLC refractory to second-generation ALK inhibitors. J Thorac Oncol 2020;15:258–65. https://doi.org/10.1016/j.jtho.2019.10.014.

[16] Solomon BJ, Besse B, Bauer TM, Felip E, Soo RA, Camidge DR, Chiari R, Bearz A, Lin CC, Gadgeel SM, Riely GJ, Tan EH, Seto T, James LP, Clancy JS, Abbattista A, Martini JF, Chen J, Peltz G, Thurm H, Ignatius Ou SH, Shaw AT. Lorlatinib in patients with ALK-positive non-small-cell lung cancer: results from a global phase 2 study. Lancet Oncol 2018;19:1654–67. https://doi.org/10.1016/S1470-2045(18)30649-1.

[17] Shaw AT, Solomon BJ, Besse B, Bauer TM, Lin C-C, Soo RA, Riely GJ, Ou S-HI, Clancy JS, Li S, Abbattista A, Thurm H, Satouchi M, Camidge DR, Kao S, Chiari R, Gadgeel SM, Felip E, Martini J-F. ALK resistance mutations and efficacy of lorlatinib in advanced anaplastic lymphoma kinase-positive non-small-cell lung cancer. J Clin Oncol 2019;37:1370–9. https://doi.org/10.1200/JCO.18.02236.

[18] Shaw AT, Yeap BY, Mino-Kenudson M, Digumarthy SR, Costa DB, Heist RS, Solomon B, Stubbs H, Admane S, McDermott U, Settleman J, Kobayashi S, Mark EJ, Rodig SJ, Chirieac LR, Kwak EL, Lynch TJ, Iafrate AJ. Clinical features and outcome of patients with non-small-cell lung cancer who harbor EML4-ALK. J Clin Oncol 2009;27:4247–53. https://doi.org/10.1200/JCO.2009.22.6993.

[19] Lee JK, Park HS, Kim DW, Kulig K, Kim TM, Lee SH, Jeon YK, Chung DH, Heo DS, Kim WH, Bang YJ. Comparative analyses of overall survival in patients with anaplastic lymphoma kinase-positive and matched wild-type advanced nonsmall cell lung cancer. Cancer 2012;118:3579–86. https://doi.org/10.1002/cncr.26668.

[20] Takeda M, Okamoto I, Sakai K, Kawakami H, Nishio K, Nakagawa K. Clinical outcome for EML4-ALK-positive patients with advanced non-small-cell lung cancer treated with first-line platinum-based chemotherapy. Ann Oncol 2012;23(11):2931–6. https://doi.org/10.1093/annonc/mds124.

[21] Camidge DR, Kono SA, Lu X, Okuyama S, Barón AE, Oton AB, Davies AM, Varella-Garcia M, Franklin W, Doebele RC. Anaplastic lymphoma kinase gene rearrangements in non-small cell lung cancer are associated with prolonged progression-free survival on pemetrexed. J Thorac Oncol 2011;6:774–80. https://doi.org/10.1097/JTO.0b013e31820cf053.

[22] Lee JO, Kim TM, Lee SH, Kim DW, Kim S, Jeon YK, Chung DH, Kim WH, Kim YT, Yang SC, Kim YW, Heo DS, Bang YJ. Anaplastic lymphoma kinase translocation: a predictive biomarker of pemetrexed in patients with non-small cell lung cancer. J Thorac Oncol 2011;6:1474–80. https://doi.org/10.1097/JTO.0b013e3182208fc2.

[23] Shaw AT, Varghese AM, Solomon BJ, Costa DB, Novello S, Mino-Kenudson M, Awad MM, Engelman JA, Riely GJ, Monica V, Yeap BY, Scagliotti GV. Pemetrexed-based chemotherapy in patients with advanced, ALK-positive non-small cell lung cancer. Ann Oncol 2013. https://doi.org/10.1093/annonc/mds242.

[24] Lee HY, Ahn HK, Jeong JY, Kwon MJ, Han JH, Sun JM, Ahn JS, Park K, La CY, Ahn MJ. Favorable clinical outcomes of pemetrexed treatment in anaplastic lymphoma kinase positive non-small-cell lung cancer. Lung Cancer 2013;79:40–5. https://doi.org/10.1016/j.lungcan.2012.10.002.

[25] Ma D, Hao X, Wang Y, Xing P, Li J. Clinical effect of pemetrexed as the first-line treatment in Chinese patients with advanced anaplastic lymphoma kinase-positive non-small cell lung cancer. Thorac Cancer 2016;7:452–8. https://doi.org/10.1111/1759-7714.12353.

[26] Assaraf YG. Molecular basis of antifolate resistance. Cancer Metastasis Rev 2007;26:153–81. https://doi.org/10.1007/s10555-007-9049-z.

[27] Sun JM, Han J, Ahn JS, Park K, Ahn MJ. Significance of thymidylate synthase and thyroid transcription factor 1 expression in patients with nonsquamous non-small cell lung cancer treated with pemetrexed-based chemotherapy. J Thorac Oncol 2011;6:1392–9. https://doi.org/10.1097/JTO.0b013e3182208ea8.

[28] Xu CW, Wang G, Wang WL, Bin GW, Han CJ, Gao JS, Zhang LY, Li Y, Wang L, Zhang YP, Tian YW, Qi DD. Association between EML4-ALK fusion gene and thymidylate synthase mRNA expression in non-small cell lung cancer tissues. Exp Ther Med 2015;9:2151–4. https://doi.org/10.3892/etm.2015.2372.

[29] Spigel DR, Schrock AB, Fabrizio D, Frampton GM, Sun J, He J, Gowen K, Johnson ML, Bauer TM, Kalemkerian GP, Raez LE, Ou S-HI, Ross JS, Stephens PJ, Miller VA, Ali SM. Total mutation burden (TMB) in lung cancer (LC) and relationship with response to PD-1/PD-L1 targeted therapies. J Clin Oncol 2016;34:9017. https://doi.org/10.1200/JCO.2016.34.15_suppl.9017.

[30] Gainor JF, Shaw AT, Sequist LV, Fu X, Azzoli CG, Piotrowska Z, Huynh TG, Zhao L, Fulton L, Schultz KR, Howe E, Farago AF, Sullivan RJ, Stone JR, Digumarthy S, Moran T, Hata AN, Yagi Y, Yeap BY, Engelman JA, Mino-Kenudson M. EGFR mutations and ALK rearrangements are associated with low response rates to -PD-1 pathway blockade in non-small cell lung cancer: a retrospective analysis. Clin Cancer Res 2016;22:4585–93. https://doi.org/10.1158/1078-0432.CCR-15-3101.

[31] Mhanna L, Guibert N, Milia J, Mazieres J. When to consider immune checkpoint inhibitors in oncogene-driven non-small cell lung cancer? Curr Treat Options Oncol 2019;20:1–11. https://doi.org/10.1007/s11864-019-0652-3.

[32] Borghaei H, Paz-Ares L, Horn L, Spigel DR, Steins M, Ready NE, Chow LQ, Vokes EE, Felip E, Holgado E, Barlesi F, Kohlhäufl M, Arrieta O, Burgio MA, Fayette J, Lena H, Poddubskaya E, Gerber DE, Gettinger SN, Rudin CM, Rizvi N, Crinò L, Blumenschein GR, Antonia SJ, Dorange C, Harbison CT, Graf Finckenstein F, Brahmer JR. Nivolumab versus docetaxel in advanced nonsquamous non-small-cell lung cancer. N Engl J Med 2015;373:1627–39. https://doi.org/10.1056/NEJMoa1507643.

[33] Herbst RS, Baas P, Kim D-W, Felip E, Pérez-Gracia JL, Han J-Y, Molina J, Kim J-H, Arvis CD, Ahn M-J, Majem M, Fidler MJ, de Castro JG, Garrido M, Lubiniecki GM, Shentu Y, Im E, Dolled-Filhart M, Garon EB. Pembrolizumab versus docetaxel for previously treated, PD-L1-positive, advanced non-small-cell lung cancer (KEYNOTE-010): a randomised controlled trial. Lancet 2016;387:1540–50. https://doi.org/10.1016/S0140-6736(15)01281-7.

[34] Fehrenbacher L, Spira A, Ballinger M, Kowanetz M, Vansteenkiste J, Mazieres J, Park K, Smith D, Artal-Cortes A, Lewanski C, Braiteh F, Waterkamp D, He P, Zou W, Chen DS, Yi J, Sandler A, Rittmeyer A. Atezolizumab versus docetaxel for patients with previously treated non-small-cell lung cancer (POPLAR): a multicentre, open-label, phase 2 randomised controlled trial. Lancet 2016;387:1837–46. https://doi.org/10.1016/S0140-6736(16)00587-0.

[35] Garassino MC, Cho BC, Kim JH, Mazières J, Vansteenkiste J, Lena H, Corral Jaime J, Gray JE, Powderly J, Chouaid C, Bidoli P, Wheatley-Price P, Park K, Soo RA, Huang Y, Wadsworth C, Dennis PA, Rizvi NA, Paz-Ares Rodriguez L, Novello S, Hiret S, Schmid P, Laack E, Califano R, Maemondo M, Kim SW, Chaft J, Vicente Baz D, Berghmans T, Kim DW, Surmont V, Reck M, Han JY, Holgado Martin E, Belda Iniesta C, Oe Y, Chella A,

Chopra A, Robinet G, Soto Parra H, Thomas M, Cheema P, Katakami N, Su WC, Kim YC, Wolf J, Lee JS, Saka H, Milella M, Ramos Garcia I, Sibille A, Yokoi T, Kang EJ, Atagi S, Spaeth-Schwalbe E, Nishio M, Imamura F, Gabrail N, Veillon R, Derijcke S, Maeda T, Zylla D, Kubiak K, Santoro A, Uy MN, Lucien Geater S, Italiano A, Kowalski D, Barlesi F, Chen YM, Spigel D, Chewaskulyong B, Garcia Gomez R, Alvarez Alvarez R, Yang CH, Hsia TC, Denis F, Sakai H, Vincent M, Goto K, Bosch-Barrera J, Weiss G, Canon JL, Scholz C, Aglietta M, Kemmotsu H, Azuma K, Bradbury P, Feld R, Chachoua A, Jassem J, Juergens R, Palmero Sanchez R, Malcolm A, Vrindavanam N, Kubota K, Waller C, Waterhouse D, Coudert B, Mark Z, Satouchi M, Chang GC, Herzmann C, Chaudhry A, Giridharan S, Hesketh P, Ikeda N, Boccia R, Iannotti N, Haigentz M, Reynolds J, Querol J, Nakagawa K, Sugawara S, Tan EH, Hirashima T, Gettinger S, Kato T, Takeda K, Juan Vidal O, Mohn-Staudner A, Panwalkar A, Daniel D, Kobayashi K, GEI L, Schulte C, Sebastian M, Cernovska M, Coupkova H, Havel L, Pauk N, Singh J, Murakami S, Csoszi T, Losonczy G, Price A, Anderson I, Iqbal M, Torri V, Juhasz E, Khanani S, Koubkova L, Levy B, Page R, Bocskei C, Crinò L, Einspahr D, Hagenstad C, Juat N, Overton L, Garrison M, Szalai Z. Durvalumab as third-line or later treatment for advanced non-small-cell lung cancer (ATLANTIC): an open-label, single-arm, phase 2 study. Lancet Oncol 2018;19:521–36. https://doi.org/10.1016/S1470-2045(18)30144-X.

[36] Antonia S, Goldberg SB, Balmanoukian A, Chaft JE, Sanborn RE, Gupta A, Narwal R, Steele K, Gu Y, Karakunnel JJ, Rizvi NA. Safety and antitumour activity of durvalumab plus tremelimumab in non-small cell lung cancer: a multicentre, phase 1b study. Lancet Oncol 2016;17:299–308. https://doi.org/10.1016/S1470-2045(15)00544-6.

[37] Spigel DR, Reynolds C, Waterhouse D, Garon EB, Chandler J, Babu S, Thurmes P, Spira A, Jotte R, Zhu J, Lin WH, Blumenschein G. Phase 1/2 study of the safety and tolerability of nivolumab plus crizotinib for the first-line treatment of anaplastic lymphoma kinase translocation—positive advanced non–small cell lung cancer (Check-Mate 370). J Thorac Oncol 2018;13:682–8. https://doi.org/10.1016/j.jtho.2018.02.022.

[38] Shaw AT, Lee S-H, Ramalingam SS, Bauer TM, Boyer MJ, Carcereny Costa E, Felip E, Han J-Y, Hida T, Hughes BGM, Kim S-W, Nishio M, Seto T, Ezeh P, Chakrabarti D, Wang J, Chang A, Fumagalli L, Solomon BJ. Avelumab (anti–PD-L1) in combination with crizotinib or lorlatinib in patients with previously treated advanced NSCLC: phase 1b results from JAVELIN lung 101. J Clin Oncol 2018;36:9008. https://doi.org/10.1200/jco.2018.36.15_suppl.9008.

[39] Felip E, de Braud FG, Maur M, Loong HH, Shaw AT, Vansteenkiste JF, John T, Liu G, Lolkema MP, Selvaggi G, Giannone V, Cazorla P, Baum J, Balbin OA, Wang L, Lau YY, Scott JW, Tan DSW. Ceritinib plus nivolumab in patients with advanced ALK-rearranged non–small cell lung cancer: results of an open-label, multicenter, phase 1B study. J Thorac Oncol 2020;15:392–403. https://doi.org/10.1016/j.jtho.2019.10.006.

[40] Kim D-W, Gadgeel SM, Gettinger SN, Riely GJ, Oxnard GR, Mekhail T, Schmid P, Dowlati A, Heist RS, Wozniak AJ, Hernandez G, Sarkar I, Mitry E, Foster P, O'Hear C, Spahn J, Ou S-HI. Safety and clinical activity results from a phase Ib study of alectinib plus atezolizumab in ALK + advanced NSCLC (aNSCLC). J Clin Oncol 2018;36:9009. https://doi.org/10.1200/jco.2018.36.15_suppl.9009.

[41] Chalmers AW, Patel S, Boucher K, Cannon L, Esplin M, Luckart J, Graves N, Van Duren T, Akerley W. Phase I trial of targeted EGFR or ALK therapy with ipilimumab in metastatic NSCLC with long-term follow-up. Target Oncol 2019;14:417–21. https://doi.org/10.1007/s11523-019-00658-0.

[42] Socinski MA, Jotte RM, Cappuzzo F, Orlandi F, Stroyakovskiy D, Nogami N, Rodríguez-Abreu D, Moro-Sibilot D, Thomas CA, Barlesi F, Finley G, Kelsch C, Lee A, Coleman S, Deng Y, Shen Y, Kowanetz M, Lopez-Chavez A, Sandler A, Reck M. Atezolizumab for first-line treatment of metastatic nonsquamous NSCLC. N Engl J Med 2018;378:2288–301. https://doi.org/10.1056/NEJMoa1716948.

[43] West H, McCleod M, Hussein M, Morabito A, Rittmeyer A, Conter HJ, Kopp H-G, Daniel D, McCune S, Mekhail T, Zer A, Reinmuth N, Sadiq A, Sandler A, Lin W, Ochi Lohmann T, Archer V, Wang L, Kowanetz M, Cappuzzo F. Atezolizumab in combination with carboplatin plus nab-paclitaxel chemotherapy compared with chemotherapy alone as first-line treatment for metastatic non-squamous non-small-cell lung cancer (IMpower130): a multicentre, randomised, open-label, phase 3 trial. Lancet Oncol 2019;20:924–37. https://doi.org/10.1016/S1470-2045(19)30167-6.

[44] Ota K, Azuma K, Kawahara A, Hattori S, Iwama E, Tanizaki J, Harada T, Matsumoto K, Takayama K, Takamori S, Kage M, Hoshino T, Nakanishi Y, Okamoto I. Induction of PD-L1 expression by the EML4-ALK oncoprotein and down stream signaling pathways in non-small cell lung cancer. Clin Cancer Res 2015;21:4014–21. https://doi.org/10.1158/1078-0432.CCR-15-0016.

[45] Koh J, Go H, Keam B, Kim MY, Nam SJ, Kim TM, Lee SH, Min HS, Kim YT, Kim DW, Jeon YK, Chung DH. Clinicopathologic analysis of programmed cell death-1 and programmed cell death-ligand 1 and 2 expressions in pulmonary adenocarcinoma: comparison with histology and driver oncogenic alteration status. Mod Pathol 2015;28:1154–66. https://doi.org/10.1038/modpathol.2015.63.

[46] Roussel H, De Guillebon E, Biard L, Mandavit M, Gibault L, Fabre E, Antoine M, Hofman P, Beau-Faller M, Blons H, Danel C, Barthes FLP, Gey A, Granier C, Wislez M, Laurent-Puig P, Oudard S, Bruneval P, Badoual C, Cadranel J, Tartour E. Composite biomarkers defined by multiparametric immunofluorescence analysis identify ALK-positive adenocarcinoma as a potential target for immunotherapy. Onco Targets Ther 2017;6. https://doi.org/10.1080/2162402X.2017.1286437.

[47] D'Incecco A, Andreozzi M, Ludovini V, Rossi E, Capodanno A, Landi L, Tibaldi C, Minuti G, Salvini J, Coppi E, Chella A, Fontanini G, Filice ME, Tornillo L, Incensati RM, Sani S, Crinò L, Terracciano L, Cappuzzo F. PD-1 and PD-L1 expression in molecularly selected non-small-cell lung cancer patients. Br J Cancer 2015;112:95–102. https://doi.org/10.1038/bjc.2014.555.

[48] Zhang M, Li G, Wang Y, Zhao S, Haihong P, Zhao H, Wang Y. Expression in lung cancer and its correlation with driver mutations: a meta-analysis. Sci Rep 2017;7:10255. https://doi.org/10.1038/s41598-017-10925-7.

[49] Lan B, Ma C, Zhang C, Chai S, Wang P, Ding L, Wang K. Association between PD-L1 expression and driver gene status in nonsmall- cell lung cancer: a meta-analysis. Oncotarget 2018;9:7684–99. https://doi.org/10.18632/oncotarget.23969.

[50] Rittmeyer A, Barlesi F, Waterkamp D, Park K, Ciardiello F, von Pawel J, Gadgeel SM, Hida T, Kowalski DM, Dols MC, Cortinovis DL, Leach J, Polikoff J, Barrios C, Kabbinavar F, Frontera OA, De Marinis F, Turna H, Lee J-S, Ballinger M, Kowanetz M, He P, Chen DS, Sandler A, Gandara DR, OAK Study Group. Atezolizumab versus docetaxel in patients with previously treated non-small-cell lung cancer (OAK): a phase 3, open-label, multicentre randomised controlled trial. Lancet (London, England) 2017;389:255–65. https://doi.org/10.1016/S0140-6736(16)32517-X.

[51] Anon. Crizotinib plus pembrolizumab in alk-positive advanced non small cell lung cancer patients. ClinicalTrials.gov; 2017. https://clinicaltrials.gov/ct2/show/NCT02511184. [Accessed 11 March 2020].

[52] Shaw AT, Felip E, Bauer TM, Besse B, Navarro A, Postel-Vinay S, Gainor JF, Johnson M, Dietrich J, James LP, Clancy JS, Chen J, Martini JF, Abbattista A, Solomon BJ. Lorlatinib in non-small-cell lung cancer with ALK or ROS1 rearrangement: an international, multicentre, open-label, single-arm first-in-man phase 1 trial. Lancet Oncol 2017;18:1590–9. https://doi.org/10.1016/S1470-2045(17)30680-0.

[53] Cappuzzo F, Mc Cleod M, Hussein M, Morabito A, Rittmeyer A, Conter HJ, Kopp H-G, Daniel D, Mc Cune S, Mekhail T, Zer A, Reinmuth N, Sadiq A, Archer V, Ochi Lohmann T, Wang L, Kowanetz M, Lin W, Sandler A, West H. IMpower130: progression-free survival (PFS) and safety analysis from a randomised phase III study of carboplatin + nab-paclitaxel (CnP) with or without atezolizumab (atezo) as first-line (1L) therapy in advanced non-squamous NSCLC. Ann Oncol 2018;29. https://doi.org/10.1093/annonc/mdy424.065. viii742–3.

[54] Socinski MA, Jotte RM, Cappuzzo F, Orlandi FJ, Stroyakoversuskiy D, Nogami N, Rodriguez-Abreu D, Moro-Sibilot D, Thomas CA, Barlesi F, Finley GG, Kelsch C, Lee A, Coleman S, Shen Y, Kowanetz M, Lopez-Chavez A, Sandler A, Reck M. Overall survival (OS) analysis of IMpower150, a randomized Ph 3 study of atezolizumab (atezo) + chemotherapy (chemo) ± bevacizumab (bev) versus chemo + bev in 1L nonsquamous (NSQ) NSCLC. J Clin Oncol 2018;36:9002. https://doi.org/10.1200/jco.2018.36.15_suppl.9002.

[55] Reck M, TSK M, Nishio M, Jotte RM, Cappuzzo F, Orlandi F, Stroyakoversuskiy D, Nogami N, Rodríguez-Abreu D, Moro-Sibilot D, Thomas CA, Barlesi F, Finley G, Lee A, Coleman S, Deng Y, Kowanetz M, Shankar G, Lin W, Socinski MA, Mok TS. Atezolizumab plus bevacizumab and chemotherapy in non-small-cell lung cancer (IMpower150): key subgroup analyses of patients with EGFR mutations or baseline liver metastases in a randomised, open-label phase 3 trial. Lancet Respir Med 2019;7:387–401. https://doi.org/10.1016/S2213-2600(19)30084-0.

[56] Anon. A study evaluating platinum-pemetrexed-atezolizumab (+/-Bevacizumab) for patients with stage IIIB/IV non-squamous non-small cell lung cancer with EGFR Mutations, ALK rearrangement or ROS1 fusion progressing after targeted therapies. ClinicalTrials.gov; 2019. https://clinicaltrials.gov/ct2/show/NCT04042558?cond=alk+pemetrexed&draw=2&rank=1. [Accessed 11 March 2020].

[57] Anon. Pembrolizumab in combination with platinum-based doublet chemotherapy in patients with EGFR mutation and ALK positive NSCLC (non-small cell lung cancer) with progressive disease following prior tyrosine kinase inhibitors (TKIs). ClinicalTrials.gov; 2017. https://clinicaltrials.gov/ct2/show/NCT03242915?cond=alk+platinum&draw=2&rank=4. [Accessed 11 March 2020].

[58] Jain RK. Normalizing tumor vasculature with anti-angiogenic therapy: a new paradigm for combination therapy. Nat Med 2001;7:987–9. https://doi.org/10.1038/nm0901-987.

[59] Ettinger DS, Wood DE, Aggarwal C, Aisner DL, Akerley W, Bauman JR, Bharat A, Bruno DS, Chang JY, Chirieac LR, D'Amico TA, Dilling TJ, Dobelbower M, Gettinger S, Govindan R, Gubens MA, Hennon M, Horn L, Lackner RP, Lanuti M, Leal TA, Lin J, Loo Jr BW, Martins RG, Otterson GA, Patel SP, Reckamp KL, Riely GJ, Schild SE, Shapiro TA, Stevenson J, Swanson SJ, Tauer KW, Yang SC, Gregory K, Hughes M. NCCN guidelines insights: non–small cell lung cancer, version 1.2020. J Natl Compr Cancer Netw 2019;17:1464–72. https://doi.org/10.6004/jnccn.2019.0059.

[60] Dejean E, Renalier MH, Foisseau M, Agirre X, Joseph N, De Paiva GR, Al Saati T, Soulier J, Desjobert C, Lamant L, Prósper F, Felsher DW, Cavaillé J, Prats H, Delsol G, Giuriato S, Meggetto F. Hypoxia-microRNA-16 downregulation induces VEGF expression in anaplastic lymphoma kinase (ALK)-positive anaplastic large-cell lymphomas. Leukemia 2011;25:1882–90. https://doi.org/10.1038/leu.2011.168.

[61] Martinengo C, Poggio T, Menotti M, Scalzo MS, Mastini C, Ambrogio C, Pellegrino E, Riera L, Piva R, Ribatti D, Pastorino F, Perri P, Ponzoni M, Wang Q, Voena C, Chiarle R. ALK-dependent control of hypoxia-inducible factors mediates tumor growth and metastasis. Cancer Res 2014;74:6094–106. https://doi.org/10.1158/0008-5472.CAN-14-0268.

[62] Chen Y, Ma G, Su C, Wu P, Wang H, Song X, Yu Q, Zeng A, Zhou S. Apatinib reverses alectinib resistance by targeting vascular endothelial growth factor receptor 2 and attenuating the oncogenic signaling pathway in echinoderm microtubule-associated protein-like 4-anaplastic lymphoma kinase fusion gene-positive lung cancer cell lines. Anticancer Drugs 2018;29:935–43. https://doi.org/10.1097/CAD.0000000000000667.

[63] Habib S, Delourme J, Dhalluin X, Petyt G, Tacelli N, Scherpereel A, Lafitte JJ, Cortot AB. Bevacizumab and weekly paclitaxel for non-squamous non small cell lung cancer patients: a retrospective study. Lung Cancer 2013;80:197–202. https://doi.org/10.1016/j.lungcan.2013.01.015.

[64] Wang HY, Zhu H, Kong L, Yu JM. Efficacy of cisplatin/pemetrexed with bevacizumab to treat advanced lung adenocarcinoma with different drive genes: case report and literature review. Onco Targets Ther 2016;9:4639–44. https://doi.org/10.2147/OTT.S101241.

[65] Liu Z, Bao Y, Li B, Sun X, Wang L. Does ALK-rearrangement predict favorable response to the therapy of bevacizumab plus pemetrexed in advanced non-small-cell lung cancer? Case report and literature review. Clin Transl Med 2018;7:1. https://doi.org/10.1186/s40169-017-0178-x.

[66] Liang Y, Wakelee HA, Neal JW. Relationship of driver oncogenes to long-term pemetrexed response in non-small-cell lung cancer. Clin Lung Cancer 2015;16:366–73. https://doi.org/10.1016/j.cllc.2014.12.009.

[67] Nonoguchi N, Miyatake S-I, Fukumoto MM, Furuse M, Hiramatsu R, Kawabata S, Kuroiwa T, Tsuji M, Fukumoto MM, Ono K. The distribution of vascular endothelial growth factor-producing cells in clinical radiation necrosis of the brain: pathological consideration of their potential roles. J Neurooncol 2011;105:423–31. https://doi.org/10.1007/s11060-011-0610-9.

[68] Tanigawa K, Mizuno K, Kamenohara Y, Unoki T, Misono S, Inoue H. Effect of bevacizumab on brain radiation necrosis in anaplastic lymphoma kinase-positive lung cancer. Respirol Case Rep 2019;7. https://doi.org/10.1002/rcr2.454.

[69] Nakasuka T, Ichihara E, Makimoto G, Maeda Y, Kiura K. Primary resistance to alectinib was lost after bevacizumab combined chemotherapy in ALK-rearranged lung adenocarcinoma. J Thorac Oncol 2019;14:e168–9. https://doi.org/10.1016/j.jtho.2019.03.009.

[70] Watanabe H, Ichihara E, Kayatani H, Higo H, Makimoto G, Kano H, Nishii K, Hara N, Ninomiya K, Kubo T, Ohashi K, Rai K, Hotta K, Tabata M, Maeda Y, Kiura K. Abstract 2131: significant combination benefit of anti-VEGFR antibody and oncogene-targeted agents in EGFR or ALK mutant NSCLC cells. Cancer Res 2019;79:2131. https://doi.org/10.1158/1538-7445.am2019-2131.

[71] Yang B, Cui Z, Meng X, Huang Z, Hu Y. Crizotinib with bevacizumab as first-line therapy in patients with advanced non-small-cell lung cancer harboring EML4-ALK fusion variant mutation: a prospective exploratory study. J Clin Oncol 2018;36:e21186. https://doi.org/10.1200/jco.2018.36.15_suppl.e21186.

[72] Anon. Phase I/II trial of alectinib and bevacizumab in patients with advanced, anaplastic lymphoma kinase (ALK)-positive, non-small cell lung cancer. ClinicalTrials.gov; 2020. https://clinicaltrials.gov/ct2/show/NCT02521051. [Accessed 15 March 2020].

[73] Anon. Brigatinib and bevacizumab for the treatment of ALK-rearranged locally advanced, metastatic, or recurrent non-small cell lung cancer. ClinicalTrials.gov; 2020. https://clinicaltrials.gov/ct2/show/NCT04227028. [Accessed 15 March 2020].

Management of ALK positive patients with tumors other than lung cancer

Charlotte Rigaud[a] and Marie-Emilie Dourthe[b]

[a]Department of Pediatric Oncology, Gustave Roussy Cancer Campus, Villejuif, France [b]Department of Pediatric Hematology, Robert Debré University Hospital, AP-HP, Paris, France

Abstract

The *ALK* gene was first described in the 1990s in a subset of non-Hodgkin lymphoma, anaplastic large-cell lymphoma (ALCL). Two decades later, a small proportion of non-small cell lung cancers (NSCLCs) were found to carry an activating rearrangement of *ALK*. This discovery drove to the development of small-molecule ALK inhibitors that demonstrated dramatic response rates in patients with ALK rearranged advanced NSCLC patients and increased 6 months the PFS. Therefore, ALK emerged has a relevant biomarker and a validated therapeutic target in ALK rearranged malignancies even in non-NCSLC, and especially in pediatric malignancies harboring ALK alterations, either as a translocation in ALCL and inflammatory myofibroblastic tumor or as activating mutations in neuroblastoma. The results of ALK inhibition by small molecules alone or in combination, especially in a relapse setting, are encouraging despite the emergence of resistance to ALK inhibitors. The introduction of ALK inhibitors in the therapeutic schedule of these patients follows two main goals: first to improve response and cure rates in these high-risk patients and second to decrease the burden of treatment and, thus, the long-term sequelae in this pediatric population. The right use and positioning of these molecules is still in process especially in ALCL and neuroblastoma and efforts have to be done to better understand the resistance mechanisms to ALK inhibitors in these malignancies.

Abbreviations

ALCL	anaplastic large-cell lymphoma
ALK	anaplastic lymphoma kinase
CNS	central nervous system
COG	Children's Oncology Group
CR	complete remission
CR2	second complete remission
EFS	event free survival
EICNHL	European Intergroup for Children with Non-Hodgkin Lymphoma
FISH	fluorescence in situ hybridation

HSCT	hematopoietic stem cell transplantation
IMT	inflammatory myofibroblastic tumors
MDD	minimal disseminated disease
MRD	minimal residual disease
NHL	non-Hodgkin lymphoma
NSCLC	non-small cell lung cancer
ORR	overall response rate
OS	overall survival
PFS	progression free survival
PR	partial response
RT-PCR	reverse transcriptase polymerase chain reaction

Conflict of interest

No potential conflicts of interest were disclosed.

Introduction

The anaplastic lymphoma kinase (ALK) is a tyrosine kinase encoded on chromosome 2 that performs a physiologic role in early brain development. Its expression is restricted to developing neural tissues and absent from other normal tissues [1, 2]. The *ALK* gene was first described in a subset of non-Hodgkin lymphoma and anaplastic large-cell lymphoma (ALCL). It was identified as a gene on chromosome 2 fused to the nucleophosmin (*NPM1*) gene in the t (2;5) translocation [3]. ALK activation by translocation, amplification, point mutations have since been identified in other malignancies. Indeed, more than 2 decades later, 3–5% of non-small cell lung cancers (NSCLCs) were found to carry an activating rearrangement of ALK [4]. This discovery drove to the development of small-molecule ALK inhibitors. The first in class, crizotinib showed dramatic response rates in patients with EML4-ALK advanced NSCLC patients [5]. Therefore, ALK has emerged a relevant biomarker and a validated therapeutic target in ALK rearranged malignancies.

We review here the management of three, mostly pediatric, malignancies harboring activating ALK alterations: anaplastic large cell lymphoma, inflammatory myofibroblastic tumor (IMT) and neuroblastoma, and the arising place of ALK targeting in these diseases.

Anaplastic large cell lymphoma

Anaplastic large cell lymphoma (ALCL) is a rare subtype of peripheral T-cell non-Hodgkin lymphoma (NHL) thought to derive from cytotoxic T cells. It is mostly considered as a pediatric disease as it accounts for 10–15% of all NHLs in children and 1–2% in adults. The annual incidence of ALCL ranges from 1.2 per million in children <15 years to ~2 per million in young adults (25–34 years) [6]. Two different histological and clinical entities of ALCL have been described: cutaneous and systemic ALCLs. Primary cutaneous ALCL is confined to the skin and could generally be treated by surgical excision without systemic therapy with an excellent prognosis.

Systemic ALCL is characterized by lymph nodes involvement, peripheral, abdominal or mediastinal frequently associated with extranodal involvement and "B symptoms" at diagnosis [7]. The most commonly involved extranodal sites include skin, bone, soft tissues, lung and liver.

More than 90% of ALCL cases in children and 50–60% of ALCL in adults show an expression of the anaplastic lymphoma kinase (ALK) on tumor immunostaining indicating the presence of a translocation involving the *ALK* gene and different partners [8–10]. The most common partner is nucleophosmin (NPM1); resulting from the t(2;5) translocation [11].

Fusion transcripts involving an *ALK* partner other than NPM1 represent about 10% of ALK-positive ALCL and various ALK partners have been described including: tropomyosin 3 (TPM3), tropomyosin 4 (TPM4), TRK fused gene (*TFG*), 5¢aminoimidazole-4-carboxiamide ribonucleotideformyltransferase/IMP cyclohydrolase (ATIC), clathrin like heavy chain 1 (CLTC), ALK lymphoma oligomerisation partner on chromosome 17 (ALO17), moesin (MSN) and myosin heavy chain (MYH9) [12–14]. It has been demonstrated that ALK immunostaining of a tumor depends on the ALK fusion partner, with different phenotypes according to the transcript [15].

ALK-positive ALCL is considered as a particular histological entity. A diverse morphological spectrum is described, with three main distinctive morphological types: common, small cell and lymphohistiocytic, sometimes associated in the same tumor sample. Other presentations such as Hodgkin-like patterns could rarely be seen. However, all morphological variants shared the presence of large cells with a highly characteristic morphology referred to hallmark cells. Concerning their immunophenotype, the tumor cells are positive for CD30, and in most cases, for EMA. One or more T-cell markers are generally expressed but in some cases the hallmark cells may have an apparent "null cell" phenotype. Cytotoxic molecules such as TIA-1, granzyme B, or perforin are found positive on immunostaining in a large majority of cases [10].

ALK-positive ALCL is a chemosensitive disease with a complete remission (CR) rate ranging from 66% to 100% whatever the drugs used but 20–40% of the patients will relapse. Event free survival (EFS) and overall survival (OS) of pediatric ALK-positive ALCL are quite stable over the last decades, 65–75% and 70–90%, respectively, and among different collaborative groups using various chemotherapy backbone with different treatment duration ranging from few months to more than a year [16].

Several factors have proven to be associated with a higher risk of treatment failure in children/adolescents: clinical factors such as the presence of mediastinal disease, visceral or cutaneous involvement [17], high risk histological subtype defined by the presence of a lymphohistiocytic or small cell component [18], positive PCR for NPM-ALK in peripheral blood and/or bone marrow at diagnosis (minimal disseminated disease, MDD) [19, 20], low anti-ALK antibody titers at diagnosis [21], and detection of minimal residual disease (MRD) by PCR for NPM1-ALK in the blood after the first course of chemotherapy [22]. Interestingly, Damm-Welk and colleagues demonstrated that ALK-positive ALCL can be classified in three prognostic groups: patients with no detectable minimal disseminated disease (MDD), patients with positive MDD and no detectable minimal residual disease (MRD) 4 weeks after the beginning of chemotherapy and patients with both positive MDD and MRD. The outcome in this last subgroup of patients accounting for 25–30% of all ALCL patients is clearly bad with 3-year EFS below 20%, with some patients experiencing multiple relapses.

There is still no gold standard for the treatment of relapse. Several retrospective studies reported a 5-year overall survival of about 70% for patients with relapsed ALCL. In these studies, various therapeutic approaches including a wide variety of chemotherapy regimens were used to achieve second complete remission (CR2) and in most publications, patients underwent consolidation regimens by high dose chemotherapy followed by either autologous or allogeneic hematopoietic stem cell transplantation (HSCT). The main factor associated with failure is less than 12 months delay from end of frontline therapy to relapse. Impact of CD3 antibody positivity on immunostaining at diagnosis was also suggested but is still debated [23–25]. Although it has been well demonstrated at diagnosis, little is known about MDD and MRD status as prognostic factors at relapse.

In children, the efficacy of a risk-adapted strategy for first ALCL relapses has been shown through a prospective trial run by the European Intergroup for Childhood non-Hodgkin Lymphoma (EICNHL) on a cohort of 118 patients reported by Ruf and colleagues in 2015 [26]. High-risk relapses (relapses during front line treatment or CD3 positive, 40% of the cases) were eligible for allogeneic HSCT after reinduction of chemotherapy. This strategy was efficient, leading to a 3-year EFS after relapse in more than 65% of patients. On the other hand, patients with low risk relapses (CD3 negative, >12 months after diagnosis, 20% of the cases) were treated with weekly vinblastine and achieved a 3-year EFS for 85% of them. Following these results, allogeneic HSCT was recommended for all high-risk relapse patients. Autologous stem cell transplant is no longer recommended in this indication.

In adults, prognosis of relapsed ALCL seems to be poorer than in children. In a retrospective study performed by the LYSA group on relapsed/refractory ALK-positive ALCL with a relapse treatment comprising autologous HSCT after reinduction therapy, the median progression free survival (PFS) and OS were 5 and 12 months, respectively, with a 10-year overall survival inferior to 15% [27].

Clearly, the burden of treatment at relapse for ALK-positive ALCL could be heavy, especially for patients with multiple relapses especially considering the major place of allograft in the relapse setting. Therefore, during the last years, other strategies were raised, first to improve outcome of patients with ALCL at diagnosis and at relapse, and second to limit long-term treatment related morbidity, with a clear need for less toxic therapeutics.

New therapeutic strategies for ALK-positive ALCL

Brentuximab vedotin

Brentuximab vedotin (BV) is an anti-CD30 antiboby conjugated to an anti-microtubule agent, monomethyle auristatin E. In a phase 2 study with 58 ALCL including 16 ALK-positive ALCL, the overall response and CR rates in AKL+ ALCL were 81% and 69%, respectively, with prolonged responses. These results led to the approval of BV by the FDA and EMA for the treatment of relapsed ALCL in adults following failure of at least one multi-agent chemotherapy protocol [28]. Given the neurologic toxicity of brentuximab in adults, prolonged treatment may be difficult to manage. This drug is, thus, mostly employed as a bridge to transplant in relapsed patients.

ALK inhibitors

Although the *ALK* gene was first described in 1994 as part of a genetic rearrangement resulting in an oncoprotein with high kinase activity in this peculiar pathology, strategy of ALK inhibition was developed decades after in ALK-positive ALCLs only after its development in NSCLC, mainly because of the globally favorable survival rates of these patients, but also its orphan disease status [3]. Since then, several ALK inhibitors have been tested in this indication.

Crizotinib, the first ALK inhibitor ever approved, is an orally available ALK and multikinase inhibitor that has been shown to induce high response rates in relapsed/refractory ALK+ ALCL: 6/8 ALCL patients achieved a CR in a pediatric phase 1 trial and all 9 ALCL patients in a retrospective report in adults [29, 30]. Only few progressions have been described for ALCL during crizotinib treatment so far, all occurring within only 2–5 months of treatment.

Even though it induces CR in most cases, crizotinib has not yet proven curative since abrupt relapses following crizotinib discontinuation have been described [31], and no successful reported cases of continuous CR after discontinuation of treatment have been reported yet. Thus, crizotinib is currently used to induce CR2 in relapsed/refractory ALK-positive ALCL patients before allogeneic or autologous HSCT or as a life-long therapy.

As seen with NSCLC, next generation ALK inhibitors have been developed in ALK-positive ALCL.

In the pediatric phase 1 of ceritinib, a 2nd generation potent ALK inhibitor, 2/2 ALK-positive.

ALCLs showed CR [32]. After a dose-escalation of ceretinib, an expansion phase is ongoing in children/adolescents. However, similarly to crizotinib, no successful discontinuation has been reported so far and, thus, ceritinib is also mostly used to induce remission in relapsed/refractory ALK+ ALCL patients before allogeneic HSCT. In adult, the efficacy of ceritinib could also be shown with long lasting responses in all 3 ALK-positive ALCL patients included in the phase I trial ASCEND-1 [33].

The next generation ALK inhibitor alectinib has also been evaluated in ALK-positive ALCLs in pediatrics and adults in a phase 1/2 trial taking place in Japan [34]. Definitive results have not been published yet but the first report released in the 2018 ASH meeting showed a good safety profile and an overall response rate of 80%, with 8 out of 10 patients with rapid CR (6) or PR (2) [35]. Interestingly, the two patients who experienced progressive disease never responded to the treatment, emphasizing the fact that resistance to ALK inhibitors in ALK-positive ALCLs is mainly primary resistance. Recently, an Italian study looking into resistance to ALK inhibitors demonstrated that an excess of NPM-ALK activation and signaling induces apoptosis via oncogenic stress responses. The conclusion of the authors was that suspension the ALK TKI treatment could represent a therapeutic option in cells that became resistant by NPM-ALK amplification [36]. To date, no large study is available regarding the mechanism of resistance to ALK inhibitors in patients with ALK-positive ALCL.

Brigatinib is also currently under evaluation in ALK-positive ALCL in adults (NCT03719898) with no available results yet. To date, no trial evaluates lorlatinib in ALK-positive ALCL neither in pediatrics nor in adults.

Despite extremely good results with crizotinib in ALK-positive ALCLs, next generation ALK inhibitors in addition to be more potent and active on some ALK resistance mutations, present the advantage of a good central nervous system penetration, which could be valuable in a disease as ALCL with a particular meningeal tropism. Indeed, some isolated CNS relapses have been reported while on crizotinib with dramatic responses to ceritinib, lorlatinib and alectinib [37]. Even though all patients achieved complete remission, they may not be cured since abrupt relapses have been reported after the discontinuation of ALK inhibitors, even after several years of treatment. A prolonged treatment or a consolidation either with allogeneic stem cell transplantation, brain radiotherapy or conventional chemotherapy should be discussed in these patients. The optimal duration of treatment with ALK inhibitors has still to be defined.

In Europe, the EICNHL group is designing a trial with an ALK inhibitor in combination with the ALCL99 standard frontline chemotherapy backbone as frontline treatment in a phase 1 safety study. Unfortunately, no ALK inhibitor has been selected yet.

Some other combinations with chemotherapy are under evaluation. Crizotinib in combination with chemotherapy has already been tested in a phase 1 trial in children with ALK-related malignancies (NCT01606878), and final safety data from the phase 2 trial of crizotinib administered in combination with multi-agent chemotherapy running by the Children's Oncology Group (COG) (NCT01979536) will be released soon.

ALK-positive ALCL and check-point inhibitors

Accumulating evidence indicates that the immune system plays a major role in both the pathogenesis and final control of ALK-positive ALCL. Antibodies against ALK and cytotoxic T-cell and CD4 T-helper responses to ALK have been observed in patients with ALK-positive ALCL both at diagnosis and at remission with a significant inverse correlation between ALK-antibody titers and the incidence of relapses. Moreover, vaccination using truncated ALK has been reported to induce potent and long lasting protection from local and systemic lymphoma growth. However, it has also been shown that ALK-positive ALCL cells strongly express the immunosuppressive cell-surface protein PD-L1, as determined at the mRNA and protein levels in ALK-positive ALCL cell lines. Furthermore, results of PD-L1 immunostaining on all patient tissue samples reported so far showed a strong PD-L1 expression [38]. Analysis revealed that PD-L1 expression is induced by the chimeric NPM1/ALK tyrosine kinase, by activating STAT3, confirming a unique function for NPM/ALK as a promoter of immune evasion by inducing PD-L1. In clinics, three case-reports clearly demonstrated that prolonged response can be achieved with anti-PD1 therapy in patients with ALK-positive ALCL refractory to chemotherapy and/or ALK inhibitors [39–41]. These three cases along with the unique immunogenic properties of ALCL including intrinsic PD-L1 induction and the good tolerance profile of PD1 inhibitors advocated for the development of clinical trials evaluating the efficacy of PD-1 inhibitors in refractory/relapsed ALCL. This hypothesis is currently under evaluation in the NIVOALCL trial testing nivolumab monotherapy either in case of progression after targeted therapies or as consolidative immunotherapy in patients in CR after ALK inhibitors or brentuximab vedotin.

To summarize, ALK-positive ALCL is a rare disease with quite good outcome with frontline standard chemotherapy and accessible to multiple targeted agents. However, some challenges remain (1) in the management of high-risk patient at diagnosis and at relapse to increase the cure rate and decrease the global treatment burden and (2) for the investigators in the optimal positioning of all these available novel agents.

Inflammatory myofibroblastic tumor

Inflammatory myofibroblastic tumor is a very rare mesenchymal tumor. Historically this disease was included in the group of "inflammatory pseudotumor" which comprises a wide range of neoplastic and non-neoplastic (reactive) lesions [37]. Over the past decades, IMT was characterized as a specific entity with particular clinical, pathological and molecular features. IMT can occur at any age but is clearly more frequent in the two first decades of life, especially in children and adolescents. This tumor can be uni- or multifocal tumor. It typically arises in the abdomen, pelvis, lung, head, or neck, but can present anywhere in the body [42–44]. Although IMT can be detected as an incidental finding, i.e. on a routine chest X-ray. Patients often present with general symptoms (fever, weight loss or fatigue) or symptoms depending on tumor location (palpable mass, abdominal pain or gastrointestinal complaints for intra-abdominal lesions, and cough, chest pain, or, less often, hemoptysis for pulmonary tumors). Signs of biological inflammation are usually present in patients presenting general symptoms [42, 45]. Histologically, IMTs are composed of a myofibroblastic spindle cell stroma with accumulation of leukocytes and plasma cells, with occasional admixed eosinophils and neutrophils [45]. Mitotic activity is generally low and vascular invasion infrequent [46–50]. Approximately 50% of IMT show positive ALK staining by immunohistochemistry (IHC) [51]. ALK expression in IMT is prominent in younger patients but not limited to this population. After the discovery of ALK rearrangement in ALCL, IMT was the first solid tumor in which ALK translocation was described [52]. Similarly to ALCL, ALK expression in IMT signs the presence of an ALK gene rearrangement. Several different translocation partners were usually correlated with the specific gene fusion by immunostaining, i.e., diffuse cytoplasmic staining with TPM3, TPM4, CARS, or ATIC, genes; nuclear membrane staining with RANBP2, etc. This last translocation is associated with a particular subset of IMT, epithelioid inflammatory myofibroblastic sarcoma that seems to have a more aggressive clinical behavior [53]. The translocation can be detected by FISH and/or RT-PCR. Of note, some ALK gene's fusion partners are associated with negative ALK expression by IHC, pleading for the systematic incorporation of molecular diagnostic tools for the diagnosis of IMT [54, 55]. The clinical course of IMT is usually relatively indolent but with a trend for local recurrence and a small risk of distant metastasis (less than 5%) [42, 43, 53]. The most commons site of metastases are lung and brain. Treatment of IMT varies according to the location of the tumor, the size and the extent of the disease. Complete surgical resection is the mainstay of curative treatment, and recurrence is very infrequent in complete resection of a solitary lesion. However, local relapse rate increase in some anatomical locations for whom obtaining complete surgical resection is very challenging. For patients with unresectable or metastatic disease, systemic strategies such as steroids or other anti-inflammatory agents and cytotoxic chemotherapy have been used with limited success.

The discovery of activating ALK rearrangements in up to half of all cases of IMT and emergence of ALK targeting in NSCLC in the 2000s gave rise to new strategic options for the treatment of ALK rearranged and not completely resectable IMTs [5, 46]. In 2010, Butrynski and colleagues reported a sustained partial response to the ALK inhibitor crizotinib in a patient with ALK-rearranged IMT, compared with the absence of activity of crizotinib in another patient with IMT without ALK translocation. This first report confirmed the dependence on ALK signaling in ALK-translocated IMT [56]. More recently, the COG published encouraging results in 14 pediatric IMT patients treated with crizotinib. This phase 1/2 study showed promising activity in metastatic or unresectable IMT with an overall response rate (ORR) of 83%. Complete response was observed in 6 out of 14 patients (36%). Lastly, a review compiled the data of 30 advanced or metastatic ALK-positive IMT treated with ALK inhibitors, mostly case-reports in addition to the 14 patients from the COG trial. The overall response rate (ORR) was 80% with 40% of CR [57]. In both publications, responders' outcome after crizotinib discontinuation was not mentioned, and the rate of patients who could finally achieve complete resection or, on the contrary, who experienced abrupt relapses after stopping ALK inhibitor is not known. Indeed, as in ALCL, relapse at ALK inhibitor ceritinib discontinuation could occur even after CR achievement and prolonged treatment as shown in a recent case-report. Nevertheless, the reintroduction of ceritinib treatment in this patient led to CR2 [58]. This demonstrates than CR2 can be reached by resuming ALK inhibitor and that, as in ALCL optimal treatment duration with ALK inhibitors to avoid relapses is not known yet. Another major issue with ALK inhibitors in ALK-positive malignancies is acquired resistance mechanisms by tumor cells. Resistance mechanisms in IMT seem to be similar to NSCLC. Several case reports illustrated a drug-resistance than can be overcome in most cases by the introduction of next generation ALK inhibitors. To date, multiple next generation ALK inhibitors have been used successfully in IMT. Notably, ALK inhibitors with good CNS penetration (brigatinib, ceritinib, alectinib and lorlatinib) are of particular interest in the rare patients with brain metastasis [57–59]. As described above in ALCL, ALK fusion could confers a particular immunogenic status to IMT and indeed particular immunologic status of IMT is demonstrated by the prominent tumor-associated inflammatory infiltrate and the constitutional symptoms, including fever. A recent pathological study, demonstrated in 35 IMT samples from 28 patients a high prevalence of PD-L1 expression on both tumor and infiltrating immune cells, with a 80% positivity for PD-L1 expression. Another publication showed that PD-L1 expression and CD8+ tumor-infiltrating lymphocytes are associated with ALK rearrangement in IMT. These results suggest that immune checkpoint inhibitors may be a novel option for treating patients with advanced IMT [60, 61].

To summarize, although IMT is considered by the World Health Organization to be of "intermediate malignancy," curative treatment can be sometimes very difficult because of the impossibility to perform complete surgical resection without mutilation. Moreover, in rare cases, patients present metastatic disease at diagnosis. When standard surgical treatment is not feasible, ALK inhibitors may play an important role in reducing tumor burden and volume. The place of immune checkpoint inhibition has to be explored in this immunogenic disease.

Neuroblastoma

Neuroblastoma is the most common solid extracranial malignant tumor in children, accounting for 8–10% of childhood cancers [62, 63]. It is an embryonal malignancy arising from neural crest stem cells and the neuroblastic tumors present along the parasympathic system from neck to pelvis or most commonly in the adrenal gland. The median age at diagnosis of neuroblastoma is around 2 years. Its presentation at diagnosis is very heterogenous.

Indeed, the natural history of the disease could be quite benign in infants with spontaneous resolution even in patients presenting with disseminated disease, whereas in older patients the disease could be biologically aggressive with widespread metastases and refractory to treatments. Considering this population, despite major progresses in neuroblastoma's therapeutic management during the last decades, neuroblastoma still account for 15% of childhood cancer deaths. To improve outcome and to overcome this complexity due to heterogenous biological and clinical behavior, a risk-tailored approach has been developed over the years. Prognostic factors used to stratify patients include age at diagnosis, localized or disseminated disease, MYCN amplification, segmental chromosomal alteration, histological type. Whereas localized mature neuroblastic tumors or neonatal tumors with numeric chromosomal alteration are highly curable with a 5-year overall survival exceeding 90%, high risk patients (>12 months with metastasis or/and MYCN amplification) accounting for almost 50% of the patients have a poorer outcome with frequent refractory disease and relapse with a 5-year survival of 40–50% [64]. For high-risk patients the therapeutic strategy became more complex over decades. By now, the treatment schedule comprises dose-intense induction with cytotoxic chemotherapy, local treatment of primary tumors by surgery plus radiotherapy, consolidation by high-dose chemotherapy followed by autologous stem cell transplantation and maintenance therapy consisting in courses of retinoic acid and immunotherapy with anti-GD2 [65, 66]. This already long and intense schedule is continuously evolving. Questions are still rising in both the European SIOPEN and American COG groups. Single vs tandem high dose consolidation regimens is currently under evaluation in both groups as the role and place of 131 I-metaiodobenzylguanidine (131I-MIBG) therapy for consolidation [67, 68]. The optimal timing of anti-GD2 therapy in frontline is still debated, and the use of GD2 target in adoptive T-cell therapy programs is currently being evaluated, with various results [69, 70]. Overall, despite better understanding in neuroblastoma biology, refinement of risk stratification, and intensive and prolonged frontline therapy providing long-term side effects, patients with high-risk neuroblastoma still suffer from frequent relapses [71]. Outcome of patients experiencing relapse is dismal with a 5-year OS of about 20% [72, 73].

ALK in neuroblastoma

In the last decade, the advent of next generation sequencing opened the way for better genomic characterization and identification of tumor-specific targets. Indeed, considering their neural crest origin, ALK full length is not surprisingly expressed in neuroblastoma cells [2]. Single-base missense germline ALK mutations were found to be the genetic origin of familial neuroblastoma and the same mutations were also detected at a somatic level in sporadic neuroblastomas, confirming ALK's special place as the only mutated oncogene suitable as a therapeutic target in this disease [74]. In 2014, Bresler and colleagues reported than the ALK tyrosine kinase domain mutations occurred in 8% of neuroblastoma. Three recurrent mutations R1275, F1174, and F1245, consistent with other studies, accounted for about 85% of ALK mutations and 13 other minor site mutations were identified. They correlated significantly with poorer survival in high- and intermediate-risk groups [75]. Some studies comparing somatic genome analysis of neuroblastoma samples at diagnosis and at relapses demonstrated that ALK mutations were enriched at relapse. Deep sequencing revealed that some of these mutations where already present at diagnosis at a subclonal level at a very low allele frequency whereas others were acquired de novo [76–78]. These observations emphasizing the role of ALK as a driver oncogene and given the restricted distribution in normal tissue and frequent expression of ALK in neuroblastoma, ALK became an attractive target in this disease.

ALK inhibition in neuroblastoma

Preclinical studies

In preclinical studies, knockdown of ALK expression in NB cells lines resulted in growth inhibition but was found more effective in cells with ALK alterations than in those with wild-type ALK. Several data suggest that not all mutations reported in neuroblastoma could be considered as oncogenic drivers. On the other hand, some of the proven activating ALK mutations lead to primary resistance to the first in class ALK inhibitor crizotinib [79]. As demonstrated by Bresler and colleagues there are differential sensitivity to crizotinib based on the type of mutation. Indeed, neublastoma with ALK R1275Q mutations were more sensitive to crizotinib, whereas those with ALK F1174L and F1245C exhibited a relative resistance [75]. Efficacy of the next generation ALK inhibitors, such as ceritinib, alectinib, lorlatinib and brigatinib has also been tested in neuroblastoma. Ceritinib demonstrated a 20-fold increased in potency over crizotinib in ALK enzymatic assays and proved efficacy in both crizotinib-naïve and resistant NSCLC. However, ceritinib did not overcome crizotinib-resistant in ALK F1174C mutant neuroblastoma [80]. Alectinib showed greater affinity than crizotinib for the ATP-binding site and thereby improved potency against the ALK-kinase. Unlike ceritinib, alectinib was proven to inhibit growth in ALK-wild type neuroblastoma cell-lines as well as in ALK F1174L mutants [81]. Brigatinib is one of the most recently developed second generation ALK inhibitors and is a highly potent and selective ALK inhibitor. In a recent study, brigatinib maintained substantial activity against 17 secondary ALK mutants, including F1174L, tested in cellular assays and demonstrated a superior inhibition profile compared with crizotinib, ceritinib, and alectinib [82]. Third generation ALK inhibitor lorlatinib was found having high potency across ALK variants and inhibited ALK more effectively than crizotinib in vitro. It has also been able to induce complete tumor regression in crizotinib-resistant xenograft mouse models of neuroblastoma, as well as in patient-derived xenografts harboring the crizotinib-resistant F1174L mutation [83]. In addition to the single drug approach and the development of next generation ALK inhibitors, several combinations have been preclinically tested to enhance crizotinib antitumor activity or to overcome resistance. For example, chemotherapy agents and small molecules inhibitors of downstream pathways such as mTOR have been combined with small molecules ALK inhibitors and anti-ALK antibodies with some successes [79].

Clinical trials

Crizotinib was the first assessed and approved ALK inhibitor in ALK and ROS1 rearranged NSCLC with promising results. These results along with the identification of ALK as an oncogenic driver in neuroblastoma led to the development of trials and clinical strategies to target ALK in neuroblastoma and in other ALK rearranged pediatric malignancies, such as IMT and ALCL.

Crizotinib was first evaluated in a cohort of 79 children and young adults (<22 years) with relapsed/refractory solid tumors, or ALCL in a phase 1–2 trial launched by the COG in 2009 [29]. Crizotinib in this population was proven safe and well tolerated and as expected, showed similar pharmacokinetics profile as in adults [84]. Among this cohort, 34 neuroblastoma patients were enrolled, with 11 of them known to harbor ALK activating mutations. Despite an objective response rate of nearly 70% in children with other ALK rearranged tumors, results were first discouraging in neuroblastoma patients. Indeed, only one out of 11 achieved complete remission and three patients experienced stable disease. When looking closer at the ALK alterations presented in these patients' tumors, three of them harbored a mutation at the R1275 locus, associated with a germline mutation in two. The last patient with stable disease presented a somatic F1174L mutation. The other seven patients with ALK mutations had progressive disease including three patients with F1174L somatic mutations. Among the other 23 patients with unknown ALK alteration status, one patient achieved a CR and five had a stable disease, prolonged for 39 cycles in one [29]. These results in neuroblastoma were clearly insufficient to claim for a curative effect but they suggested an anti-tumor activity in neuroblastoma and confirmed the differential efficacy profile according to the locus of the ALK mutation.

To enhance the results obtained in monotherapy, and assessed safety and tolerability of the addition of crizotinib to chemotherapy agents, the next trial designed by the COG associated crizotinib with either cyclophosphamide/topotecan or vincristine/doxorubicine in patients with refractory or relapsed ALCL or solid tumors. Limiting toxicities were dehydration, diarrhea and prolonged QTc and according to the authors could be due to the poor palatability of the oral solution of crizotinib [85]. Results from this trial, especially concerning efficacy are not yet published but seemed encouraging as crizotinib in association with standard therapy is currently evaluated in a COG phase III trial designed for high-risk neuroblastoma with ALK mutations (NCT03126916).

Additionally, next generation ALK inhibitors such as ceritinib, ensartinib, lorlatinib and entrectinib are currently under clinical trial evaluation for high risk or relapsed/refractory neuroblastoma, alone or in combination with either chemotherapy agents or other targeted small molecules (NCT01742286, NCT02780128, NCT02650401, NCT03213652, NCT03107988). Ceritinib is evaluated in neuroblastoma in two pediatric trials: 1/single agent in an international pharma sponsored phase 1–2 trial with a neuroblastoma dedicated cohort (NCT01742286) 2/in combination with the small molecule CDK4/6 inhibitor ribociclib as part of a molecular profiling program for neuroblastoma (The NEPENTHE Trial; NCT02780128). Interestingly, in the ceritinib monotherapy trial, one patient with relapsed neuroblastoma with an ALK F1174L mutation experienced partial response on a retroperitoneal mass but the disease concurrently progressed in central nervous CNS, suggesting that other ALK inhibitors with better CNS penetration could be used to achieve adequate concentrations in neuroblastoma sanctuary sites such CNS [32]. It could also suggest a spatial and subclonal evolution of the disease. Ensartinib has been chosen as the ALK and ROS1 inhibitor of the Pediatric MATCH (Molecular Analysis for Therapy Choice) phase 2 trial, stratifying patients' treatment according to molecular alterations and including neuroblastoma among other pediatric relapsed or refractory pediatric malignancies (NCT03213652). Lorlatinib is under evaluation in addition to the chemotherapy backbone cyclophosphamide plus topotecan in newly diagnosed high-risk neuroblastoma patient in a phase 1 trial designed by the NANT (New Approaches to Neuroblastoma Therapy) (NCT03107988). Lastly, ALK is targeted along with other neuroblastoma alterations by the pan- TRK, ROS1 and ALK entrectinib in the dose finding trial with a neuroblastoma specific cohort (NCT02650401). Completion and results of these trials evaluating next generation ALK will be crucial.

To summarize, neuroblastoma is a highly heterogenous disease in its clinical features as well as on a biological and molecular bases, with clear unmet medical needs in the management of high-risk patients who still suffer from dismal outcome. With promising activity in preclinical data and encouraging signals in the first clinical trials, ALK inhibition will definitely play a role in the management of the nearly 10% of high-risk neuroblastoma patients harboring ALK alterations. Nevertheless, the completion and results of the ongoing trials evaluating next generation ALK inhibitors and combinations will be crucial to (1) increase the knowledge on resistance to ALK inhibition in neuroblastoma patients and (2) ideally position ALK inhibition in the already dense treatment schedule of high-risk patients.

Conclusion

During the past years, ALK targeting has clearly became a weapon of choice in pediatric malignancies harboring ALK rearrangements, either translocation as in ALCL and IMT or activating mutations as in neuroblastoma. Results of ALK inhibition by small molecules alone or in combination, especially in a relapse setting, are encouraging despite the emergence of resistance to ALK inhibitors. The introduction of ALK inhibitors in the therapeutic schedule of these patients follows two main goals: first to improve response and cure rates in the high-risk patients and second to decrease the burden of treatment and thus the long-term sequelae in this pediatric population. The right use and positioning of these molecules are still in progress especially in ALCL and neuroblastoma and efforts have to be done to better understand the resistance mechanisms to ALK inhibitors in these malignancies.

References

[1] Carpenter EL, Haglund EA, Mace EM, Deng D, Martinez D, Wood AC, et al. Antibody targeting of anaplastic lymphoma kinase induces cytotoxicity of human neuroblastoma. Oncogene 2012;31(46):4859–67.
[2] Lamant L, Pulford K, Bischof D, Morris SW, Mason DY, Delsol G, et al. Expression of the ALK tyrosine kinase gene in neuroblastoma. Am J Pathol 2000;156(5):1711–21.
[3] Morris SW, Kirstein MN, Valentine MB, Dittmer KG, Shapiro DN, Saltman DL, et al. Fusion of a kinase gene, ALK, to a nucleolar protein gene, NPM, in non-Hodgkin's lymphoma. Science 1994;263(5151):1281–4.
[4] Soda M, Choi YL, Enomoto M, Takada S, Yamashita Y, Ishikawa S, et al. Identification of the transforming EML4-ALK fusion gene in non-small-cell lung cancer. Nature 2007;448(7153):561–6.
[5] Kwak EL, Bang Y-J, Camidge DR, Shaw AT, Solomon B, Maki RG, et al. Anaplastic lymphoma kinase inhibition in non-small-cell lung cancer. N Engl J Med 2010;363(18):1693–703.
[6] Alessandri AJ, Pritchard SL, Schultz KR, Massing BG. A population-based study of pediatric anaplastic large cell lymphoma. Cancer 2002;94(6):1830–5.
[7] Minard-Colin V, Brugières L, Reiter A, Cairo MS, Gross TG, Woessmann W, et al. Non-Hodgkin lymphoma in children and adolescents: progress through effective collaboration, current knowledge, and challenges ahead. J Clin Oncol Off J Am Soc Clin Oncol 2015;33(27):2963–74.

[8] Brugières L, Deley MC, Pacquement H, Meguerian-Bedoyan Z, Terrier-Lacombe MJ, Robert A, et al. CD30(+) anaplastic large-cell lymphoma in children: analysis of 82 patients enrolled in two consecutive studies of the French Society of Pediatric Oncology. Blood 1998;92(10):3591–8.

[9] Sibon D, Fournier M, Brière J, Lamant L, Haioun C, Coiffier B, et al. Long-term outcome of adults with systemic anaplastic large-cell lymphoma treated within the Groupe d'Etude des Lymphomes de l'Adulte trials. J Clin Oncol Off J Am Soc Clin Oncol 2012;30(32):3939–46.

[10] Swerdlow SH, Campo E, Pileri SA, Harris NL, Stein H, Siebert R, et al. The 2016 revision of the World Health Organization classification of lymphoid neoplasms. Blood 2016;127(20):2375–90.

[11] Lamant L, Meggetto F, al Saati T, Brugières L, de Paillerets BB, Dastugue N, et al. High incidence of the t (2, 5) (p 23; q35) translocation in anaplastic large cell lymphoma and its lack of detection in Hodgkin's disease. Comparison of cytogenetic analysis, reverse transcriptase-polymerase chain reaction, and P-80 immunostaining. Blood 1996;87(1):284–91.

[12] Stein H, Foss HD, Dürkop H, Marafioti T, Delsol G, Pulford K, et al. CD30(+) anaplastic large cell lymphoma: a review of its histopathologic, genetic, and clinical features. Blood 2000;96(12):3681–95.

[13] Duyster J, Bai RY, Morris SW. Translocations involving anaplastic lymphoma kinase (ALK). Oncogene 2001;20 (40):5623–37.

[14] Pulford K, Lamant L, Espinos E, Jiang Q, Xue L, Turturro F, et al. The emerging normal and disease-related roles of anaplastic lymphoma kinase. Cell Mol Life Sci 2004;61(23):2939–53.

[15] Damm-Welk C, Klapper W, Oschlies I, Gesk S, Röttgers S, Bradtke J, et al. Distribution of NPM1-ALK and X-ALK fusion transcripts in paediatric anaplastic large cell lymphoma: a molecular-histological correlation. Br J Haematol 2009;146(3):306–9.

[16] Prokoph N, Larose H, Lim MS, Burke GAA, Turner SD. Treatment options for paediatric anaplastic large cell lymphoma (ALCL): current standard and beyond. Cancers 2018;10(4). https://doi.org/10.3390/cancers10040099.

[17] Le Deley M-C, Reiter A, Williams D, Delsol G, Oschlies I, McCarthy K, et al. Prognostic factors in childhood anaplastic large cell lymphoma: results of a large European intergroup study. Blood 2008;111(3):1560–6.

[18] Lamant L, McCarthy K, d'Amore E, Klapper W, Nakagawa A, Fraga M, et al. Prognostic impact of morphologic and phenotypic features of childhood ALK-positive anaplastic large-cell lymphoma: results of the ALCL99 study. J Clin Oncol Off J Am Soc Clin Oncol 2011;29(35):4669–76.

[19] Damm-Welk C, Busch K, Burkhardt B, Schieferstein J, Viehmann S, Oschlies I, et al. Prognostic significance of circulating tumor cells in bone marrow or peripheral blood as detected by qualitative and quantitative PCR in pediatric NPM-ALK-positive anaplastic large-cell lymphoma. Blood 2007;110(2):670–7.

[20] Mussolin L, Pillon M, d'Amore ES, Santoro N, Lombardi A, Fagioli F, et al. Prevalence and clinical implications of bone marrow involvement in pediatric anaplastic large cell lymphoma. Leukemia 2005;19(9):1643–7.

[21] Ait-Tahar K, Damm-Welk C, Burkhardt B, Zimmermann M, Klapper W, Reiter A, et al. Correlation of the autoantibody response to the ALK oncoantigen in pediatric anaplastic lymphoma kinase-positive anaplastic large cell lymphoma with tumor dissemination and relapse risk. Blood 2010;115(16):3314–9.

[22] Damm-Welk C, Mussolin L, Zimmermann M, Pillon M, Klapper W, Oschlies I, et al. Early assessment of minimal residual disease identifies patients at very high relapse risk in NPM-ALK-positive anaplastic large-cell lymphoma. Blood 2014;123(3):334–7.

[23] Brugières L, Quartier P, Le Deley MC, Pacquement H, Perel Y, Bergeron C, et al. Relapses of childhood anaplastic large-cell lymphoma: treatment results in a series of 41 children—a report from the French Society of Pediatric Oncology. Ann Oncol Off J Eur Soc Med Oncol 2000;11(1):53–8.

[24] Mori T, Takimoto T, Katano N, Kikuchi A, Tabuchi K, Kobayashi R, et al. Recurrent childhood anaplastic large cell lymphoma: a retrospective analysis of registered cases in Japan. Br J Haematol 2006;132(5):594–7.

[25] Woessmann W, Zimmermann M, Lenhard M, Burkhardt B, Rossig C, Kremens B, et al. Relapsed or refractory anaplastic large-cell lymphoma in children and adolescents after Berlin-Frankfurt-Muenster (BFM)-type first-line therapy: a BFM-group study. J Clin Oncol Off J Am Soc Clin Oncol 2011;29(22):3065–71.

[26] Ruf R, Brugieres L, Pillon M, Zimmermann M, Attarbaschi A, Melgrenn K, Williams D, Uyttebroeck A, Wrobel G, Reiter A, Woessman W. Risk-adapted therapy for patients with relapsed or refractory ALCL – final report of the prospective ALCL-relapse Trial of the EICNHL. In: 5th Int Symp Child Adolesc Young Adult NHL Br J Haematol Varese Italy; 2015.

[27] Sibon D, Morschhauser F, Resche-Rigon M, Ghez D, Dupuis J, Marçais A, et al. Single or tandem autologous stem-cell transplantation for first-relapsed or refractory Hodgkin lymphoma: 10-year follow-up of the prospective H96 trial by the LYSA/SFGM-TC study group. Haematologica 2016;101(4):474–81.

[28] Pro B, Advani R, Brice P, Bartlett NL, Rosenblatt JD, Illidge T, et al. Five-year results of brentuximab vedotin in patients with relapsed or refractory systemic anaplastic large cell lymphoma. Blood 2017;130(25):2709–17.

[29] Mossé YP, Lim MS, Voss SD, Wilner K, Ruffner K, Laliberte J, et al. Safety and activity of crizotinib for paediatric patients with refractory solid tumours or anaplastic large-cell lymphoma: a Children's Oncology Group phase 1 consortium study. Lancet Oncol 2013;14(6):472–80.

[30] Gambacorti-Passerini C, Messa C, Pogliani EM. Crizotinib in anaplastic large-cell lymphoma. N Engl J Med 2011;364(8):775–6.

[31] Gambacorti-Passerini C, Mussolin L, Brugieres L. Abrupt relapse of ALK-positive lymphoma after discontinuation of crizotinib. N Engl J Med 2016;374(1):95–6.

[32] Geoerger B, Schulte J, Zwaan CM, Casanova M, Fischer M, Moreno L, et al. Phase I study of ceritinib in pediatric patients (Pts) with malignancies harboring a genetic alteration in ALK (ALK+): safety, pharmacokinetic (PK), and efficacy results. J Clin Oncol 2015;33(15_suppl):10005.

[33] Richly H, Kim TM, Schuler M, Kim D-W, Harrison SJ, Shaw AT, et al. Ceritinib in patients with advanced anaplastic lymphoma kinase-rearranged anaplastic large-cell lymphoma. Blood 2015;126(10):1257–8.

[34] Nagai H, Fukano R, Sekimizu M, Kada A, Saito AM, Asada R, et al. Phase II trial of CH5424802 (alectinib hydrochloride) for recurrent or refractory ALK-positive anaplastic large cell lymphoma: study protocol for a nonrandomized non-controlled trial. Nagoya J Med Sci 2017;79(3):407–13.

[35] Sekimizu M, Fukano R, Choi I, Kada A, Saito A, Asada R, et al. Phase II trial of CH5424802 (alectinib hydrochloride) for recurrent or refractory ALK-positive anaplastic large cell lymphoma. Blood 2018;132(Suppl 1):2924.

[36] Ceccon M, Merlo MEB, Mologni L, Poggio T, Varesio LM, Menotti M, et al. Excess of NPM-ALK oncogenic signaling promotes cellular apoptosis and drug dependency. Oncogene 2016;35(29):3854–65.

[37] Rigaud C, Abbas R, Simonin M, Le Mouel L, Pereira V, Geoerger B, et al. Profound and sustained response with ALK inhibitors in patients with relapsed or progressive ALK positive anaplastic large cell lymphoma with CNS involvement. In: SIOP 2019 Congr; 2019.

[38] Marzec M, Zhang Q, Goradia A, Raghunath PN, Liu X, Paessler M, et al. Oncogenic kinase NPM/ALK induces through STAT3 expression of immunosuppressive protein CD274 (PD-L1, B7-H1). Proc Natl Acad Sci U S A 2008;105(52):20852–7.

[39] Rigaud C, Abbou S, Minard-Colin V, Geoerger B, Scoazec JY, Vassal G, et al. Efficacy of nivolumab in a patient with systemic refractory ALK+ anaplastic large cell lymphoma. Pediatr Blood Cancer 2018;65(4). https://doi.org/10.1002/pbc.26902.

[40] Chan TSY, Khong P-L, Kwong Y-L. Pembrolizumab for relapsed anaplastic large cell lymphoma after allogeneic haematopoietic stem cell transplantation: efficacy and safety. Ann Hematol 2016;95(11):1913–5.

[41] Hebart H, Lang P, Woessmann W. Nivolumab for refractory anaplastic large cell lymphoma: a case report. Ann Intern Med 2016;165(8):607–8.

[42] Coffin CM, Watterson J, Priest JR, Dehner LP. Extrapulmonary inflammatory myofibroblastic tumor (inflammatory pseudotumor). A clinicopathologic and immunohistochemical study of 84 cases. Am J Surg Pathol 1995;19(8):859–72.

[43] Janik JS, Janik JP, Lovell MA, Hendrickson RJ, Bensard DD, Greffe BS. Recurrent inflammatory pseudotumors in children. J Pediatr Surg 2003;38(10):1491–5.

[44] Buccoliero AM, Ghionzoli M, Castiglione F, Paglierani M, Baroni G, Messineo A, et al. Inflammatory myofibroblastic tumor: clinical, morphological, immunohistochemical and molecular features of a pediatric case. Pathol Res Pract 2014;210(12):1152–5.

[45] Pettinato G, Manivel JC, De Rosa N, Dehner LP. Inflammatory myofibroblastic tumor (plasma cell granuloma). Clinicopathologic study of 20 cases with immunohistochemical and ultrastructural observations. Am J Clin Pathol 1990;94(5):538–46.

[46] Coffin CM, Hornick JL, Fletcher CDM. Inflammatory myofibroblastic tumor: comparison of clinicopathologic, histologic, and immunohistochemical features including ALK expression in atypical and aggressive cases. Am J Surg Pathol 2007;31(4):509–20.

[47] Hussong JW, Brown M, Perkins SL, Dehner LP, Coffin CM. Comparison of DNA ploidy, histologic, and immunohistochemical findings with clinical outcome in inflammatory myofibroblastic tumors. Mod Pathol Off J U S Can Acad Pathol Inc 1999;12(3):279–86.

[48] Ramachandra S, Hollowood K, Bisceglia M, Fletcher CD. Inflammatory pseudotumour of soft tissues: a clinicopathological and immunohistochemical analysis of 18 cases. Histopathology 1995;27(4):313–23.

[49] Warter A, Satge D, Roeslin N. Angioinvasive plasma cell granulomas of the lung. Cancer 1987;59(3):435–43.

[50] Yamamoto H, Oda Y, Saito T, Sakamoto A, Miyajima K, Tamiya S, et al. p53 mutation and MDM2 amplification in inflammatory myofibroblastic tumours. Histopathology 2003;42(5):431–9.

[51] Cook JR, Dehner LP, Collins MH, Ma Z, Morris SW, Coffin CM, et al. Anaplastic lymphoma kinase (ALK) expression in the inflammatory myofibroblastic tumor: a comparative immunohistochemical study. Am J Surg Pathol 2001;25(11):1364–71.

[52] Griffin CA, Hawkins AL, Dvorak C, Henkle C, Ellingham T, Perlman EJ. Recurrent involvement of 2p23 in inflammatory myofibroblastic tumors. Cancer Res 1999;59(12):2776–80.

[53] Mariño-Enríquez A, Wang W-L, Roy A, Lopez-Terrada D, Lazar AJF, Fletcher CDM, et al. Epithelioid inflammatory myofibroblastic sarcoma: an aggressive intra-abdominal variant of inflammatory myofibroblastic tumor with nuclear membrane or perinuclear ALK. Am J Surg Pathol 2011;35(1):135–44.

[54] Takeuchi K, Soda M, Togashi Y, Sugawara E, Hatano S, Asaka R, et al. Pulmonary inflammatory myofibroblastic tumor expressing a novel fusion, PPFIBP1-ALK: reappraisal of anti-ALK immunohistochemistry as a tool for novel ALK fusion identification. Clin Cancer Res Off J Am Assoc Cancer Res 2011;17(10):3341–8.

[55] Lovly CM, Gupta A, Lipson D, Otto G, Brennan T, Chung CT, et al. Inflammatory myofibroblastic tumors harbor multiple potentially actionable kinase fusions. Cancer Discov 2014;4(8):889–95.

[56] Butrynski JE, D'Adamo DR, Hornick JL, Dal Cin P, Antonescu CR, Jhanwar SC, et al. Crizotinib in ALK-rearranged inflammatory myofibroblastic tumor. N Engl J Med 2010;363(18):1727–33.

[57] Theilen T-M, Soerensen J, Bochennek K, Becker M, Schwabe D, Rolle U, et al. Crizotinib in ALK+ inflammatory myofibroblastic tumors—current experience and future perspectives. Pediatr Blood Cancer 2018;65(4). https://doi.org/10.1002/pbc.26920.

[58] Brivio E, Zwaan CM. ALK inhibition in two emblematic cases of pediatric inflammatory myofibroblastic tumor: efficacy and side effects. Pediatr Blood Cancer 2019;66(5), e27645.

[59] Parker BM, Parker JV, Lymperopoulos A, Konda V. A case report: pharmacology and resistance patterns of three generations of ALK inhibitors in metastatic inflammatory myofibroblastic sarcoma. J Oncol Pharm Pract Off Publ Int Soc Oncol Pharm Pract 2019;25(5):1226–30.

[60] Cha YJ, Shim HS. PD-L1 expression and CD8+ tumor-infiltrating lymphocytes are associated with ALK rearrangement and clinicopathological features in inflammatory myofibroblastic tumors. Oncotarget 2017;8(52):89465–74.

[61] Cottrell TR, Duong AT, Gocke CD, Xu H, Ogurtsova A, Taube JM, et al. PD-L1 expression in inflammatory myofibroblastic tumors. Mod Pathol Off J U S Can Acad Pathol Inc 2018;31(7):1155–63.

[62] Li J, Thompson TD, Miller JW, Pollack LA, Stewart SL. Cancer incidence among children and adolescents in the United States, 2001–2003. Pediatrics 2008;121(6):e1470–7.

[63] Spix C, Pastore G, Sankila R, Stiller CA, Steliarova-Foucher E. Neuroblastoma incidence and survival in European children (1978–1997): report from the Automated Childhood Cancer Information System project. Eur J Cancer Oxf Engl 1990 2006;42(13):2081–91.

[64] Cohn SL, Pearson ADJ, London WB, Monclair T, Ambros PF, Brodeur GM, et al. The International Neuroblastoma Risk Group (INRG) classification system: an INRG Task Force report. J Clin Oncol Off J Am Soc Clin Oncol 2009;27(2):289–97.

[65] Matthay KK, Villablanca JG, Seeger RC, Stram DO, Harris RE, Ramsay NK, et al. Treatment of high-risk neuroblastoma with intensive chemotherapy, radiotherapy, autologous bone marrow transplantation, and 13-cis-retinoic acid. Children's Cancer Group. N Engl J Med 1999;341(16):1165–73.

[66] Yu AL, Gilman AL, Ozkaynak MF, London WB, Kreissman SG, Chen HX, et al. Anti-GD2 antibody with GM-CSF, interleukin-2, and isotretinoin for neuroblastoma. N Engl J Med 2010;363(14):1324–34.

[67] Park JR, Kreissman SG, London WB, Naranjo A, Cohn SL, Hogarty MD, et al. Effect of tandem autologous stem cell transplant vs single transplant on event-free survival in patients with high-risk neuroblastoma: a randomized clinical trial. JAMA 2019;322(8):746–55.

[68] French S, DuBois SG, Horn B, Granger M, Hawkins R, Pass A, et al. 131I-MIBG followed by consolidation with busulfan, melphalan and autologous stem cell transplantation for refractory neuroblastoma. Pediatr Blood Cancer 2013;60(5):879–84.

[69] Heczey A, Louis CU, Savoldo B, Dakhova O, Durett A, Grilley B, et al. CAR T cells administered in combination with lymphodepletion and PD-1 inhibition to patients with neuroblastoma. Mol Ther J Am Soc Gene Ther 2017;25(9):2214–24.

[70] Louis CU, Savoldo B, Dotti G, Pule M, Yvon E, Myers GD, et al. Antitumor activity and long-term fate of chimeric antigen receptor-positive T cells in patients with neuroblastoma. Blood 2011;118(23):6050–6.

[71] Oeffinger KC, Mertens AC, Sklar CA, Kawashima T, Hudson MM, Meadows AT, et al. Chronic health conditions in adult survivors of childhood cancer. N Engl J Med 2006;355(15):1572–82.

[72] London WB, Bagatell R, Weigel BJ, Fox E, Guo D, Van Ryn C, et al. Historical time to disease progression and progression-free survival in patients with recurrent/refractory neuroblastoma treated in the modern era on Children's Oncology Group early-phase trials. Cancer 2017;123(24):4914–23.

[73] Moreno L, Rubie H, Varo A, Le Deley MC, Amoroso L, Chevance A, et al. Outcome of children with relapsed or refractory neuroblastoma: a meta-analysis of ITCC/SIOPEN European phase II clinical trials. Pediatr Blood Cancer 2017;64(1):25–31.

[74] Janoueix-Lerosey I, Lequin D, Brugières L, Ribeiro A, de Pontual L, Combaret V, et al. Somatic and germline activating mutations of the ALK kinase receptor in neuroblastoma. Nature 2008;455(7215):967–70.

[75] Bresler SC, Weiser DA, Huwe PJ, Park JH, Krytska K, Ryles H, et al. ALK mutations confer differential oncogenic activation and sensitivity to ALK inhibition therapy in neuroblastoma. Cancer Cell 2014;26(5):682–94.

[76] Schleiermacher G, Javanmardi N, Bernard V, Leroy Q, Cappo J, Rio Frio T, et al. Emergence of new ALK mutations at relapse of neuroblastoma. J Clin Oncol Off J Am Soc Clin Oncol 2014;32(25):2727–34.

[77] Bellini A, Bernard V, Leroy Q, Rio Frio T, Pierron G, Combaret V, et al. Deep sequencing reveals occurrence of subclonal ALK mutations in neuroblastoma at diagnosis. Clin Cancer Res Off J Am Assoc Cancer Res 2015;21 (21):4913–21.

[78] Trigg RM, Turner SD. ALK in neuroblastoma: biological and therapeutic implications. Cancers 2018;10(4). https://doi.org/10.3390/cancers10040113.

[79] Chen Y, Takita J, Choi YL, Kato M, Ohira M, Sanada M, et al. Oncogenic mutations of ALK kinase in neuroblastoma. Nature 2008;455(7215):971–4.

[80] Friboulet L, Li N, Katayama R, Lee CC, Gainor JF, Crystal AS, et al. The ALK inhibitor ceritinib overcomes crizotinib resistance in non-small cell lung cancer. Cancer Discov 2014;4(6):662–73.

[81] Lu J, Guan S, Zhao Y, Yu Y, Woodfield SE, Zhang H, et al. The second-generation ALK inhibitor alectinib effectively induces apoptosis in human neuroblastoma cells and inhibits tumor growth in a TH-MYCN transgenic neuroblastoma mouse model. Cancer Lett 2017;400:61–8.

[82] Zhang S, Anjum R, Squillace R, Nadworny S, Zhou T, Keats J, et al. The potent ALK inhibitor brigatinib (AP26113) overcomes mechanisms of resistance to first- and second-generation ALK inhibitors in preclinical models. Clin Cancer Res Off J Am Assoc Cancer Res 2016;22(22):5527–38.

[83] Infarinato NR, Park JH, Krytska K, Ryles HT, Sano R, Szigety KM, et al. The ALK/ROS1 inhibitor PF-06463922 overcomes primary resistance to crizotinib in ALK-driven neuroblastoma. Cancer Discov 2016;6(1):96–107.

[84] Balis FM, Thompson PA, Mosse YP, Blaney SM, Minard CG, Weigel BJ, et al. First-dose and steady-state pharmacokinetics of orally administered crizotinib in children with solid tumors: a report on ADVL0912 from the Children's Oncology Group Phase 1/Pilot Consortium. Cancer Chemother Pharmacol 2017;79(1):181–7.

[85] Greengard EG, Mosse YP, Liu X, Ahern CH, Minard C, Blaney S, et al. Safety and tolerability of crizotinib in combination with chemotherapy for relapsed or refractory solid tumors or anaplastic large cell lymphoma: a Children's Oncology Group phase I consortium study. J Clin Oncol 2015;33(15_suppl):10058.

Resistance to anaplastic lymphoma kinase (ALK) tyrosine kinase inhibitors (TKIs) in patients with lung cancer: Single mutations, compound mutations, and other mechanisms of drug resistance

Ryohei Katayama[a,b]

[a]Division of Experimental Chemotherapy, Cancer Chemotherapy Center, Japanese Foundation for Cancer Research, Tokyo, Japan [b]Department of Computational Biology and Medical Sciences, Graduate School of Frontier Sciences, The University of Tokyo, Tokyo, Japan

Abstract

Lung cancer is a leading cause of the deaths due to cancer, and the development of effective treatments has been an urgent issue. Over the last 15 years, EGFR activating mutations and fusion genes such as *ALK* or *ROS1* have been discovered, and a number of molecular targeting drugs have been developed, that results in a re- markable antitumor effect and great extension of the survival period for the advanced lung cancer patients. In ALK rearranged cancer, many ALK tyrosine kinase inhibitors (ALK-TKIs) have been developed and 5 ALK- TKIs were approved and used in clinical setting. However, the major problem is the tumor recurrence due to the acquired resistances, and that make it difficult to achieve complete cure. The mechanism of acquired re- sistance varies not only from case to case, but also multiple resistance mechanisms exist in each patient with high diversity. This chapter focuses on the drug resistance mechanisms to ALK-TKIs in ALK-rearranged lung cancer, and the potential therapeutic strategies to overcome resistance.

Abbreviations

ALK	anaplastic lymphoma kinase
TKI	tyrosine kinase inhibitor
NSCLC	non-small cell lung cancer
ROS1	c-ros oncogene 1
EGFR	epidermal growth factor receptor

Conflict of interest

R. Katayama reports research grants from Chugai, TAKEDA, TOPPAN Printing, Daiichi-Sankyo, and lecture's fee from Pfizer, outside the submitted work.

Introduction

ALK (anaplastic lymphoma kinase) is expressed focally in neuronal cells and in the small intestine. Although the precise physiological functions of ALK remain unclear, it is believed to play a role in promoting normal development and function of the nervous system. Interestingly, no apparent dysfunction was observed in the *Alk* gene-deleted mice [1,2], and no universally-recognized adverse events have been associated with the administration of ALK-specific tyrosine kinase inhibitors (TKIs). However, there are numerous reports documenting the function of ALK in model organisms, including *Drosophila* and nematodes. In *Drosophila*, ALK binds to its ligand, Jeb (Jelly belly) and promotes synapse formation and gastrointestinal development [3,4].

By contrast, the function of ALK tyrosine kinase and its role in promoting malignant disease have been well studied, most notably in lung cancer associated with ALK gene rearrangements [5]. Upon formation of fusion genes, including EML4-ALK or NPM-ALK, the fusion protein is expressed constitutively, and ALK tyrosine kinase is activated in response to oligomerization of the fusion partner protein; this results in constitutive activation of cell growth signaling and the induction of tumorigenesis [5,6] (Fig. 1). To inhibit the pro-oncogenic signaling promoted by the ALK fusion protein, ALK TKIs have been developed. Administration of the inhibitors resulted in marked tumor shrinkage in vitro, in vivo, and in clinical trials. To date, five ALK inhibitors have been approved and have been introduced for the treatment of non-small-cell lung cancer (NSCLC) with *ALK*-gene rearrangements. The use of potent ALK inhibitors has resulted in significant improvements with respect to the prognosis of ALK-rearranged NSCLC; survival has been prolonged remarkably. Currently, the average survival time for patients diagnosed with NSCLC with ALK gene rearrangements is over 4 years, even among those with advanced disease. Unfortunately, despite these initial remarkable responses, relapse is inevitable due to acquired resistance. Drug resistance mechanisms can be categorized into three groups, including (1) secondary mutations that develop in ALK (including compound mutations), (2) activation of bypass pathways, and (3) other mechanisms including transformation to small cell lung cancer and other cancer phenotypes [7]. This chapter provides a comprehensive insight into our current understanding of resistance mechanisms to ALK-TKIs. Therapeutic strategies that have the potential to overcome this resistance are also briefly reviewed.

ALK fusion genes

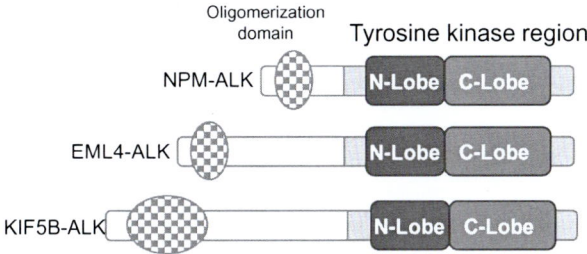

Reported ALK fusion genes
(including multiple cancer types)

- NPM-ALK	- MSN-ALK
- EML4-ALK	- PPFIBP1-ALK
- ATIC-ALK	- PRKAR1A-ALK
- C2orf44-ALK	- RANBP2-ALK
- CARS-ALK	- SEC31A-ALK
- CLTC-ALK	- SQSTM1-ALK
- CLTCL1-ALK	- STRN-ALK
- FN1-ALK	- TFG-ALK
- HIP1-ALK	- TPM3-ALK
- KIF5B-ALK	- TPM4-ALK
- KLC1-ALK	- VCL-ALK
- LMNA-ALK	

FIG. 1   Reported ALK fusion oncogenes including the fusion genes identified in multiple cancer types.

Mechanisms of resistance to ALK-TKIs

Point mutations in the ALK kinase domain mediate resistance to ALK-TKI

The first-generation ALK-TKI, crizotinib, reduced tumors in more than 60% of patients diagnosed with ALK-positive lung cancer, although the median progression-free survival of patients treated with the agent was about 10 months [8–10]. This result suggested that acquired resistance emerged in more than half of the patients treated with crizotinib within one year. The first reported mechanism associated with crizotinib resistance involved the generation of two independent mutations in ALK; specifically, the mutations C1156Y and L1196M were identified in cells in a pleural effusion from a patient that had relapsed while on crizotinib treatment [11]. The ALK-L1196M mutation is at a "gatekeeper site" and is analogous to epidermal growth factor (EGFR)-T790M, the most frequently observed resistance-associated mutation identified in first- or second-generation EGFR-TKI treated patients; it is also analogous to the T315I mutant of ABL kinase that has been observed in relapsed chronic myeloid leukemia in patients that had undergone treatment with the BCR-ABL inhibitors, imatinib, or dasatinib. The gatekeeper residue is situated within a hydrophobic pocket located at the hinge region at the site of ATP binding to ALK tyrosine kinase. The gatekeeper residue plays an important role with respect to the interactions between kinases and kinase inhibitors and can serve to increase the specificity of specific TKIs. For example, in the case of ABL kinase, the gatekeeper residue, T315, interacts directly to a methyl group of the phenyl ring of imatinib and thereby increases the potency and specificity of imatinib; the T315I mutation impairs the affinity of the ABL kinase for imatinib. Likewise, in the case of the EGFR-T790M, this mutation is believed to increase the affinity of EGFR for ATP, thereby promoting a slight decrease in its affinity for first- or second- generation EGFR-TKIs; in other words, this results in the relative decrease in the affinity of EGFR for the TKI, given the overall abundance of ATP within the target cells. Similarly, the ALK-L1196M mutation has been reported to promote an increase in kinase activity along with a diminished affinity for crizotinib [12].

On the other hand, the ALK-C1156Y mutation is situated in the N-lobe of ALK and may change the conformation of the alpha-C helix and as such it may have an impact on the conformation of the crizotinib binding region; however, a detailed mechanism has not yet been clarified.

A number of additional mutations in ALK have been identified from cells isolated from patients who underwent relapse while under treatment with crizotinib; among these are a 1151T-insertion, L1152R, I1171T, F1174L, G1202R, S1206Y, F1245V, and G1269A (Fig. 2). The L1196M and G1269A mutations have been reported at relatively high frequency [13,14]. The G1269 residue is located immediately adjacent to an Asp-Phe-Gly (DFG) motif that has been recognized within the catalytic cores of protein kinases. The G1269A mutation induces a slight structural change. However crizotinib binds at a site near to the G1269 residue, and as such, this point mutation may largely affect its binding affinity. G1202R and S1206Y mutations are "solvent front" mutations; G1202 and S1206 are situated on the outer

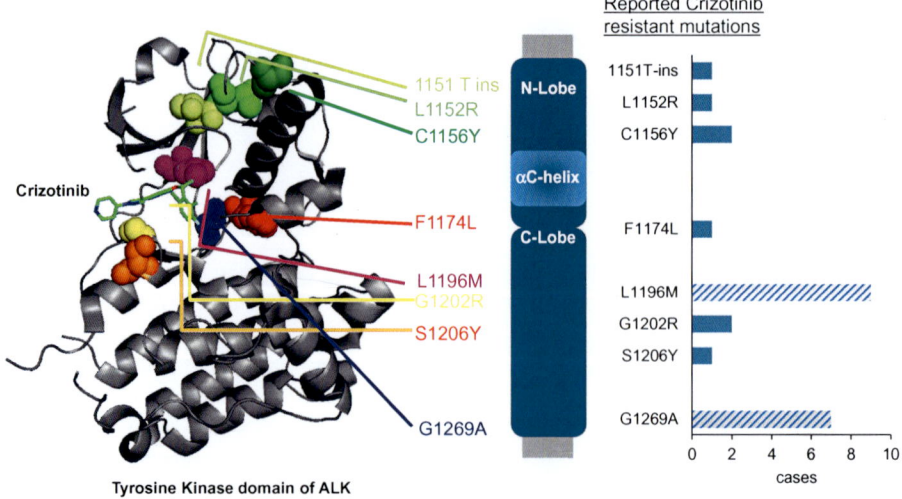

FIG. 2   Major crizotinib resistant mutations. Mutated amino acid residues in ALK kinase domain were pointed in ribbon structure of ALK (PDB-ID: 2XP2). *Adapted from Katayama et al. Clin Cancer Res 2015.*

face of the ALK and are in contact with solvent. It is critical to recognize that these two mutations replace relatively small amino acids (G and S) with bulky (Y) and charged (R) amino acids, that could provide steric hindrance to crizotinib binding. Similarly, the 1151T insertion and the L1152R mutations are located close to C1156 in N-lobe of ALK; as such, they may also have a negative impact on the structure of alpha-C helix that contributes to the active protein conformation [13]. Interestingly, ALK mutations F1174L and F1245V were initially identified as activating mutations in cases of neuroblastoma [12,15]; as such, they have the potential to induce constitutive activation of full length ALK. Specifically, the mutations introduced at either F1174 or F1245 destroy the π-π interactions between their two phenyl rings that are critical in order to maintain ALK in an inactive conformation. This conformation prohibits auto-phosphorylation of ALK at Y1278, Y1282 and Y1283 residues found in the activation loop (A-loop).

The second-generation ALK inhibitors, alectinib and ceritinib, were reported to have the capacity to overcome crizotinib resistance both in vitro and in vivo. Indeed, both alectinib and ceritinib exhibited marked anti-tumor effects in patients who underwent previous treatment with crizotinib as well as in those who are ALK-TKI treatment-naïve. Alectinib was first reported as a potent and selective ALK-TKI with the capacity to overcome the L1196M gatekeeper mutation [16]. ALKs with other mutations associated with crizotinib resistance, including C1156Y, F1174I/L, and S1206Y were also sensitive to alectinib; these results suggest that clinical resistance to crizotinib may be overcome by alectinib. However, analysis of specimens from patients with refractory disease together with experimental data from cell line models has revealed that ALK mutations including I1171N/S/T, V1180L, and G1202R all promote resistance to alectinib [17]. Similarly, albeit controversial, the L1196M gatekeeper mutation associated with resistance to crizotinib was also identified in tumor cells from patients demonstrating resistance to alectinib [18]. However, it is important to recognize that the IC_{50} of alectinib for Ba/F3 cells that express mutant EML4-ALK-L1196M is about 10-fold higher than reported in cells expressing EML4-ALK-WT; in other words, serum concentrations of alectinib that can be reasonably achieved may not be high enough to inhibit the activity of ALK-L1196M in some cases.

Ceritinib is another potent and selective ALK-TKI that has the capacity to overcome several of the characterized mutations associated with crizotinib resistance. Indeed, cells with multiple crizotinib-resistant mutations, including L1196M and G1269A, were shown to be highly sensitive to administration of ceritinib [19,20]. Furthermore, the I1171T and V1180L ALK mutations associated with alectinib resistance could be readily overcome by treatment with ceritinib; in fact, mutated EML4-ALK-V1180L was found to be more sensitive to ceritinib than was the wild-type EML4-ALK [17]. Ceritinib resistant mutations have also been identified, including F1174V and G1202R [20]. Among these, the G1202R mutation confers resistance to crizotinib, alectinib and ceritinib, while several characterized ALK mutations, including I1171T/N/S, V1180L, and F1174V, can be overcome by any of the first or second-generation ALK-TKIs. Interestingly, IC_{50} of ceritinib when used to treat cells with the EML4-ALK-F1174V mutations in only a few-fold higher than the wild-type; nonetheless, this "activating" mutation was observed frequently among ceritinib resistant patients. The mechanisms underlying this observation have not yet been clarified.

Brigatinib is another potent, second-generation ALK-TKI that can overcome the gatekeeper L1196M mutation [21,22]. In addition, brigatinib was shown to be effective against ALK with multiple single mutations including L1152R, C1156Y, V1180L, and I1171T/S/N. In vitro findings have shown that brigatinib can effectively inhibit ALK-G1202R with an

IC_{50} of approximately 100 nM. However, the G1202R mutation was also observed in samples from 3 of 7 patients who had undergone treatment with brigatinib; these results implied that brigatinib may have only marginal activity against cells harboring the ALK G1202R mutation. Several ALK mutations were reported in cells of patients who relapsed while on brigatinib treatment, including the single mutations D1203N and G1202R and compound mutations including E1210K+S1206C and E1210K+D1203N [18].

Lorlatinib is a third generation ALK-TKI with high potency and high selectivity for both ALK and the related receptor tyrosine kinase, ROS1, in studies carried out in vitro. Lorlatinib was shown to be effective in overcoming the impact of all single point mutations detected in cells from crizotinib, alectinib and ceritinib-refractory patients [23]. Although the IC_{50} for lorlatinib-mediated inhibition of for ALK-G1202R was much higher than that reported for wild-type ALK, it was measured at ~50–100 nM. Experiments performed in vivo also clearly showed that treatment with lorlatinib of 7.5 mg/kg or higher resulted in tumor shrinkage; the lorlatinib concentration that resulted in complete inhibition of ALK-G1202R was lower than that detected in free plasma [23]. Lorlatinib also showed activity against NSCLC cells expressing EML4-ALK-G1202R in vivo, and patients with tumors that harbored the G1202R mutation responded to lorlatinib therapy in clinical trials [24,25]. Taken together, these results suggest that lorlatinib is active against all the single ALK-TKI resistant mutations. However, relapse due to acquired resistance to lorlatinib was observed in tumor cells with compound ALK mutations, including C1156Y+L1198F, L1196M+G1202R, I1171N+ L1198F, I1171N+1203N, G1202R+G1269A, I1171S+G1269A, L1196M+D1203N, F1174L+ G1202R, and C1156Y+G1269A [26–28]. Mutagenesis screening resulted in the identification of a number of lorlatinib-resistant compound mutations; many of these were also identified in specimens from lorlatinib-resistant patients. Interestingly, ALK with the compound I1171N+ L1256F mutation was extremely resistant to lorlatinib but remained sensitive to alectinib [29] (Fig. 3). ALK with a single L1256F mutation was also found to be tremendously resistant

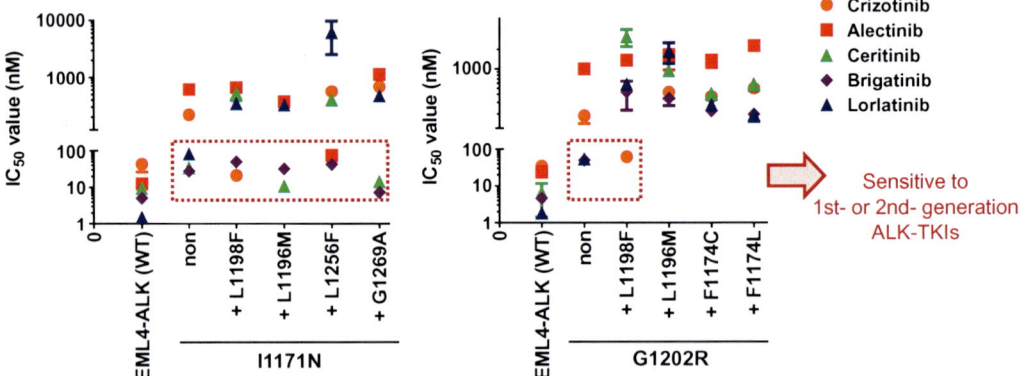

FIG. 3   Compound mutations such as G1202R+L1196M in ALK induced marked increase of IC50 to lorlatinib. Some of the lorlatinib resistant mutants were sensitive to first or second generation ALK inhibitors. *Adapted from Okada et al. EBioMed 2019.*

to lorlatinib, but remained quite sensitive to alectinib. Furthermore, the EML4-ALK-L1256F single mutation was associated with reduced oncogenic activity, although ALK with an I1171N + L1256F compound mutation maintained tyrosine kinase activity [29]. Interestingly, Yoda et al. identified triple compound mutations (G1202R + L1204V + G1269A or E1210K + D1203N + G1269A) in cells from lorlatinib-resistant patients [27]. Taken together, these results suggest that while lorlatinib is active against single point mutations that confer resistance to other ALK-TKIs, this drug will no longer be effective once two or more ALK mutations develop simultaneously; this has been found to be the case even when the compound mutations include single mutations that would otherwise be susceptible to lorlatinib treatment.

Shaw et al. reported a case of a patient diagnosed with NSCLC with an ALK rearrangement that had undergone sequential treatment with crizotinib, alectinib, ceritinib, and lorlatinib. After completing lorlatinib treatment, the tumor eventually relapsed with an ALK C1156Y + L1198F compound mutation. Of interest, only the C1156Y mutation had been identified after the first round of treatment; addition of the L1198F mutation resulted in re-sensitization of ALK to crizotinib [26]. Given this observation, the patient underwent a second round of treatment with crizotinib after developing lorlatinib resistance; tumor shrinkage was confirmed for a period of approximately half a year. Structural analysis provided an explanation for the ALK L1198F mutation and its role in promoting resistance to lorlatinib with re-sensitization to crizotinib. Lorlatinib has a macrocyclic structure similar to that of crizotinib and includes a specific side chain modification. This side chain derivative was optimized to increase selectivity for and inhibition of ALK, and to reduce activity against neurotrophic tropomyosin-related kinase (NTRK) proteins. In the sequences of NTRKs, the amino acid corresponding to L1198 residue in ALK is an F residue. To increase the selectivity of lorlatinib for ALK alone, a nitrile group was added in a portion near the site of interaction with L1198. As such, the ALK-L1198F mutation is resistant to lorlatinib due to a marked decrease in binding affinity notably at or near F1198. However, the ALK L1198F mutation increased the binding affinity of crizotinib as determined by analysis of its crystal structure [26]. Nonetheless, compound ALK mutations including L1196M + G1202R, F1174I + G1202R, I1171N + L1198H are resistant to all clinically-available ALK-TKIs. Okada and colleagues have shown that adaphostin and its analogs have inhibitory activity against ALK L1196M + G1202R mutants at relatively low concentrations [29]. Currently, sequential treatment with ALK inhibitors is a widely accepted practice, as most of the single ALK mutations conferring resistance to TKIs can be overcome with lorlatinib or brigatinib treatment; furthermore, as shown, earlier generation ALK-TKIs might be able to overcome resistance to specific compound mutations. As such, identification of resistance patterns and their underlying mechanisms could be very important in order to decide optimal strategy with respect to treatment with ALK-TKIs.

Comparison of resistance-associated ALK mutations to those of other tyrosine kinases

As noted in the previous section, many secondary mutations in ALK have been identified that promote resistance to TKIs; the mutations in ALK that have been associated with drug resistance are more complex and varied than those promoting TKI resistance in EGFR.

For example, the ALK-L1196M gatekeeper mutation is in a position analogous to that of EGFR-T790M, Abl-T315I, ROS1-L2026M, and RET-V804M. Likewise, the ALK-G1202R solvent front mutation corresponds to NTRK1-G595R [30] and ROS1-G2032R [31], which is the most common of the crizotinib-resistant mutations in cancers with a ROS1 rearrangement. Similarly, the ALK-G1269A mutation correspond to NTRK-G667C [30]; other parallels include ALK-L1256F to ROS1-L2086F [32], ALK-C1156Y to ROS1-L1986F, and F1174I/L/V to ROS1-F2004V. As indicated here, most of these mutations are commonly observed among the characterized oncogenic tyrosine kinases.

Computational simulation to predict the resistance and effective drugs

There are a number of reports that focus on in silico computational molecular dynamic simulations of drug binding and affinity estimation. We previously performed computational simulations focused on ALK and using MP-CAFEE and successfully quantified the binding affinity of the ALK-TKI alectinib to several ALK-TKI resistant mutants [17,29,33]. In order to provide a strict assessment of the prediction accuracy, we applied computational thermodynamic simulation with MP-CAFEE to the interactions of multiple ALK-TKIs (crizotinib, alectinib or ceritinib) and specific resistant mutants and compared the calculated binding free energies (ΔG) with the experimental IC_{50} values measured in in vitro cell viability assays targeting Ba/F3 cells expressing the EML4-ALK mutants. A clear, linear correlation was identified when comparing the experimental IC_{50} values and the calculated ΔGs for all three TKIs [29]. This result suggests that the free energy estimation using MP-CAFEE can correctly predict the impact of each ALK-TKI resistant mutation on drug binding. Similar results were obtained for calculated ΔGs predicted for several of the lorlatinib-resistant ALK double mutations and their experimental IC_{50} values; however, the prediction failed for several lorlatinib-resistant ALK compound mutations including I1171N + L1256F. This may relate to the fact that the L1256F mutation added a phenyl ring with the potential to contribute to an increase in π-π electric interactions; these interactions cannot be formally evaluated by molecular dynamic simulations such as MP-CAFEE, but might be amenable to evaluation by quantum chemistry [29]. Multiple strategies will be needed to improve current in silico simulations so that they might be capable of providing accurate estimations of drug binding affinities [34].

ALK-TKI resistance mechanisms other than secondary mutations

In addition to the secondary mutations within the ALK kinase domain, amplification of the ALK fusion gene has been identified as another significant mechanism underlying crizotinib resistance [21]. However, it is not clear that ALK gene amplification is involved in the resistance observed after treatment with second or third generation ALK-TKIs. Likewise, other mechanisms that promote growth activation, including those involving EGFR, may induce the resistance to ALK-TKIs. This form of resistance occurs via what is known as "bypass pathway" activation (Fig. 4). For example, specimens from patients that have developed crizotinib resistance have included amplification of the cKIT gene and elevated stromal levels of expression of its ligand, stem cell factor [13] (Fig. 5). Likewise, EGFR and insulin-like growth factor

Resistance mechanisms in ALK rearranged cancer

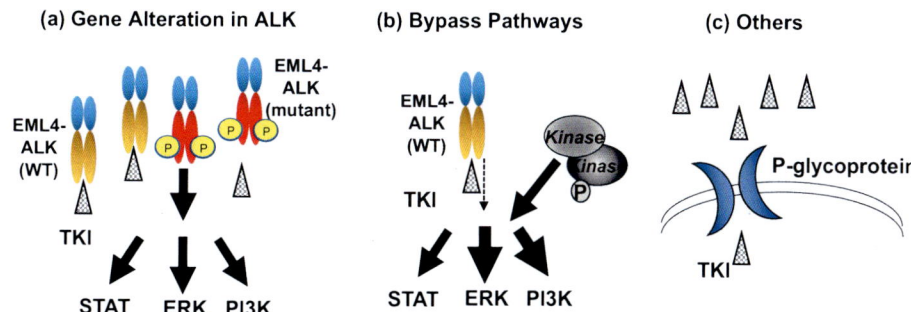

	(a) Gene Alteration in ALK	(b) Bypass Pathways	(c) Others
Crizotinib resistance	L1196M, G1269A, C1156Y, G1202R, I1151T ins, etc...	EGFR or IGF1R activation, cKIT amplification with SCF, Src family activation, etc...	- P-glycoprotein upregulation - Microenvironment (?) - Transformation to Squamous or SCLC - Leptomeningeal metastasis
Alectinib Resistance	I1171T/N/S, G1202R, V1180L	cMET amplification, Ligand mediated activation (EGFR, cMET, etc...)	
Ceritinib Resistance	F1174C/V, G1202R,	MEK active mutation	

FIG. 4   ALK-TKIs (crizotinib, alectinib and ceritinib) resistance mechanisms. (A) Mutations in ALK is observed in approximately half of the resistant patients. In other cases, (B) Bypass pathway activation, such as other receptor tyrosine kinase activation or growth signaling pathway activation, induced ALK-TKI resistances are observed. As other mechanisms, (C) P-glycoprotein upregulation (to crizotinib and ceritinib), or small cell lung cancer transformation were observed. Multiple mechanism can be found even in one patient.

Bypass Pathway mediated resistance mechanisms

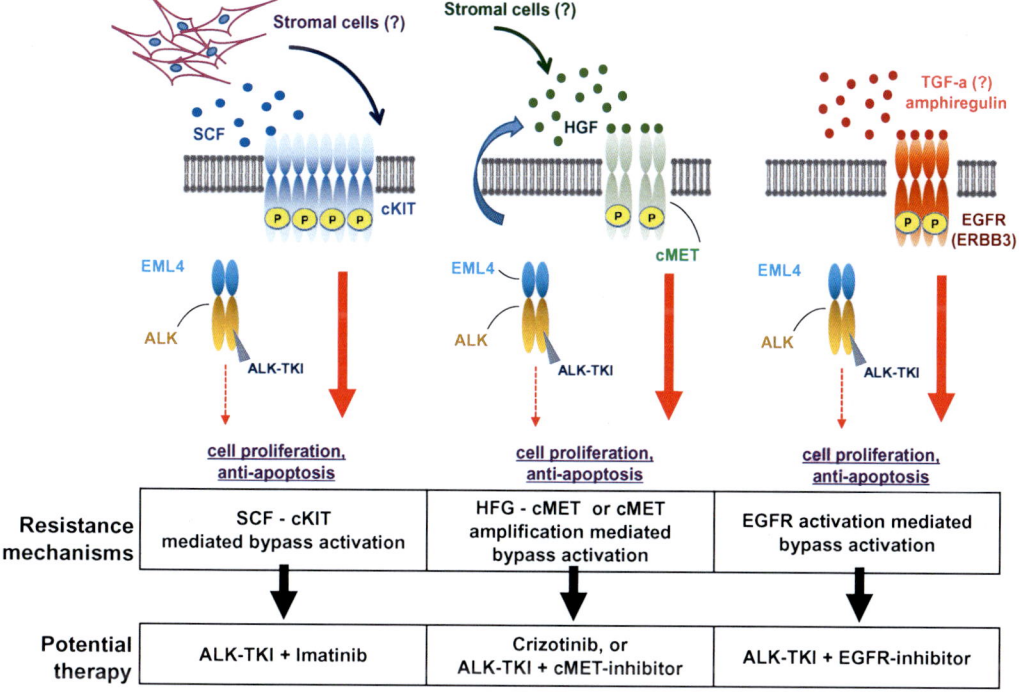

Resistance mechanisms	SCF - cKIT mediated bypass activation	HFG - cMET or cMET amplification mediated bypass activation	EGFR activation mediated bypass activation
Potential therapy	ALK-TKI + Imatinib	Crizotinib, or ALK-TKI + cMET-inhibitor	ALK-TKI + EGFR-inhibitor

FIG. 5   Typical example of the bypass pathway-mediated resistance mechanisms and the potential combination therapies to overcome these resistances are shown.

(IGF)1R activation, upregulation of ligands of ERBB receptor family, and KRAS- activating mutations were all identified as mechanisms underlying acquisition of crizotinib resistance [35–38]. Furthermore, amplification of *cMET* and/or upregulation its ligand, HGF, MEK activating mutations, and activating mutations of the RAS/Src signaling pathways have been associated with resistance to alectinib and ceritinib [36,38–40]. Of note, since crizotinib has a potent cMET inhibiting activity (as crizotinib was originally developed and clinically evaluated as a cMET inhibitor), no cases of cMET amplification have been reported in association with resistance to crizotinib, although this aberration has been identified at significant frequencies in association with resistance to alectinib or ceritinib. Experimental studies have revealed that bypass pathway mediated resistance can be overcome by combining an inhibitor that suppresses the activated bypass pathway with an ALK inhibitor, such as alectinib. However, combination therapy has not yet been established; at this time, there are no combination therapies proven to be safe and effective in a clinical setting.

Among other resistance mechanisms, overexpression of P-glycoprotein/ ABCB1, a drug efflux membrane transporter, was observed in association with ceritinib resistance [41]. In this report, crizotinib and ceritinib, but not alectinib or lorlatinib, were effectively exported by P-glycoprotein. To overcome resistance secondary to P-glycoprotein overexpression, monotherapy with either alectinib or lorlatinib or the combination therapy with ceritinib and a P-glycoprotein inhibitor such as MS209 has been shown to be effective. Of note, the blood brain barrier expresses high levels of P-glycoprotein, which contributes to drug export from cerebrospinal fluid. Lorlatinib, but not crizotinib, reached high concentration in the CNS; as such, lorlatinib was shown to be effective for treating brain and/or leptomeningeal metastasis in patients with lung cancer with ALK rearrangements. Recently, Arai et al. reported that cells from metastatic leptomeningeal cancer with an ALK rearrangement secreted amphiregulin and activated the EGFR/ERBB3 signaling pathway [42]. Experimental studies revealed that the combination of osimertinib (an inhibitor of amphiregulin-mediated EGFR/ERBB3 activation) with any of the ALK inhibitors served to control leptomeningeal metastases. Higher levels of amphiregulin were detected in the medullary cavity of patients with ALK-TKI resistant tumors.

Results from recent studies have suggested that the ALK fusion protein is stabilized by heat shock protein (Hsp90); stability of the ALK fusion protein was reduced in response to treatment with an Hsp90 inhibitor, resulting in a dramatic reduction in the cellular levels of the ALK fusion protein. Clinical trials featuring Hsp90 inhibitors for the treatment of cancers with ALK rearrangements revealed their significant clinical efficacy [43].

Other resistance mechanisms

In addition to the ALK-TKI resistance identified by analysis of clinical specimens, various mechanisms underlying ALK-TKI resistance have been reported from experimental studies. For example, the mediator complex subunit-12 (MED-12) was identified from an RNAi screening study; knockdown of MED-12 activated the transforming growth factor (TGF)-β receptor signaling and MEK/ERK signaling, resulting in resistance of ALK-TKIs [44]. In addition, resistance-associated with activation of yes-associated protein (YAP)1 has been reported; treatment with an ALK-TKI together with an inhibitor of YAP1 may result in prolonged suppression of tumor re-growth and clinical relapse [45]. However, the detailed molecular mechanisms underlying the crosstalk between ALK-mediated signaling and the YAP1-associated Hippo pathway remain unclear. It has also been shown that activation of the signaling via the Axl receptor may induce ALK-TKI resistance.

Conclusion

Summary, discoveries, and future challenges

ALK-TKIs have been under active development over the past 10 years; five drugs are currently available and several are undergoing testing in clinical trials. Various resistance mechanisms have been identified through genetic analysis of clinical specimens from ALK-TKI- resistant patients and via experiments carried out in cultured cell lines. Although resistance due to primary ALK mutations frequently arises in response to treatment with first-line ALK-TKIs, this can be overcome by at least one of the approved second-generation ALK-TKIs. Further resistance emerges due to the acquisition of secondary compound mutations in ALK; most of these are resistant to all the approved ALK-TKIs except for several compound mutations, such as C1156Y+L1198F, I1171N+L1256F, and I1171N+G1269A. TKI resistance can also emerge in response to bypass pathway activation. Experimental studies in animals have revealed that combination therapy consisting of an inhibitor of the bypass pathway together with an ALK-TKI typically overcomes this resistance; however, there is no clinical evidence at this time that shows that combination therapy is safe and that it is effective at overcoming bypass pathway-mediated resistance. In addition, there are also many cases in which the mechanisms underlying drug resistance are unclear; elucidation of these unknown mechanisms remains an important issue. Other issues include clarification of the mechanism by which the ALK-G1202R promotes TKI resistance in EML4-ALK variant type 3 (a short variant in which exon 6 of EML4 is fused to exon 20 of ALK) but not in EML4-ALK variant 1 (a longer variant in which exon13 of EML4 is fused to exon20 of ALK). There is as yet no detailed whole structural analysis that explains how the EML4-ALK fusion proteins are activated via the dimerization or trimerization of the EML4 domains. There are several reports that suggest that next-generation sequencing of cell-free (cf) DNA may be able to capture resistant mutations and resistance mechanisms in advance, personalized treatment strategies might be developed to maximize the clinical efficacy of ALK-TKIs. However, ongoing effort will be needed in order to uncover critical problems and to improve our understanding of the inevitable emergence of resistance to TKI therapeutics. Additional studies may provide insights into how the emergence of resistance might be delayed or even counteracted as a means to improve the prognosis of NSCLC.

References

[1] Bilsland JG, et al. Behavioral and neurochemical alterations in mice deficient in anaplastic lymphoma kinase suggest therapeutic potential for psychiatric indications. Neuropsychopharmacology 2008;33:685–700. https://doi.org/10.1038/sj.npp.1301446.

[2] Weiss JB, et al. Anaplastic lymphoma kinase and leukocyte tyrosine kinase: functions and genetic interactions in learning, memory and adult neurogenesis. Pharmacol Biochem Behav 2012;100:566–74. https://doi.org/10.1016/j.pbb.2011.10.024.

[3] Englund C, et al. Jeb signals through the Alk receptor tyrosine kinase to drive visceral muscle fusion. Nature 2003;425:512–6. https://doi.org/10.1038/nature01950.

[4] Lee HH, Norris A, Weiss JB, Frasch M. Jelly belly protein activates the receptor tyrosine kinase Alk to specify visceral muscle pioneers. Nature 2003;425:507–12. https://doi.org/10.1038/nature01916.

[5] Soda M, et al. Identification of the transforming EML4-ALK fusion gene in non-small-cell lung cancer. Nature 2007;448:561–6.

[6] Morris SW, et al. Fusion of a kinase gene, ALK, to a nucleolar protein gene, NPM, in non-Hodgkin's lymphoma. Science 1994;263:1281–4.

[7] Katayama R. Drug resistance in anaplastic lymphoma kinase-rearranged lung cancer. Cancer Sci 2018;109: 572–80. https://doi.org/10.1111/cas.13504.

[8] Camidge DR, et al. Clinical activity of crizotinib (PF-02341066), in ALK-positive patients with advanced non-small cell lung cancer. Ann Oncol 2010;21:123. suppl 8.

[9] Kwak EL, et al. Anaplastic lymphoma kinase inhibition in non-small-cell lung cancer. N Engl J Med 2010;363:1693–703.

[10] Shaw AT, et al. Effect of crizotinib on overall survival in patients with advanced non-small-cell lung cancer harbouring ALK gene rearrangement: a retrospective analysis. Lancet Oncol 2011;12:1004–12. https://doi.org/10.1016/S1470-2045(11)70232-7.

[11] Choi YL, et al. EML4-ALK mutations in lung cancer that confer resistance to ALK inhibitors. N Engl J Med 2010;363:1734–9. https://doi.org/10.1056/NEJMoa1007478.

[12] Bresler SC, et al. ALK mutations confer differential oncogenic activation and sensitivity to ALK inhibition therapy in neuroblastoma. Cancer Cell 2014;26:682–94. https://doi.org/10.1016/j.ccell.2014.09.019.

[13] Katayama R, et al. Mechanisms of acquired crizotinib resistance in ALK-rearranged lung cancers. Sci Transl Med 2012;4, 120ra117.

[14] Lovly CM, Pao W. Escaping ALK inhibition: mechanisms of and strategies to overcome resistance. Sci Transl Med 2012;4, 120ps2.

[15] Chen Y, et al. Oncogenic mutations of ALK kinase in neuroblastoma. Nature 2008;455:971–4.

[16] Sakamoto H, et al. CH5424802, a selective ALK inhibitor capable of blocking the resistant gatekeeper mutant. Cancer Cell 2011;19:679–90.

[17] Katayama R, et al. Two novel ALK mutations mediate acquired resistance to the next-generation ALK inhibitor alectinib. Clin Cancer Res 2014;20:5686–96. https://doi.org/10.1158/1078-0432.CCR-14-1511.

[18] Gainor JF, et al. Molecular mechanisms of resistance to first- and second-generation ALK inhibitors in ALK-rearranged lung cancer. Cancer Discov 2016. https://doi.org/10.1158/2159-8290.CD-16-0596.

[19] Marsilje TH, et al. Synthesis, structure-activity relationships, and in vivo efficacy of the novel potent and selective anaplastic lymphoma kinase (ALK) inhibitor 5-chloro-N2-(2-isopropoxy-5-methyl-4-(piperidin-4-yl)phenyl)-N4-(2-(isopropylsulf onyl)phenyl)pyrimidine-2,4-diamine (LDK378) currently in phase 1 and phase 2 clinical trials. J Med Chem 2013;56:5675–90.

[20] Friboulet L, et al. The ALK inhibitor ceritinib overcomes crizotinib resistance in non-small cell lung cancer. Cancer Discov 2014;4:662–73. https://doi.org/10.1158/2159-8290.CD-13-0846.

[21] Katayama R, et al. Therapeutic strategies to overcome crizotinib resistance in non-small cell lung cancers harboring the fusion oncogene EML4-ALK. Proc Natl Acad Sci U S A 2011;108:7535–40.

[22] Huang WS, et al. Discovery of brigatinib (AP26113), a phosphine oxide-containing, potent, orally active inhibitor of anaplastic lymphoma kinase. J Med Chem 2016;59:4948–64. https://doi.org/10.1021/acs.jmedchem.6b00306.

[23] Zou HY, et al. PF-06463922, an ALK/ROS1 inhibitor, overcomes resistance to first and second generation ALK inhibitors in preclinical models. Cancer Cell 2015. https://doi.org/10.1016/j.ccell.2015.05.010.

[24] Shaw AT, et al. Lorlatinib in non-small-cell lung cancer with ALK or ROS1 rearrangement: an international, multicentre, open-label, single-arm first-in-man phase 1 trial. Lancet Oncol 2017;18(12). https://doi.org/10.1016/S1470-2045(17)30680-0.

[25] Solomon BJ, et al. Lorlatinib in patients with ALK-positive non-small-cell lung cancer: results from a global phase 2 study. Lancet Oncol 2018;19:1654–67. https://doi.org/10.1016/S1470-2045(18)30649-1.

[26] Shaw AT, et al. Resensitization to crizotinib by the lorlatinib ALK resistance mutation L1198F. N Engl J Med 2016;374:54–61. https://doi.org/10.1056/NEJMoa1508887.

[27] Yoda S, et al. Sequential ALK inhibitors can select for lorlatinib-resistant compound ALK mutations in ALK-positive lung cancer. Cancer Discov 2018;8:714–29. https://doi.org/10.1158/2159-8290.CD-17-1256.

[28] Takahashi K, et al. Overcoming resistance by ALK compound mutation (I1171S + G1269A) after sequential treatment of multiple ALK inhibitors in non-small cell lung cancer. Thorac Cancer 2020;11:581–7. https://doi.org/10.1111/1759-7714.13299.

[29] Okada K, et al. Prediction of ALK mutations mediating ALK-TKIs resistance and drug re-purposing to overcome the resistance. EBioMedicine 2019;41:105–19. https://doi.org/10.1016/j.ebiom.2019.01.019.

[30] Russo M, et al. Acquired resistance to the TRK inhibitor entrectinib in colorectal cancer. Cancer Discov 2016;6: 36–44. https://doi.org/10.1158/2159-8290.CD-15-0940.

[31] Awad MM, et al. Acquired resistance to crizotinib from a mutation in CD74-ROS1. N Engl J Med 2013;368: 2395–401.

[32] Papadopoulos KP, et al. U.S. Phase I first-in-human study of taletrectinib (DS-6051b/AB-106), a ROS1/TRK inhibitor, in patients with advanced solid tumors. Clin Cancer Res 2020. https://doi.org/10.1158/1078-0432.CCR-20-1630.

[33] Fujitani H, Tanida Y, Matsuura A. Massively parallel computation of absolute binding free energy with well-equilibrated states. Phys Rev E Stat Nonlin Soft Matter Phys 2009;79, 021914.

[34] Ono F, et al. Improvement in predicting drug sensitivity changes associated with protein mutations using a molecular dynamics based alchemical mutation method. Sci Rep 2020;10:2161. https://doi.org/10.1038/s41598-020-58877-9.

[35] Sasaki T, et al. A novel ALK secondary mutation and EGFR signaling cause resistance to ALK kinase inhibitors. Cancer Res 2011;71:6051–60.

[36] Doebele RC, et al. Mechanisms of resistance to crizotinib in patients with ALK gene rearranged non-small cell lung cancer. Clin Cancer Res 2012;18:1472–82.

[37] Lovly CM, et al. Rationale for co-targeting IGF-1R and ALK in ALK fusion-positive lung cancer. Nat Med 2014;20:1027–34. https://doi.org/10.1038/nm.3667.

[38] Tanimoto A, et al. Receptor ligand-triggered resistance to alectinib and its circumvention by Hsp90 inhibition in EML4-ALK lung cancer cells. Oncotarget 2014;5:4920–8.

[39] Crystal AS, et al. Patient-derived models of acquired resistance can identify effective drug combinations for cancer. Science 2014;346:1480–6. https://doi.org/10.1126/science.1254721.

[40] Yanagitani N, et al. Drug resistance mechanisms in Japanese anaplastic lymphoma kinase-positive non-small cell lung cancer and the clinical responses based on the resistant mechanisms. Cancer Sci 2020;111:932–9. https://doi.org/10.1111/cas.14314.

[41] Katayama R, et al. P-glycoprotein mediates ceritinib resistance in anaplastic lymphoma kinase-rearranged non-small cell lung cancer. EBioMedicine 2016;3:54–66. https://doi.org/10.1016/j.ebiom.2015.12.009.

[42] Arai S, et al. Osimertinib overcomes alectinib resistance caused by amphiregulin in a leptomeningeal carcinomatosis model of ALK-rearranged lung cancer. J Thorac Oncol 2020;15:752–65. https://doi.org/10.1016/j.jtho.2020.01.001.

[43] Sequist LV, et al. Activity of IPI-504, a novel heat-shock protein 90 inhibitor, in patients with molecularly defined non-small-cell lung cancer. J Clin Oncol 2010;28:4953–60.

[44] Huang S, et al. MED12 controls the response to multiple cancer drugs through regulation of TGF-beta receptor signaling. Cell 2012;151:937–50. https://doi.org/10.1016/j.cell.2012.10.035.

[45] Tsuji T, et al. YAP1 mediates survival of ALK-rearranged lung cancer cells treated with alectinib via pro-apoptotic protein regulation. Nat Commun 2020;11:74. https://doi.org/10.1038/s41467-019-13771-5.

Resistance mechanisms to ALK TKIs in tumors other than lung cancer

Luca Mologni

School of Medicine and Surgery, University of Milano-Bicocca, Monza, Italy

Abstract

The Anaplastic Lymphoma Kinase (ALK) was identified in 1994 as the C-terminal fusion partner in the chimeric protein arising from the recurrent t(2;5) translocation found in the anaplastic large-cell lymphoma (ALCL), a rare subgroup of T-cell malignancy. Since then, ALK has been involved in several additional tumors, both solid and hematological, through chromosomal rearrangements and activating point mutations. Following a spectacular success of targeted therapy in another fusion gene-driven cancer (chronic myeloid leukemia) a significant research effort has led to the development of several fairly selective ALK inhibitors. The use of such drugs is still investigational/compassionate (i.e., off-label) in ALK-dependent tumors other than lung cancer. Nevertheless, the activity of ALK inhibitors in ALK-positive ALCLs has been well documented, displaying high rates of long-term disease control. Despite excellent clinical activity, however, a significant fraction of patients still relapses after response, via both ALK-dependent (drug-resistant mutant clones) and ALK-independent (activation of alternative pathways) mechanisms. Moreover, different ALK-positive cancers show different responses to ALK inhibitors, likely reflecting different tumor biology, variable heterogeneity of the cancer cell population, as well as different co-dependencies to additional survival pathways, that will need to be targeted to achieve better therapeutic effects. Here, I reviewed the current literature on the involvement of ALK, the clinical activity of ALK inhibitors and the mechanisms of resistance to ALK inhibition in tumors other than non-small cell lung cancer.

Abbreviations

ALK	anaplastic lymphoma kinase
ALCL	anaplastic large-cell lymphoma
CHOP	cyclophosphamide-doxorubicin-vincristine-prednisone
CR	complete response
DLBCL	diffuse large B-cell lymphoma
HDAC	histone deacetylase
IC50	50% inhibitory concentration
IMT	inflammatory myofibroblastic tumor
MAPK	mitogen-activated protein kinase

mTOR	mammalian target of rapamycin
NPM	nucleophosmin
NSCLC	non-small cell lung cancer
ORR	overall response rate
OS	overall survival
PR	partial response
SD	stable disease

Conflict of interest

No potential conflicts of interest were disclosed.

Introduction

Although ALK has attracted much attention and research effort mostly after its discovery as a driver of lung cancer oncogenesis, the first ever description of ALK involvement in human cancer occurred in a rare T-cell lymphoma. This finding determined its official name, that is, Anaplastic Lymphoma Kinase. Ever since, ALK has been implicated in a variety of tumors, of different origins, all of which have in common the aberrant activation of this receptor tyrosine kinase. This chapter will describe the role of ALK alterations in cancers excluding lung carcinoma, their targeting and the associated mechanisms of drug resistance in the context of these tumors.

Mechanisms of ALK-mediated transformation in tumors other than lung cancer

In 1994, Morris et al. identified a recurrent fusion of ALK to the nucleophosmin (*NPM1*) gene in anaplastic large-cell lymphoma (ALCL) cases associated with t(2;5) chromosomal rearrangement [1] and later described the full-length ALK as a neural receptor tyrosine kinase [2]. Between 50% and 80% of ALCLs have been reported to carry ALK rearrangements and this led to the definition of ALK-positive (ALK+) ALCL as a new disease entity within non-Hodgkin's lymphomas [3]. About 80% of ALK+ ALCLs harbor the t(2;5) rearrangement encoding for the NPM-ALK fusion protein. Indeed, transgenic expression of NPM-ALK in CD4+ T lymphocytes induces lymphomagenesis [4]. Subsequently, several groups independently identified novel ALK fusions in a minority of ALCLs [5–12]. ALCL accounts for approximately 15% of pediatric non-Hodgkin lymphomas [13,14], while it is rarer in adults (2–3% of all cases). Standard treatment is based on multidrug cytotoxic combinations such as CHOP (Cyclophosphamide, Doxorubicin, Vincristine, Prednisone) or similar, in adults [15]; for children, the ALCL-99 protocol, which includes high-dose methotrexate, has been shown to achieve better results [16,17]. Responses to chemotherapy are generally higher in ALK+ compared to ALK- negative ALCL, resulting in significantly better prognosis [14].

ALK rearrangements were later identified in approximately 50% of patients affected by the inflammatory myofibroblastic tumor (IMT), a rare soft tissue cancer [12,18–27], suggesting a broader oncogenic role for the ALK kinase, not restricted to ALCL. As expected, the use of next generation sequencing technologies led to the identification of ALK involvement in several additional cancer types, both hematological and solid: besides non-small cell lung cancer (see Chapter 2) ALK fusions were found to be oncogenic in rare cases of diffuse large B-cell lymphoma (DLBCL) [28–31], spitz tumors [32], anaplastic and medullary thyroid cancer [33,34], leiomyosarcoma [35], mesothelioma [36], squamous cell carcinoma of the esophagus [37], renal cancer [38–42], myeloid leukemia [43–45], glioma [46], melanoma [47], breast carcinoma [48], ovarian cancer [49] and colorectal tumors [48,50–52] (Table 1).

Whatever the type of rearrangement, the common themes among all ALK fusions are: (i) high aberrant expression driven by the partner gene promoter; (ii) an amino-terminal domain, derived from the fusion partner, which dictates subcellular localization and provides an oligomerization interface that mimics physiological ligand-induced receptor dimerization; and (iii) a carboxy-terminal tyrosine kinase region derived from the *ALK* gene, which is constitutively active as a result of constitutive homotypic interaction.

Besides gene fusions, an alternative mechanism of ALK oncogenic activation in cancer is represented by activating point mutations. The majority of familial neuroblastoma patients and up to 10% of sporadic cases harbor mutations in the catalytic domain of the full-length ALK that activate the receptor in a ligand-independent manner [73–75]. In a subset of patients, wild-type ALK is overexpressed, often as a result of gene amplification, again resulting in excessive kinase activity [76,77]. A third group of neuroblastoma patients expresses ligand-dependent ALK mutants, whose role as drivers of oncogenesis is doubtful [78]. In general, ALK point mutations have been validated as therapeutic targets in neural cancers. As ALK is normally expressed (albeit at low levels) in neural tissue, these cells simply hijack an existing signaling circuitry to gain a selective growth advantage. Point mutations were also identified in thyroid cancers [103] and myeloid neoplasms [104], although their oncogenic role in these cases is not well established. Again, it was suggested that some of these mutants may be ligand-dependent [105]. Furthermore, deletion mutants have also been reported to be oncogenic: aberrant activation of the ALK kinase is caused by genomic amplification of a truncated form of the receptor in neuroblastoma cell lines and primary tumors [79,80]. Similarly, an in-frame deletion of exons 2–17 was identified in an ALCL patient [53]. In both cases, the extracellular domain was partially lost and the protein was predicted to retain the transmembrane region. However, truncated ALK localized to the endoplasmic reticulum, in contrast to plasma membrane-bound full-length receptor. How these deletion variants undergo constitutive activation is unclear, possibly the extracellular portion carries some auto-inhibitory properties. In a significant fraction of melanoma patients, an aberrant ALK transcript variant was identified, due to an alternative transcription start site that excludes the whole extracellular domain sequence, leading to expression of an N-terminally truncated constitutively active form of the kinase [100]. The transcribed region corresponds to the ALK portion that is retained in fusions observed in other malignancies. Similarly, a Δ2-17 deletion variant was found in synovial sarcoma cells, again losing the extracellular domain [101]. Various other exon deletions were reported in Ewing sarcoma, whose significance is unclear [102]. In several soft tissue sarcomas, ALK is found overexpressed through gene copy number gain [101,102,106]. In alveolar rhabdomyosarcoma, ALK gene copy gain was detected in 88% of specimens by in situ hybridization and correlated with poor survival [107]. An alternative mechanism of ALK activation was reported in glioblastoma multiforme, whereby the co-overexpression of pleiotrophin and ALK led to an indirect ligand-dependent hyper-activation of the wild-type ALK receptor, without gene rearrangement or amplification [88–90].

Oncogenic ALK signaling has mostly been studied in ALCL models [108]. The most relevant signals activated by the NPM-ALK fusion include the JAK/STAT, RAS/MAPK and

TABLE 1 Summary of ALK + diseases other than NSCLC.

Disease	Aberrations	Reference	Clinical results: Drug – outcome (patients)	Reference	Acquired resistance	Reference
ALCL	Fusion; Truncation	[1–12,53]	Crizotinib – CR (2/2) Crizotinib – CR (7/9); PR (1/9); SD (1/9) Crizotinib – CR (21/26); PR (2/26); SD (3/26) Crizotinib – CR (9/9) Crizotinib – CR (8/16); PR (1/16); SD (3/16) Ceritinib – CR (2/3); PR (1/3)	[54] [55] [56] [57] [58] [59]	*Crizotinib:* L1196M, L1196Q, I1171N/T/S, R1192P, G1269A, overexpression, bypass (IGF-1R) *ASP3026:* G1128S, C1156F, I1171N/T, F1174I, N1178H, E1210K, C1156F+ D1203N *Brigatinib:* F1174V + L1198F, L1122V + L1196M, L1122V + L1196M+G1202R, S1206C, gene amplification. *Ceritinib:* overexpression, F1174L, T1151M *Alectinib:* F1174C *Lorlatinib:* G1202R+ G1269A, C1156F+ L1198F, N1178H+ G1269A, I1171T+ E1210K, L1196M, N1178H, G1202R, G1269A, I1171T, bypass (PI3K, RAS/MAPK)	[57,60–64] [65] [66] [63,67,68] [62] [69]
DLBCL	Fusion	[28–31]	Crizotinib – PR (1/2); SD (1/2) Crizotinib – SD (1/1) Crizotinib – PR (3 cases)	[57] [58] [70–72]	–	–
Neuroblastoma	Mutation; Amplification; Truncation	[73–80]	Crizotinib – CR (1/11); SD (2/11) Crizotinib – SD (1/2) Ceritinib – PR (1/1) ASP3026 – SD (1/1) Alectinib – CR (1/1) Alectinib – PR (1/1) Lorlatinib – CR (1/1)	[55] [58] [81] [82] [83] [84] [85]	*Ceritinib:* bypass (AXL, MYCN) *Lorlatinib:* bypass (EGFR, HER4, RAS/MAPK)	[86,87] [69]
Glioma/ Glioblastoma	Fusion; Amplification; Activation	[46,88–90]	–	–	*Alectinib:* bypass (c-Myc, RAS/MAPK)	[91]

Cancer	Mechanism	Ref.	Clinical results	Ref.	*Crizotinib*: F1174L, G1128A	Ref.
IMT	Fusion	[12,18–27]	Crizotinib – PR (1/2) Crizotinib – PR (3/7); SD (4/7) Crizotinib – CR (5/14); PR (7/14) Crizotinib – CR (1/9); PR (5/9); SD (3/9) Ceritinib – PR (1/1) Brigatinib – PR (1/2); SD (1/2) Entrectinib – CR (1/1)	[92] [55] [56] [58] [93] [94] [95]	*Crizotinib*: F1174L, G1128A	[96,97]
Colorectal Cancer	Fusion	[48,50–52]	Crizotinib – PR (1/1) Brigatinib – SD (1/1) Entrectinib – PR (1/1)	[58] [94] [95]	–	–
Esophageal Cancer	Fusion	[37]	–	–	–	–
Thyroid Cancer	Fusion; Mutation	[33,34]	Crizotinib – PR (1/1) Crizotinib – PR (1/1) Ceritinib/brigatinib – PR (1/1)	[58] [98] [99]	–	–
Renal Cancer	Fusion; Mutation	[38–42]	Entrectinib – PR (1/1)	[95]	–	–
Breast Cancer	Fusion	[48]	Crizotinib – No response (1/1)	[58]	–	–
Ovarian Cancer	Fusion; Amplification	[49]	–	–	–	–
Melanoma	Fusion; Truncation	[47,100]	Entrectinib – No response	[47]	–	–
Spitz Tumors	Fusion	[32]	–	–	–	–
Myeloid Leukemia	Fusion; Mutation	[43–45]	–	–	–	–
Mesothelioma	Fusion	[36]	–	–	–	–
Sarcoma	Fusion; Amplification; Truncation	[35,101,102]	Crizotinib – No response (1/1)	[58]	–	–

Mechanisms of ALK activation, clinical results and mechanisms of acquired resistance to ALK inhibitors are indicated, along with references to relevant papers cited in the text.

PI3K/AKT pathways [108–111]. In particular, STAT3 appears as a dominant mediator of cells survival in ALCL, both ALK fusion-positive (through NPM-ALK [112]) and ALK-negative (via JAK1 or STAT3 activating mutations [113]). Moreover, PLCγ has been shown to transduce the oncogenic signaling through direct interaction with NPM-ALK [114]. Also, β-catenin was found to be constitutively activated in ALK+ ALCL; interestingly, STAT3 was among its transcriptional targets in these cells [115]. In addition, cytoskeletal regulators such as the Rho family GTPases Cdc42 and Rac1 [116–118] and the Wiskott-Aldrich syndrome protein (WASP [119,120]) have been involved in the intricate signaling network triggered by ALK fusions in ALCL. Finally, tyrosine phosphatases have been involved in the regulation of ALK- driven pathways [121,122]. Some of the identified downstream pathways are likely to be cell type-specific. In particular, cytoskeletal proteins governing cell shape are in line with the anaplastic nature of ALCL, but are not expected to have a significant role in other ALK-dependent neoplasms. Similarly, the JAK/STAT signaling is typical of lymphoid malignancies and ALK fusions simply take advantage of its role in T cell biology. In contrast, the RAS/MAPK pathway seems to be physiologically activated by wild-type ALK [123] and it is also a major effector of ALK oncogenes in NSCLC [124] and neuroblastoma [78].

Tyrosine kinase inhibitors in ALK+ tumors other than lung cancer

First-in-class: Crizotinib

The identification of the NPM-ALK fusion protein in ALCL suggested that specific inhibition of its kinase activity may be a viable therapeutic option, similar to BCR-ABL1 inhibitors in chronic myeloid leukemia. Biological validation of NPM-ALK as a therapeutic target was obtained by siRNA-mediated silencing. NPM- ALK knock-down in ALCL cells led to inhibition of downstream signaling, cell growth in vitro, and tumor growth in vivo, and synergized with chemotherapeutic agents [125,126]. The first description of an ALK small molecule inhibitor was reported in 2007 by Galkin et al. [127]. The compound was identified through a cell screening, where NPM-ALK transformed Ba/F3 cells where used as a target and parental IL3-dependent Ba/F3 cells as a non-target reference. This led to the discovery of a phenylamino-pyrimidine compound named **NVP-TAE684**, which showed fairly specific inhibition of ALK kinase, in vitro growth inhibition and in vivo regression of ALK+ ALCL and DLBCL xenografts [127,128]. Unfortunately, TAE684 was not further developed into clinical use due to unexpected toxicity issues. Soon after, serendipitous discovery of potent anti-ALK activity by a MET inhibitor opened the era of clinical use of ALK inhibitors: **crizotinib** (initially known as PF-2341066, as an anti-angiogenic drug [129]) showed nanomolar inhibition of NPM/ALK-dependent signaling and cell growth, and exhibited strong anti-tumor efficacy in ALCL xenografts [130]. Preclinical evaluation of crizotinib in ALK-mutated neuroblastoma showed that different mutants have differential sensitivity to treatment, introducing the issue of drug resistance: crizotinib inhibited the growth of cells carrying an R1275Q ALK mutation, or an amplified wild-type ALK, while cells with the F1174L mutation were more resistant [131]. On the basis of excellent preclinical results, crizotinib entered clinical investigation in NSCLC (see Chapter 2). Within the same initial trial, crizotinib was tested in two IMT patients, one of whom was RANBP2/ALK+ [92]. Remarkably, only the ALK+ patient showed a response to the drug. At about the same time, crizotinib was used off-label in two ALK+ ALCL patients with excellent outcomes [54]. Interestingly, the value of targeted ALK inhibition in ALCL was also shown by rapid relapses upon crizotinib discontinuation, which could be successfully brought to a new remission when treatment was resumed [132]. These case reports indicated that ALK+ tumors (other than NSCLC) are sensitive to ALK inhibition and paved the way for larger investigations. A dose finding phase I trial evaluated pediatric patients with documented genetic alterations in ALK, including 9 ALCL, 11 neuroblastoma and 7 IMT cases, recurrent or refractory to chemotherapy [55]. Seven out of nine ALCLs had a complete response (CR), one achieved a partial response (PR) and one showed stable disease (SD), achieving 100% disease control (Table 1). In 7 IMT patients, 3 PRs were observed while 4 patients had SD. These data confirmed clinical activity of crizotinib in ALK+ tumors. However, in 11 ALK-mutated neuroblastoma patients, only one CR and two SD were observed, in line with poor activity on mutated ALK. A more recent update of this study reported 88% and 86% overall response rate (ORR) in 26 ALCL and 14 IMT patients, respectively [56]. Further investigation of crizotinib activity in lymphoma patients demonstrated impressive response rates, with 9/9 ALCL cases obtaining a CR [57]. The cohort included 2 DLBCL patients, with limited clinical benefit (1 PR and 1 SD). Overall survival (OS) at 2 years was 73%. A larger phase Ib study evaluating long-term effects of crizotinib in ALK + tumors, excluding NSCLC (PROFILE 1013; ClinicalTrials.gov Identifier: NCT01121588), reported the results in ALK+ lymphoma, IMT and other tumors, with a median follow-up of 3 years [58]. ORR was 56% in ALCL patients (3-year OS, 72%), 67% in the IMT group (3-year OS, 67%) and only 12% (median OS, 8.3 months) in other ALK+ tumors, which included 4 nasopharyngeal cancers, 2 neuroblastomas and single cases of colon, thyroid, pancreatic, and other cancers. Poor responses in this group may be related to several factors, such as tumor type, genetics, concurrent mutations in other oncogenes, drug-resistant ALK mutants, which were not evaluated. This study confirmed marked antitumor activity of crizotinib in ALCL and IMT. Interestingly, in ALCL patients, the progression free survival (PFS) curve stabilized after early relapses, i.e., all relapsing patients progressed within 3 months. After that, no relapses were observed. These data strongly suggest that ALK inhibition induces durable responses and should be considered as first-line treatment indication in ALCL and IMT tumors.

Several trials have recently started to fulfill this above need. A phase II trial in patients with ALK+ ALCL resistant or refractory to standard cytotoxic treatment is reported (NCT02419287) [133]. Another phase I/II study of crizotinib in relapsed ALK+ ALCL and neuroblastoma is currently recruiting patients in Japan (UMIN000028075). A larger cross-tumoral phase II study is enrolling any ALK+ patient, including renal, sarcoma, IMT and ALCL cases (NCT01524926). By contrast, DLBCL patients seem to respond poorly compared to ALCL, although there is scarce literature concerning the use of crizotinib in this rare subtype of lymphoma: in the PROFILE 1013 study, one ALK+ DLBCL patient only achieved a SD [58]; few other reports described cases of very short-lived symptoms alleviation [70–72]. Further studies are needed to determine the clinical activity of ALK inhibitors in other tumors. A patient with anaplastic thyroid carcinoma harboring an ALK rearrangement showed a remarkable and durable response to crizotinib [98]. A new trial will evaluate any ALK inhibitor in colorectal cancer patients with ALK mutation (NCT03792568).

Resistance to crizotinib

As tumors are characterized by high genetic heterogeneity and instability, they adapt to environmental changes and evolve under selective pressure. Therefore, despite excellent results, the selection of TKI- resistant disease can be observed in several cases. An in vitro study analyzed the effects of introducing mutations in key residues, based on homology with other tyrosine kinases. The authors found that substitution of the gatekeeper leucine to a bulkier methionine (L1196M) caused marked resistance against two unrelated compounds [60]. Indeed, the gatekeeper was one of the first described crizotinib-resistant mutants in NSCLC patients [134]. Analysis of drug-resistant ALCL clones selected in vitro revealed a different gatekeeper substitution (L1196Q) and a novel I1171N mutation [61]. The I1171N mutant was later observed also in an ALCL patient relapsing on crizotinib [57]. The structural effect of a larger gatekeeper amino acid on drug binding is well known from previous experience with other TKIs: a bulkier side chain reduces the space that would be occupied by the drug, thus creating steric hindrance to inhibitor binding [135]. The reason for resistance mediated by I1171N is less obvious. This residue belongs to the αC helix and is part of the R-spine, two key structural elements of tyrosine kinases, governing enzyme activation [136]. Molecular modeling studies suggested that mutations at I1171 alter enzyme kinetics, destabilizing drug binding. The I1171N mutation was also insensitive to compound TAE684. Additional in vitro analyses of ALCL cell lines selected by crizotinib identified the emergence of I1171T and I1171S substitutions [62,63]. These mutants showed moderate resistance to crizotinib and cross-resistance to other inhibitors. A close amino acid, F1174, situated at the end of the αC helix, was found mutated in an IMT patient who developed resistance to crizotinib [96]. The authors showed that a F1174L mutant has higher kinase activity compared with wild- type ALK and accelerates the growth of RANBP2-ALK transformed cells. Interestingly, this mutation is also found in neuroblastoma patients at onset, conferring primary resistance to crizotinib [131]. In some cases, mutations at F1174 appear in neuroblastoma patients after relapse from chemotherapy, leading to a more aggressive disease [137]. This residue is located at the center of a hydrophobic cluster of phenylalanine residues involved in DFG positioning and kinase autoinhibition [138]. Computational analysis confirmed that the F1174L mutant has higher flexibility within the hinge region and displays weaker crizotinib binding [139]. Moreover, ALK F1174L shows higher affinity for ATP, which contributes to its reduced sensitivity to crizotinib [131]. Another case of IMT with acquired resistance to crizotinib was found to carry a G1128A mutation [97]. Glycine 1128 is part of the glycine-rich nucleotide-binding loop; mutations at this position likely alter ATP binding kinetics. The patient was then shifted to ceritinib with success.

Besides ALK point mutations, the inadvertent activation of alternative survival pathways, bypassing ALK inhibition, can lead to drug resistance. A bypass mechanism was identified by Li et al. in crizotinib-resistant ALCL cells harboring a G1269A mutation [64]. Thus, these cells developed a dual resistance, with both ALK-dependent and -independent coexisting mechanisms. These data tell about the complexity of the drug resistance problem.

Second- and third-generation ALK inhibitors

The appearance of crizotinib-resistant disease in several settings fostered the search for alternative drugs, collectively referred to as second-generation ALK inhibitors. These include alectinib, ceritinib, brigatinib and lorlatinib, that have all reached clinical use in ALK+ tumors; and other compounds such as ASP3026, giltertinib, ensartinib, entrectinib, repotrectinib, and others, which are still investigational, or have been approved for different indications.

Ceritinib, brigatinib and ASP3026 all have a chemical structure that is very similar to NVP-TAE684, with little modifications to improve their toxicity profiles. **Ceritinib** inhibits several crizotinib-resistant mutants, including I1171T, L1196M and G1269A, and was shown to suppress Karpas299 xenograft growth in rats [140,141]. Among over 300 patients treated with ceritinib in the ASCEND-1 trial, 3 chemoresistant ALK+ ALCL patients were evaluated: two had a CR and one achieved a PR (95% reduction). These responses were still ongoing at the time of reporting, with durations >20 months [59]. These results confirm excellent long-term responses in ALCL by ALK inhibition, as described with crizotinib. Efficacy in other ALK+ tumors remains anecdotal: in a Japanese study, one ALK+ IMT patient, previously treated with ASP3026, achieved a PR with ceritinib [93]. One case of ALK I1171T neuroblastoma achieved a complete response [81]. Recently, an ALK fusion positive thyroid cancer patient resistant to crizotinib was reported to respond to ceritinib and brigatinib [99].

Brigatinib has been shown to block most drug-resistant ALK mutants at clinically relevant concentrations (i.e., below mean plasma levels) [142] and showed superior in vivo efficacy in ALK-mutated neuroblastoma compared with crizotinib [143]. In a large phase I-II trial of brigatinib, 8 patients with malignancies other than NSCLC were evaluated [94]: four of them, with confirmed rearranged ALK, achieved a PR and two had SD (one ALK+ IMT and one colon cancer).

ASP3026 eradicated NPM/ALK+ ALCL xenografts and showed activity against the L1196M mutant [144,145]. A phase Ib study enrolled one neuroblastoma patient, carrying an ALK F1174L mutation, who achieved stable disease as best response [82]. Clinical development of the drug was subsequently discontinued by Astellas Pharma, for commercial reasons.

Alectinib, designed by Chugai Pharmaceutical with a completely novel structure, showed potent antitumor activity against several crizotinib-resistant mutants, including the F1174L and R1275Q neuroblastoma hotspots, and induced full regression of ALCL and neuroblastoma tumors in vivo [146–148]. Trials were only run with NSCLC patients, so reports of clinical activity in other cancers are sporadic; in one case of chemotherapy resistant ALCL, alectinib was successfully given as a bridge to bone marrow transplantation, leading to a durable complete remission of the disease [83]. Another group reported a case of refractory ALK F1245C neuroblastoma showing good response to alectinib [84]. In Japan, a non-randomized non- controlled phase II trial of alectinib is actively recruiting ALCL patients [149].

Lorlatinib was the last ALK inhibitor to be registered for clinical use in ALK+ NSCLC. It is usually referred to as third-generation compound. Preclinical evaluation of lorlatinib indicated sub-nanomolar inhibition and in vivo antitumor activity against most drug-resistant ALK mutants, including the solvent front refractory G1202R variant [150]. In non-NSCLC setting, lorlatinib showed superior in vitro and in vivo inhibitory activity against neuroblastoma-associated ALK variants, compared with crizotinib [151,152]. In TKI-resistant ALCL cells harboring various mutant subclones, lorlatinib suppressed all clones at nontoxic doses [65]. A monocentric phase II clinical trial in ALCL patients is ongoing (NCT03505554) and results are estimated to be ready by the end of 2020. In a patient with metastatic neuroblastoma carrying an unusual Y1278S ALK mutation, who progressed on crizotinib, lorlatinib induced complete resolution of symptoms and a confirmed PR [85].

Entrectinib is a multitarget pan-TRK, ROS1 and ALK inhibitor endowed with potent antitumor activity in ALK+ ALCL and neuroblastoma models [153,154]. Clinical efficacy was evaluated in few non-NSCLC patients enrolled in two phase I trials. Three patients with diverse ALK fusion positive tumors (one renal, one colorectal, one of unknown origin) were eligible for phase II evaluation: two PR and one SD were observed. In addition, one neuroblastoma patient (ALK F1245V) achieved a PR [95]. However, only TKI-naïve patients showed responses. Clinical benefit was then observed in a DCTN1/ALK+ IMT patient achieving a durable CR [155].

Ensartinib is a close analogue of crizotinib with 10-fold increased potency against ALK [156]. It was significantly more effective than crizotinib in subcutaneous neuroblastoma models [157]. First-in-human results were recently disclosed in a NSCLC cohort, but no data are available in other ALK+ tumors, yet. An ongoing trial (NCT03213652) will evaluate activity in pediatric ALK+ patients after molecular analysis, within the larger NCI-MATCH study [158].

TSR-011 is a highly selective benzimidazole ALK inhibitor, which induced complete regression of Karpas-299 xenografts in mice and retained activity against the gatekeeper mutant [159]. However, due to initial clinical results that showed lower response rates compared to approved inhibitors, TSR-011 development has been discontinued.

Repotrectinib is a macrocyclic compound with some structural analogies to lorlatinib, endowed with activity against the ALK G1202R solvent front mutant [160]. This drug was shown to inhibit a range of ALK variants found in neuroblastoma and exhibited strong antitumor activity in neuroblastoma murine models [161].

Resistance to second- and third-generation drugs

Although ALK inhibitors have not reached formal approval by drug agencies for non-NSCLC tumors, it is clear from the available data that they will eventually have a role in front-line treatment in these cancers. Nevertheless, drug resistance will inevitably limit their efficacy, similarly to crizotinib. The impact of this phenomenon will depend on tumor type, heterogeneity, and stage. Preclinical studies have already identified a number of new mutants that confer resistance to second-generation inhibitors. ALCL cells selected under pressure by ASP3026 developed new variants in the P-loop, αC helix and ATP pocket, including a compound C1156F/D1203N mutation that was refractory to all ALK inhibitors [65]. Cross-resistance analysis of the selected mutants suggested that I1171N/T and E1210K mutants would also be resistant to alectinib. Several of these variants were later found in NSCLC patients, indicating that in vitro studies can predict clinical occurrences, and that resistance profiles of different ALK fusions are similar, as demonstrated by Fontana et al. [162]. Interestingly, in a heterogeneous ASP3026-resistant population in which several mutant clones co-existed, further treatment with second-generation drugs, brigatinib, alectinib and ceritinib, led to differential clonal selection according to the relative drug sensitivity of the various mutations, strongly indicating that the precise knowledge of drug-mutation interactions can allow better treatment choices [65]. Selection in vitro by brigatinib led to the emergence of two different mechanisms of resistance in ALCL cell lines: SUPM2 cells acquired various double mutants (Table 1) and even a low frequency triple mutation (L1122V/L1196M/G1202R), reflecting high potency of this compound that is active against most crizotinib-resistant single mutants; in contrast, Karpas-299 cells evolved a very high expression of the NPM/ALK fusion oncogene, that provided full cross resistance to all tested inhibitors [66]. Intriguingly, ALK kinase overexpression led to a paradoxical phenomenon: cells appeared to be not only resistant, but even addicted to the presence of the inhibitor, and massively died upon drug removal [163]. This was explained by toxicity induced by an excess of signaling, unleashed by drug withdrawal. Indeed, cells re-selected in the absence of inhibitor lost NPM/ALK gene amplification and re-gained sensitivity to the drug. The same effects were also described in crizotinib- and ceritinib-resistant ALCL cells by Amin and colleagues, who then exploited this feature in vivo, by applying a drug holiday schedule leading to long-term control of tumor growth [67]. Further analysis of drug addiction in lorlatinib-resistant ALCL cells revealed a mitochondrial stress response triggered by fusion kinase signaling overflow [164]. More recently, a phosphoproteomic analysis suggested the involvement of STAT1 hyper-activation in toxicity induced by drug withdrawal in ceritinib-addicted ALCL cells and confirmed that intermittent dosing can significantly delay tumor relapse [68]. These studies may lead to a new approach in selected patients who develop ALK gene amplification. However, caution should be taken before translating these results in human subjects. It is interesting to note that such mechanism is not restricted to ALK + cancer [164–166]. Encouragingly, anecdotal clinical remissions upon drug discontinuation have been documented in melanoma patients [167].

Alectinib-resistant ALCL cells developed an F1174C mutation [62], while F1174L and T1151M mutants were selected by exposure to ceritinib in vitro [63]. Similarly, SH-SY5Y neuroblastoma cells carrying the F1174L variant showed modest sensitivity to ceritinib (IC50, 150–180 nM) if compared to cells with wild-type ALK. Nonetheless, further resistance to ceritinib occurred in these cells through overexpression of AXL kinase, which led to ALK-independence and full insensitivity to ceritinib (IC50 > 1 μM) [86]. AXL silencing by siRNA or inhibition by small molecules resensitized these cells to ALK inhibition. Mechanisms of target-independent resistance are well documented in targeted therapies, including ALK-driven NSCLC. Therefore, it is not surprising that it also occurs in other ALK + tumors. In another example, neuroblastoma xenografts resistant to ceritinib were found to upregulate MYCN [87]. To add further complexity, an in vivo study of resistance to lorlatinib in ALCL xenografts revealed composite mechanisms arising with concomitant ALK mutations and by-pass pathway activation [69]. The latter included PI3K/AKT and RAS/MAPK signaling. In the same paper, lorlatinib-resistant neuroblastoma cells (ALK R1275Q) were found to hyper-activate EGFR and HER4 receptors; in addition, they acquired a nonsense mutation in NF1 gene, leading to constitutive MAPK signal activation. The RAS/MEK/ERK pathway was also identified as a bypass mechanism in NSCLC cells [124,168], appearing as a common rescue pathway in ALK + cancers treated with ALK inhibitors. Therefore, drug combination strategies may be devised to block simultaneously ALK and MAPK signaling in these tumors (see below). In vitro, lorlatinib-resistant ALCL cells acquired highly overexpressed compound mutants (G1202R/G1269A and C1156F/L1198F) which rendered them able to survive at high drug concentrations, but made them again drug-dependent [69].

In ALK + glioblastoma cells, overexpression of c-Myc and activation of the MAPK pathway conferred resistance to alectinib. Hence, inhibition of these pathways by a Myc inhibitor or trametinib overcame resistance and resensitized resistant cells to alectinib [91].

Finally, a recent intriguing finding by Janostiak et al. shows how kinase signaling can be interchangeable in tumors, and that we can expect 'creative' solutions exploited by cancer cells to evolve drug resistance: while ALK is in general a therapy target, in this work the authors identified ALK as a by-pass kinase in melanoma cells treated with BRAF inhibitors [169]. On the other hand, a similar dual role (target/bypass) has been also described for EGFR in NSCLC [170].

Drug combinations in ALK+ tumors other than lung cancer

The data discussed above, altogether, show that ALK-driven tumors can rely on different mechanisms to survive ALK-targeted therapy. These include ALK point mutations, ALK overexpression and several by-pass salvage pathways. While ALK-dependent resistance can be tackled by drug switch or increase, ALK- independent mechanisms call for alternative treatments to be combined with ALK inhibition. Therefore, various studies have addressed the possibility to combine ALK inhibitors and other drugs to achieve superior efficacy. Co-targeting of ALK and mTOR showed enhanced antitumor activity in ALK-mutated/MYCN-amplified neuroblastoma mouse models [171,172]. The molecular basis for this result resides in the necessity to block two cooperating oncogenic signals (ALK and MYCN) in these cells. Along the same lines, a combination of temsirolimus with ALK inhibitors showed synergistic effects in ALCL cells in vitro and in vivo, leading to complete regression of relapsed xenografts [69]. Similar results were observed with everolimus and crizotinib [173]. Since MYCN is not involved in ALCL pathobiology, the synergy was ascribed in this case to a more profound suppression of oncogenic signaling by the combined treatment, which restrained resistance.

Mitou and colleagues found that crizotinib induces cytoprotective autophagy in ALK+ ALCL cells [174]. Therefore, targeting autophagy with chloroquine caused an autophagic switch from cytoprotection to cytotoxicity. Moreover, BCL-2 is involved in the regulation of this switch and can be targeted in combination with crizotinib to enhance cell killing [175].

Although we are in the era of personalized therapies, cytotoxic chemotherapy is still a mainstay of cancer treatment. Thus, it could be useful to explore synergism of chemotherapy with targeted drugs. Such strategy may show superior ability to prevent resistance, due to unspecific suppression of mutant clones that resist targeted drugs. Indeed, crizotinib was shown to synergize with chemotherapy in neuroblastoma mouse models, achieving long-term complete tumor remission, whereas single treatments allowed relapse [176]. The combination was also more effective in neuroblastoma xenografts carrying crizotinib-resistant mutants (e.g., F1174L). Synergism was ascribed to a potentiated p53 activation. Similarly, Wang et al. used MDM2 inhibitors in combination with ceritinib to activate p53 and improve antitumor activity [87]. This combination also overcame resistance to single-agent ceritinib.

In a drug combination screening, Wood et al. identified ribociclib, a dual CDK4/6 inhibitor, as a synergistic agent in combination with ceritinib in neuroblastoma cells. Combined ALK/CDKs inhibition caused complete regressions in xenografts with ALK F1174L and ALK F1245C mutations and prevented the emergence of resistance [177].

Histone deacetylase (HDAC) inhibitors have been recognized as promising tools to restore normal epigenetic regulation in cancer cells and sensitize them to anticancer therapy. The combination of vorinostat and alectinib was investigated in neuroblastoma cells in vitro and was shown to cause MYCN down-regulation and induce cell death. However, the combination index was mostly additive, indicating poor pharmacological interaction [178]. Similarly, in ALCL cells, we observed no synergism of ALK inhibitors with valproic acid (unpublished results). Thus, ALK/HDAC dual inhibition needs further investigation before it can be suggested in ALK+ tumors.

IGF-1R was found to interact and cooperate with NPM/ALK, thus enhancing STAT downstream signaling in ALCL cells. From there, it was an easy step for the cells to switch to IGF-1R when ALK was blocked by ASP3026. Therefore, dual blockade was needed to suppress NPM/ALK+ tumor growth in vivo [179].

As noted above, the RAS/MAPK pathway has been often involved in drug resistance. Thus, dual ALK/MEK inhibition could be a valuable strategy to prevent (or treat) resistant disease. While several data are available in ALK+ NSCLC, only one study specifically addressed this point in other ALK+ tumors: Ito et al. [180] combined shRNA-induced ALK downregulation with UO126, a relatively specific inhibitor of MEK1/2 activation. The authors found that concomitant ALK and MEK inhibition prevented tumor re-growth after treatment discontinuation in vivo, suggesting a strong synergy of this combination to abolish resistance. In lorlatinib-resistant tumors that showed evidence of MAPK activation, trametinib partially restored sensitivity to lorlatinib [69]. However, as these cells exhibited complex patterns of intracellular pathways mediating resistance, three-wise drug combinations were necessary to achieve better effects: ALK/PI3K/MEK triple inhibition in ALCL and ALK/pan-HER/MEK in neuroblastoma cells.

ALCL is characterized by high expression of the CD30 antigen, a TNFR-related transmembrane receptor. It has been identified as an attractive target for antibody-based therapy in ALCL. Brentuximab vedotin is an antibody-drug conjugate which brings the cytotoxic drug monomethyl auristatin E to CD30+ tumor cells [181]. It is approved for the treatment of relapsed or refractory Hodgkin lymphoma and ALCL. Starting from the observation that relapses occur both in brentuximab-treated and in crizotinib-treated patients, Wang et al. asked whether combining the two therapies could be an option in ALCL, to improve the outcome. They demonstrated that mice treated with the combination regimen showed better control of xenografts growth [182]. Interestingly, the MAPK pathway was identified again as a key player in this setting.

As a further confirmation of MAPK involvement in ALK+ tumor biology, downstream effector transcription factors JUN and JUNB (which are stimulated by ERK1/2 activity) have been implicated in ALCL survival. Indeed, genetic deletion of both factors impaired NPM/ALK-driven tumor growth in transgenic mice. In this setting, JUN and JUNB were found to induce the expression of PDGFRB [183]. Interestingly, the combination of the PDGFRB inhibitor imatinib with ALK inhibitors significantly reduced tumor mass and the chance of relapses. PDGFRB activation was also observed in some ALCL patients and the authors successfully treated one such PDGFRB+ patient with imatinib monotherapy. Once again, this observation strengthens the idea that precise medicine requires precise characterization of each single tumor.

Conclusion

Following the medical revolution brought by kinase inhibitors in chronic myeloid leukemia, the discovery of ALK aberrant activation in cancer prompted research efforts to design specific ALK inhibitors, which took relatively short time to be identified. Indeed, ALK inhibitors caused a paradigm change in clinical management of ALK+ patients. Unfortunately, for obvious reasons, this is mostly true for NSCLC patients, while clinical development in other ALK+ tumors, including the "original" ALK-dependent disease, is lagging behind. Hopefully, things will change soon and ALK targeted drugs will be approved within the next few years at least for ALCL and IMT patients. However, despite great success of ALK inhibitors, tumors find their way to survive therapy, thanks to rapid clonal evolution. This is particularly true in advanced, heterogeneous disease, where several mutant subclones coexist, or cells can easily turn on alternative signaling pathways and switch to ALK independence. In this regard, it is interesting to note that ALK+ ALCL patients display pretty long responses if they do not relapse in the initial three months, which happens in 30–40% of the cases. This is in contrast to NSCLC patients, who develop resistance in virtually all cases, within approximately 1 year. Understanding the reasons of this different behavior, identifying the causes of resistance, and finding clinical or genetic determinants of long-term responses will be the focus of research in the coming years and will lead to improved outcome for ALK+ cancer patients.

References

[1] Morris SW, et al. Fusion of a kinase gene, ALK, to a nucleolar protein gene, NPM, in non-Hodgkin's lymphoma. Science 1994;263(5151):1281–4.
[2] Morris SW, et al. ALK, the chromosome 2 gene locus altered by the t(2;5) in non-Hodgkin's lymphoma, encodes a novel neural receptor tyrosine kinase that is highly related to leukocyte tyrosine kinase (LTK). Oncogene 1997; 14(18):2175–88.
[3] Morris SW, et al. Alk+ CD30+ lymphomas: a distinct molecular genetic subtype of non-Hodgkin's lymphoma. Br J Haematol 2001;113(2):275–95.
[4] Chiarle R, et al. NPM-ALK transgenic mice spontaneously develop T-cell lymphomas and plasma cell tumors. Blood 2003;101(5):1919–27.
[5] Hernandez L, et al. TRK-fused gene (TFG) is a new partner of ALK in anaplastic large cell lymphoma producing two structurally different TFG-ALK translocations. Blood 1999;94(9):3265–8.
[6] Lamant L, et al. A new fusion gene TPM3-ALK in anaplastic large cell lymphoma created by a (1;2)(q25;p23) translocation. Blood 1999;93(9):3088–95.
[7] Rosenwald A, et al. t(1;2)(q21;p23) and t(2;3)(p23;q21): two novel variant translocations of the t(2;5)(p23;q35) in anaplastic large cell lymphoma. Blood 1999;94(1):362–4.
[8] Colleoni GW, et al. ATIC-ALK: a novel variant ALK gene fusion in anaplastic large cell lymphoma resulting from the recurrent cryptic chromosomal inversion, inv(2)(p23q35). Am J Pathol 2000;156(3):781–9.
[9] Tort F, et al. Molecular characterization of a new ALK translocation involving moesin (MSN-ALK) in anaplastic large cell lymphoma. Lab Invest 2001;81(3):419–26.
[10] Feldman AL, et al. Novel TRAF1-ALK fusion identified by deep RNA sequencing of anaplastic large cell lymphoma. Genes Chromosomes Cancer 2013;52(11):1097–102.
[11] Lamant L, et al. Non-muscle myosin heavy chain (MYH9): a new partner fused to ALK in anaplastic large cell lymphoma. Genes Chromosomes Cancer 2003;37(4):427–32.
[12] Cools J, et al. Identification of novel fusion partners of ALK, the anaplastic lymphoma kinase, in anaplastic large-cell lymphoma and inflammatory myofibroblastic tumor. Genes Chromosomes Cancer 2002;34(4):354–62.

[13] Turner SD, et al. Anaplastic large cell lymphoma in paediatric and young adult patients. Br J Haematol 2016;173 (4):560–72.

[14] Ferreri AJ, et al. Anaplastic large cell lymphoma, ALK-positive. Crit Rev Oncol Hematol 2012;83(2):293–302.

[15] Cederleuf H, et al. The addition of etoposide to CHOP is associated with improved outcome in ALK+ adult anaplastic large cell lymphoma: a nordic lymphoma group study. Br J Haematol 2017;178(5):739–46.

[16] Prokoph N, et al. Treatment options for paediatric anaplastic large cell lymphoma (ALCL): current standard and beyond. Cancers (Basel) 2018;10(4):99.

[17] Brugières L, et al. Impact of the methotrexate administration dose on the need for intrathecal treatment in children and adolescents with anaplastic large-cell lymphoma: results of a randomized trial of the EICNHL group. J Clin Oncol 2009;27(6):897–903.

[18] Chen ST, Lee JC. An inflammatory myofibroblastic tumor in liver with ALK and RANBP2 gene rearrangement: combination of distinct morphologic, immunohistochemical, and genetic features. Hum Pathol 2008;39 (12):1854–8.

[19] Ma Z, et al. Fusion of ALK to the Ran-binding protein 2 (RANBP2) gene in inflammatory myofibroblastic tumor. Genes Chromosomes Cancer 2003;37(1):98–105.

[20] Alaggio R, et al. Inflammatory myofibroblastic tumors in childhood: a report from the Italian cooperative group studies. Cancer 2010;116(1):216–26.

[21] Bridge JA, et al. Fusion of the ALK gene to the clathrin heavy chain gene, CLTC, in inflammatory myofibroblastic tumor. Am J Pathol 2001;159(2):411–5.

[22] Coffin CM, et al. ALK1 and p80 expression and chromosomal rearrangements involving 2p23 in inflammatory myofibroblastic tumor. Mod Pathol 2001;14(6):569–76.

[23] Debelenko LV, et al. Identification of CARS-ALK fusion in primary and metastatic lesions of an inflammatory myofibroblastic tumor. Lab Invest 2003;83(9):1255–65.

[24] Debiec-Rychter M, et al. ALK-ATIC fusion in urinary bladder inflammatory myofibroblastic tumor. Genes Chromosomes Cancer 2003;38(2):187–90.

[25] Lawrence B, et al. TPM3-ALK and TPM4-ALK oncogenes in inflammatory myofibroblastic tumors. Am J Pathol 2000;157(2):377–84.

[26] Panagopoulos I, et al. Fusion of the SEC31L1 and ALK genes in an inflammatory myofibroblastic tumor. Int J Cancer 2006;118(5):1181–6.

[27] Patel AS, et al. RANBP2 and CLTC are involved in ALK rearrangements in inflammatory myofibroblastic tumors. Cancer Genet Cytogenet 2007;176(2):107–14.

[28] Delsol G, et al. A new subtype of large B-cell lymphoma expressing the ALK kinase and lacking the 2; 5 translocation. Blood 1997;89(5):1483–90.

[29] Beltran B, et al. ALK-positive diffuse large B-cell lymphoma: report of four cases and review of the literature. J Hematol Oncol 2009;2:11.

[30] Lee SE, et al. Identification of RANBP2-ALK fusion in ALK positive diffuse large B-cell lymphoma. Hematol Oncol 2014;32(4):221–4.

[31] De Paepe P, et al. ALK activation by the CLTC-ALK fusion is a recurrent event in large B-cell lymphoma. Blood 2003;102(7):2638–41.

[32] Rand AJ, et al. Atypical ALK-positive Spitz tumors with 9p21 homozygous deletion: report of two cases and review of the literature. J Cutan Pathol 2018;45(2):136–40.

[33] Ji JH, et al. Identification of driving ALK fusion genes and genomic landscape of medullary thyroid cancer. PLoS Genet 2015;11(8), e1005467.

[34] Kelly LM, et al. Identification of the transforming STRN-ALK fusion as a potential therapeutic target in the aggressive forms of thyroid cancer. Proc Natl Acad Sci U S A 2014;111(11):4233–8.

[35] Davis LE, et al. Discovery and characterization of recurrent, targetable ALK fusions in leiomyosarcoma. Mol Cancer Res 2019;17(3):676–85.

[36] Loharamtaweethong K, et al. Anaplastic lymphoma kinase (ALK) translocation in paediatric malignant peritoneal mesothelioma: a case report of novel ALK-related tumour spectrum. Histopathology 2016;68(4):603–7.

[37] Jazii FR, et al. Identification of squamous cell carcinoma associated proteins by proteomics and loss of beta tropomyosin expression in esophageal cancer. World J Gastroenterol 2006;12(44):7104–12.

[38] Sukov WR, et al. ALK alterations in adult renal cell carcinoma: frequency, clinicopathologic features and outcome in a large series of consecutively treated patients. Mod Pathol 2012;25(11):1516–25.

[39] Debelenko LV, et al. Renal cell carcinoma with novel VCL-ALK fusion: new representative of ALK-associated tumor spectrum. Mod Pathol 2011;24(3):430–42.

[40] Hang JF, Chung HJ, Pan CC. ALK-rearranged renal cell carcinoma with a novel PLEKHA7-ALK translocation and metanephric adenoma-like morphology. Virchows Arch 2020;476:921–9.

[41] Marino-Enriquez A, et al. ALK rearrangement in sickle cell trait-associated renal medullary carcinoma. Genes Chromosomes Cancer 2011;50(3):146–53.

[42] Chen W, et al. Identification of anaplastic lymphoma kinase fusions in clear cell renal cell carcinoma. Oncol Rep 2020;43:817–26.

[43] Maesako Y, et al. inv(2)(p23q13)/RAN-binding protein 2 (RANBP2)-ALK fusion gene in myeloid leukemia that developed in an elderly woman. Int J Hematol 2014;99(2):202–7.

[44] Hayashi A, et al. Crizotinib treatment for refractory pediatric acute myeloid leukemia with RAN- binding protein 2-anaplastic lymphoma kinase fusion gene. Blood Cancer J 2016;6(8):e456.

[45] Rottgers S, et al. ALK fusion genes in children with atypical myeloproliferative leukemia. Leukemia 2010;24 (6):1197–200.

[46] Aghajan Y, et al. Novel PPP1CB-ALK fusion protein in a high-grade glioma of infancy. BMJ Case Rep 2016;2016. https://doi.org/10.1136/bcr-2016-217189.

[47] Couts KL, et al. ALK inhibitor response in melanomas expressing EML4-ALK fusions and alternate ALK isoforms. Mol Cancer Ther 2018;17(1):222–31.

[48] Lin E, et al. Exon array profiling detects EML4-ALK fusion in breast, colorectal, and non-small cell lung cancers. Mol Cancer Res 2009;7(9):1466–76.

[49] Ren H, et al. Identification of anaplastic lymphoma kinase as a potential therapeutic target in ovarian cancer. Cancer Res 2012;72(13):3312–23.

[50] Yakirevich E, et al. Oncogenic ALK fusion in rare and aggressive subtype of colorectal adenocarcinoma as a potential therapeutic target. Clin Cancer Res 2016;22(15):3831–40.

[51] Lipson D, et al. Identification of new ALK and RET gene fusions from colorectal and lung cancer biopsies. Nat Med 2012;18(3):382–4.

[52] Amatu A, et al. Novel CAD-ALK gene rearrangement is drugable by entrectinib in colorectal cancer. Br J Cancer 2015;113(12):1730–4.

[53] Fukuhara S, et al. Partial deletion of the ALK gene in ALK-positive anaplastic large cell lymphoma. Hematol Oncol 2018;36(1):150–8.

[54] Gambacorti-Passerini C, Messa C, Pogliani EM. Crizotinib in anaplastic large-cell lymphoma. N Engl J Med 2011;364(8):775–6.

[55] Mossé YP, et al. Safety and activity of crizotinib for paediatric patients with refractory solid tumours or anaplastic large-cell lymphoma: a Children's Oncology Group phase 1 consortium study. Lancet Oncol 2013;14(6): 472–80.

[56] Mosse YP, et al. Targeting ALK with crizotinib in pediatric anaplastic large cell lymphoma and inflammatory myofibroblastic tumor: a Children's Oncology Group Study. J Clin Oncol 2017;35(28):3215–21.

[57] Gambacorti Passerini C, et al. Crizotinib in advanced, chemoresistant anaplastic lymphoma kinase-positive lymphoma patients. J Natl Cancer Inst 2014;106(2):djt378.

[58] Gambacorti-Passerini C, et al. Long-term effects of crizotinib in ALK-positive tumors (excluding NSCLC): a phase 1b open-label study. Am J Hematol 2018.

[59] Richly H, et al. Ceritinib in patients with advanced anaplastic lymphoma kinase-rearranged anaplastic large-cell lymphoma. Blood 2015;126(10):1257–8.

[60] Wan W, et al. Anaplastic lymphoma kinase activity is essential for the proliferation and survival of anaplastic large-cell lymphoma cells. Blood 2006;107(4):1617–23.

[61] Ceccon M, et al. Crizotinib-resistant NPM-ALK mutants confer differential sensitivity to unrelated Alk inhibitors. Mol Cancer Res 2013;11(2):122–32.

[62] Zdzalik D, et al. Activating mutations in ALK kinase domain confer resistance to structurally unrelated ALK inhibitors in NPM-ALK-positive anaplastic large-cell lymphoma. J Cancer Res Clin Oncol 2014; 140(4):589–98.

[63] Amin AD, et al. TKI sensitivity patterns of novel kinase-domain mutations suggest therapeutic opportunities for patients with resistant ALK+ tumors. Oncotarget 2016;7(17):23715–29.

[64] Li Y, et al. Activation of IGF-1R pathway and NPM-ALK G1269A mutation confer resistance to crizotinib treatment in NPM-ALK positive lymphoma. Invest New Drugs 2019;38(3):599–609.

[65] Mologni L, et al. NPM/ALK mutants resistant to ASP3026 display variable sensitivity to alternative ALK inhibitors but succumb to the novel compound PF-06463922. Oncotarget 2015;6(8):5720–34.

[66] Ceccon M, et al. Treatment efficacy and resistance mechanisms using the second-generation ALK inhibitor AP26113 in human NPM-ALK-positive anaplastic large cell lymphoma. Mol Cancer Res 2014;13(4):775–83.

[67] Amin AD, et al. Evidence suggesting that discontinuous dosing of ALK kinase inhibitors may prolong control of ALK+ tumors. Cancer Res 2015.

[68] Rajan SS, et al. The mechanism of cancer drug addiction in ALK-positive T-cell lymphoma. Oncogene 2020; 39(10):2103–17.

[69] Redaelli S, et al. Lorlatinib treatment elicits multiple on- and off-target mechanisms of resistance in ALK-driven cancer. Cancer Res 2018;78(24):6866–80.

[70] Wass M, et al. Crizotinib in refractory ALK-positive diffuse large B-cell lymphoma: a case report with a short-term response. Eur J Haematol 2014;92(3):268–70.

[71] Li J, et al. Promising response of anaplastic lymphoma kinase-positive large B-cell lymphoma to crizotinib salvage treatment: case report and review of literature. Int J Clin Exp Med 2015;8(5):6977–85.

[72] Mehra V, et al. ALK-positive large B-cell lymphoma with strong CD30 expression; a diagnostic pitfall and resistance to brentuximab and crizotinib. Histopathology 2016;69(5):880–2.

[73] George RE, et al. Activating mutations in ALK provide a therapeutic target in neuroblastoma. Nature 2008;455 (7215):975–8.

[74] Janoueix-Lerosey I, et al. Somatic and germline activating mutations of the ALK kinase receptor in neuroblastoma. Nature 2008;455(7215):967–70.

[75] Mosse YP, et al. Identification of ALK as a major familial neuroblastoma predisposition gene. Nature 2008;455 (7215):930–5.

[76] Passoni L, et al. Mutation-independent anaplastic lymphoma kinase overexpression in poor prognosis neuroblastoma patients. Cancer Res 2009;69(18):7338–46.

[77] Subramaniam MM, et al. Aberrant copy numbers of ALK gene is a frequent genetic alteration in neuroblastomas. Hum Pathol 2009;40(11):1638–42.

[78] Chand D, et al. Cell culture and Drosophila model systems define three classes of anaplastic lymphoma kinase mutations in neuroblastoma. Dis Model Mech 2013;6(2):373–82.

[79] Okubo J, et al. Aberrant activation of ALK kinase by a novel truncated form ALK protein in neuroblastoma. Oncogene 2012;31(44):4667–76.

[80] Cazes A, et al. Characterization of rearrangements involving the ALK gene reveals a novel truncated form associated with tumor aggressiveness in neuroblastoma. Cancer Res 2013;73(1):195–204.

[81] Guan J, et al. Clinical response of the novel activating ALK-I1171T mutation in neuroblastoma to the ALK inhibitor ceritinib. Cold Spring Harb Mol Case Stud 2018;4(4), a002550.

[82] Li T, et al. First-in-human, open-label dose-escalation and dose-expansion study of the safety, pharmacokinetics, and antitumor effects of an oral ALK inhibitor ASP3026 in patients with advanced solid tumors. J Hematol Oncol 2016;9:23.

[83] Nakai R, et al. Alectinib, an anaplastic lymphoma kinase (ALK) inhibitor, as a bridge to allogeneic stem cell transplantation in a patient with ALK-positive anaplastic large-cell lymphoma refractory to chemotherapy and brentuximab vedotin. Clin Case Rep 2019;7(12):2500–4.

[84] Heath JA, et al. Good clinical response to alectinib, a second generation ALK inhibitor, in refractory neuroblastoma. Pediatr Blood Cancer 2018;65(7):e27055.

[85] Vasseur A, et al. Efficacy of lorlatinib in primary crizotinib-resistant adult neuroblastoma harboring ALK Y1278S mutation. JCO Precis Oncol 2019;(3):1–5.

[86] Debruyne DN, et al. ALK inhibitor resistance in ALK(F1174L)-driven neuroblastoma is associated with AXL activation and induction of EMT. Oncogene 2016;35(28):3681–91.

[87] Wang HQ, et al. Combined ALK and MDM2 inhibition increases antitumor activity and overcomes resistance in human. Elife 2017;6. https://doi.org/10.7554/eLife.171371.

[88] Karagkounis G, et al. Anaplastic lymphoma kinase expression and gene alterations in glioblastoma: correlations with clinical outcome. J Clin Pathol 2017;70(7):593–9.

[89] Grzelinski M, et al. Enhanced antitumorigenic effects in glioblastoma on double targeting of pleiotrophin and its receptor ALK. Neoplasia 2009;11(2):145–56.

[90] Powers C, et al. Pleiotrophin signaling through anaplastic lymphoma kinase is rate-limiting for glioblastoma growth. J Biol Chem 2002;277(16):14153–8.

[91] Berberich A, et al. cMyc and ERK activity are associated with resistance to ALK inhibitory treatment in glioblastoma. J Neurooncol 2020;146(1):9–23.

[92] Butrynski JE, et al. Crizotinib in ALK-rearranged inflammatory myofibroblastic tumor. N Engl J Med 2010;363 (18):1727–33.

[93] Nishio M, et al. Phase I study of ceritinib (LDK378) in Japanese patients with advanced, anaplastic lymphoma kinase-rearranged non-small-cell lung cancer or other tumors. J Thorac Oncol 2015;10(7):1058–66.

[94] Gettinger SN, et al. Activity and safety of brigatinib in ALK-rearranged non-small-cell lung cancer and other malignancies: a single-arm, open-label, phase 1/2 trial. Lancet Oncol 2016;17(12):1683–96.

[95] Drilon A, et al. Safety and antitumor activity of the multitargeted Pan-TRK, ROS1, and ALK inhibitor entrectinib: combined results from two phase I trials (ALKA-372-001 and STARTRK-1). Cancer Discov 2017;7(4):400–9.

[96] Sasaki T, et al. The neuroblastoma-associated F1174L ALK mutation causes resistance to an ALK kinase inhibitor in ALK-translocated cancers. Cancer Res 2010;70(24):10038–43.

[97] Tsakiri K, et al. Crizotinib failure in a TPM4-ALK–rearranged inflammatory myofibroblastic tumor with an emerging ALK kinase domain mutation. JCO Precis Oncol 2017;(1):1–7.

[98] Godbert Y, et al. Remarkable response to crizotinib in woman with anaplastic lymphoma kinase-rearranged anaplastic thyroid carcinoma. J Clin Oncol 2015;33(20):e84–7.

[99] Leroy L, et al. Remarkable response to ceritinib and brigatinib in an anaplastic lymphoma kinase-rearranged anaplastic thyroid carcinoma previously treated with crizotinib. Thyroid 2020;30(2):343–4.

[100] Wiesner T, et al. Alternative transcription initiation leads to expression of a novel ALK isoform in cancer. Nature 2015;526(7573):453–7.

[101] Fleuren EDG, et al. Phosphoproteomic profiling reveals ALK and MET as novel actionable targets across synovial sarcoma subtypes. Cancer Res 2017;77(16):4279–92.

[102] Fleuren ED, et al. Expression and clinical relevance of MET and ALK in Ewing sarcomas. Int J Cancer 2013; 133(2):427–36.

[103] Murugan AK, Xing M. Anaplastic thyroid cancers harbor novel oncogenic mutations of the ALK gene. Cancer Res 2011;71(13):4403–11.

[104] Maxson JE, et al. Therapeutically targetable ALK mutations in leukemia. Cancer Res 2015;75(11):2146–50.

[105] Guan J, et al. Anaplastic lymphoma kinase L1198F and G1201E mutations identified in anaplastic thyroid cancer patients are not ligand-independent. Oncotarget 2017;8(7):11566–78.

[106] Kelleher FC, McDermott R. The emerging pathogenic and therapeutic importance of the anaplastic lymphoma kinase gene. Eur J Cancer 2010;46(13):2357–68.

[107] van Gaal JC, et al. Anaplastic lymphoma kinase aberrations in rhabdomyosarcoma: clinical and prognostic implications. J Clin Oncol 2012;30(3):308–15.

[108] Chiarle R, et al. The anaplastic lymphoma kinase in the pathogenesis of cancer. Nat Rev Cancer 2008;8(1):11–23.

[109] Zamo A, et al. Anaplastic lymphoma kinase (ALK) activates Stat3 and protects hematopoietic cells from cell death. Oncogene 2002;21(7):1038–47.

[110] Ruchatz H, et al. Constitutive activation of Jak2 contributes to proliferation and resistance to apoptosis in NPM/ALK-transformed cells. Exp Hematol 2003;31(4):309–15.

[111] Prutsch N, et al. Dependency on the TYK2/STAT1/MCL1 axis in anaplastic large cell lymphoma. Leukemia 2019;33(3):696–709.

[112] Chiarle R, et al. Stat3 is required for ALK-mediated lymphomagenesis and provides a possible therapeutic target. Nat Med 2005;11(6):623–9.

[113] Crescenzo R, et al. Convergent mutations and kinase fusions lead to oncogenic STAT3 activation in anaplastic large cell lymphoma. Cancer Cell 2015;27(4):516–32.

[114] Bai RY, et al. Nucleophosmin-anaplastic lymphoma kinase of large-cell anaplastic lymphoma is a constitutively active tyrosine kinase that utilizes phospholipase C-gamma to mediate its mitogenicity. Mol Cell Biol 1998; 18(12):6951–61.

[115] Anand M, Lai R, Gelebart P. β-catenin is constitutively active and increases STAT3 expression/activation in anaplastic lymphoma kinase-positive anaplastic large cell lymphoma. Haematologica 2011;96(2):253–61.

[116] Ambrogio C, et al. The anaplastic lymphoma kinase controls cell shape and growth of anaplastic large cell lymphoma through Cdc42 activation. Cancer Res 2008;68(21):8899–907.

[117] Colomba A, et al. Activation of Rac1 and the exchange factor Vav3 are involved in NPM-ALK signaling in anaplastic large cell lymphomas. Oncogene 2008;27(19):2728–36.

[118] Choudhari R, et al. Redundant and nonredundant roles for Cdc42 and Rac1 in lymphomas developed in NPM-ALK transgenic mice. Blood 2016;127(10):1297–306.

[119] Menotti M, et al. Wiskott-Aldrich syndrome protein (WASP) is a tumor suppressor in T cell lymphoma. Nat Med 2019;25(1):130–40.

[120] Murga-Zamalloa CA, et al. NPM-ALK phosphorylates WASp Y102 and contributes to oncogenesis of anaplastic large cell lymphoma. Oncogene 2017;36(15):2085–94.

[121] Honorat JF, et al. SHP1 tyrosine phosphatase negatively regulates NPM-ALK tyrosine kinase signaling. Blood 2006;107(10):4130–8.

[122] Voena C, et al. The tyrosine phosphatase Shp2 interacts with NPM-ALK and regulates anaplastic lymphoma cell growth and migration. Cancer Res 2007;67(9):4278–86.

[123] Mologni L, Gambacorti-Passerini C. New developments in the treatment of anaplastic lymphoma kinase-driven malignancies. Clin Investig 2012;2(8):835–52.

[124] Hrustanovic G, et al. RAS-MAPK dependence underlies a rational polytherapy strategy in EML4-ALK-positive lung cancer. Nat Med 2015;21(9):1038–47.

[125] Piva R, et al. Ablation of oncogenic ALK is a viable therapeutic approach for anaplastic large-cell lymphomas. Blood 2006;107(2):689–97.

[126] Hsu FY, et al. Downregulation of NPM-ALK by siRNA causes anaplastic large cell lymphoma cell growth inhibition and augments the anti cancer effects of chemotherapy in vitro. Cancer Invest 2007;25(4):240–8.

[127] Galkin AV, et al. Identification of NVP-TAE684, a potent, selective, and efficacious inhibitor of NPM-ALK. Proc Natl Acad Sci U S A 2007;104(1):270–5.

[128] Cerchietti L, et al. Inhibition of anaplastic lymphoma kinase (ALK) activity provides a therapeutic approach for CLTC-ALK-positive human diffuse large B cell lymphomas. PLoS ONE 2011;6(4):e18436.

[129] Zou HY, et al. An orally available small-molecule inhibitor of c-Met, PF-2341066, exhibits cytoreductive antitumor efficacy through antiproliferative and antiangiogenic mechanisms. Cancer Res 2007;67(9):4408–17.

[130] Christensen JG, et al. Cytoreductive antitumor activity of PF-2341066, a novel inhibitor of anaplastic lymphoma kinase and c-Met, in experimental models of anaplastic large-cell lymphoma. Mol Cancer Ther 2007;6(12 Pt 1):3314–22.

[131] Bresler SC, et al. Differential inhibitor sensitivity of anaplastic lymphoma kinase variants found in neuroblastoma. Sci Transl Med 2011;3(108):108ra114.

[132] Gambacorti-Passerini C, Mussolin L, Brugieres L. Abrupt relapse of ALK-positive lymphoma after discontinuation of Crizotinib. N Engl J Med 2016;374(1):95–6.

[133] Bossi E, Aroldi A, Brioschi FA, Steidl C, Baretta S, Renso R, et al. Phase two study of crizotinib in patients with anaplastic lymphoma kinase (ALK)-positive anaplastic large cell lymphoma relapsed/refractory to chemotherapy. Am J Hematol 2020;95(12):E319–21.

[134] Choi YL, et al. EML4-ALK mutations in lung cancer that confer resistance to ALK inhibitors. N Engl J Med 2010;363(18):1734–9.

[135] Zuccotto F, et al. Through the "gatekeeper door": exploiting the active kinase conformation. J Med Chem 2010;53(7):2681–94.

[136] Meharena HS, et al. Deciphering the structural basis of eukaryotic protein kinase regulation. PLoS Biol 2013; 11(10), e1001680.

[137] Schleiermacher G, et al. Emergence of new ALK mutations at relapse of neuroblastoma. J Clin Oncol 2014; 32(25):2727–34.

[138] Bossi RT, et al. Crystal structures of anaplastic lymphoma kinase in complex with ATP competitive inhibitors. Biochemistry 2010;49(32):6813–25.

[139] Kumar A, Ramanathan K. Exploring the structural and functional impact of the ALK F1174L mutation using bioinformatics approach. J Mol Model 2014;20(7):2324.

[140] Marsilje TH, et al. Synthesis, structure-activity relationships, and in vivo efficacy of the novel potent and se-lective anaplastic lymphoma kinase (ALK) inhibitor 5-chloro-N2-(2-isopropoxy-5- methyl-4-(piperidin-4-yl) phenyl)-N4-(2-(isopropylsulf onyl)phenyl)pyrimidine-2,4-diamine (LDK378) currently in phase 1 and phase 2 clinical trials. J Med Chem 2013;56(14):5675–90.

[141] Friboulet L, et al. The ALK inhibitor ceritinib overcomes crizotinib resistance in non-small cell lung cancer. Cancer Discov 2014;4(6):662–73.

[142] Zhang S, et al. The potent ALK inhibitor brigatinib (AP26113) overcomes mechanisms of resistance to first- and second-generation ALK inhibitors in preclinical models. Clin Cancer Res 2016;22(22):5527–38.

[143] Siaw JT, et al. Brigatinib, an anaplastic lymphoma kinase inhibitor, abrogates activity and growth in ALK-positive neuroblastoma cells, Drosophila and mice. Oncotarget 2016;7(20):29011–22.

[144] George SK, et al. The ALK inhibitor ASP3026 eradicates NPM-ALK(+) T-cell anaplastic large-cell lymphoma in vitro and in a systemic xenograft lymphoma model. Oncotarget 2014;5(14):5750–63.

[145] Mori M, et al. The selective anaplastic lymphoma receptor tyrosine kinase inhibitor ASP3026 induces tumor regression and prolongs survival in non-small cell lung cancer model mice. Mol Cancer Ther 2014; 13(2):329–40.

[146] Sakamoto H, et al. CH5424802, a selective ALK inhibitor capable of blocking the resistant gatekeeper mutant. Cancer Cell 2011;19(5):679–90.

[147] Kodama T, et al. Selective ALK inhibitor alectinib with potent antitumor activity in models of crizotinib resis-tance. Cancer Lett 2014;351(2):215–21.

[148] Lu J, et al. The second-generation ALK inhibitor alectinib effectively induces apoptosis in human neuroblas-toma cells and inhibits tumor growth in a TH-MYCN transgenic neuroblastoma mouse model. Cancer Lett 2017;400:61–8.

[149] Nagai H, et al. Phase II trial of CH5424802 (alectinib hydrochloride) for recurrent or refractory ALK- positive anaplastic large cell lymphoma: study protocol for a non-randomized non-controlled trial. Nagoya J Med Sci 2017;79(3):407–13.

[150] Zou HY, et al. PF-06463922, an ALK/ROS1 inhibitor, overcomes resistance to first and second generation ALK inhibitors in preclinical models. Cancer Cell 2015;28(1):70–81.

[151] Infarinato NR, et al. The ALK/ROS1 inhibitor PF-06463922 overcomes primary resistance to crizotinib in ALK-driven neuroblastoma. Cancer Discov 2016;6(1):96–107.

[152] Guan J, et al. The ALK inhibitor PF-06463922 is effective as a single agent in neuroblastoma driven by expres-sion of ALK and MYCN. Dis Model Mech 2016;9(9):941–52.

[153] Ardini E, et al. Entrectinib, a Pan-TRK, ROS1, and ALK inhibitor with activity in multiple molecularly defined cancer indications. Mol Cancer Ther 2016;15(4):628–39.

[154] Aveic S, et al. Combating autophagy is a strategy to increase cytotoxic effects of novel ALK inhibitor entrectinib in neuroblastoma cells. Oncotarget 2016;7(5):5646–63.

[155] Ambati SR, et al. Entrectinib in two pediatric patients with inflammatory myofibroblastic tumors harboring. JCO Precis Oncol 2018;2018. https://doi.org/10.1200/PO.18.00095.

[156] Lovly CM, et al. Insights into ALK-driven cancers revealed through development of novel ALK tyrosine kinase inhibitors. Cancer Res 2011;71(14):4920–31.

[157] Di Paolo D, et al. New therapeutic strategies in neuroblastoma: combined targeting of a novel tyrosine kinase inhibitor and liposomal siRNAs against ALK. Oncotarget 2015;6(30):28774–89.

[158] Flaherty KT, et al. THE molecular analysis for therapy choice (NCI-match) trial: lessons for genomic trial design. J Natl Cancer Inst 2020. https://doi.org/10.1093/jnci/djz245.

[159] Lewis RT, et al. The discovery and optimization of a novel class of potent, selective, and orally bioavailable anaplastic lymphoma kinase (ALK) inhibitors with potential utility for the treatment of cancer. J Med Chem 2012;55(14):6523–40.

[160] Drilon A, et al. Repotrectinib (TPX-0005) is a next-generation ROS1/TRK/ALK inhibitor that potently inhibits ROS1/TRK/ALK solvent-front mutations. Cancer Discov 2018;8(10):1227–36.

[161] Cervantes-Madrid D, et al. Repotrectinib (TPX-0005), effectively reduces growth of ALK driven neuroblastoma cells. Sci Rep 2019;9(1):19353.

[162] Fontana D, et al. Activity of second-generation ALK inhibitors against crizotinib-resistant mutants in an NPM-ALK model compared to EML4-ALK. Cancer Med 2015.

[163] Ceccon M, et al. Excess of NPM-ALK oncogenic signaling promotes cellular apoptosis and drug dependency. Oncogene 2016;35(29):3854–65.

[164] Ceccon M, et al. Mitochondrial hyperactivation and enhanced ROS production are involved in toxicity induced by oncogenic kinases over-signaling. Cancers (Basel) 2018;10(12):509.

[165] Tipping AJ, et al. Restoration of sensitivity to STI571 in STI571-resistant chronic myeloid leukemia cells. Blood 2001;98(13):3864–7.

[166] Das Thakur M, et al. Modelling vemurafenib resistance in melanoma reveals a strategy to forestall drug resistance. Nature 2013;494(7436):251–5.

[167] Koop A, et al. Intermittent BRAF-inhibitor therapy is a feasible option: report of a patient with metastatic melanoma. Br J Dermatol 2014;170(1):220–2.

[168] Crystal AS, et al. Patient-derived models of acquired resistance can identify effective drug combinations for cancer. Science 2014;346(6216):1480–6.

[169] Janostiak R, Malvi P, Wajapeyee N. Anaplastic lymphoma kinase confers resistance to BRAF kinase inhibitors in melanoma. iScience 2019;16:453–67.

[170] Shaw AT, Engelman JA. ALK in lung cancer: past, present, and future. J Clin Oncol 2013;31(8):1105–11.

[171] Moore NF, et al. Molecular rationale for the use of PI3K/AKT/mTOR pathway inhibitors in combination with crizotinib in ALK-mutated neuroblastoma. Oncotarget 2014;5(18):8737–49.

[172] Berry T, et al. The ALK(F1174L) mutation potentiates the oncogenic activity of MYCN in neuroblastoma. Cancer Cell 2012;22(1):117–30.

[173] Xu W, et al. Crizotinib in combination with everolimus synergistically inhibits proliferation of anaplastic lymphoma kinase–positive anaplastic large cell lymphoma. Cancer Res Treat 2018;50(2):599–613.

[174] Mitou G, et al. Targeting autophagy enhances the anti-tumoral action of crizotinib in ALK-positive anaplastic large cell lymphoma. Oncotarget 2015;6(30):30149–64.

[175] Torossian A, et al. Blockade of crizotinib-induced BCL2 elevation in ALK-positive anaplastic large cell lymphoma triggers autophagy associated with cell death. Haematologica 2019;104(7):1428–39.

[176] Krytska K, et al. Crizotinib synergizes with chemotherapy in preclinical models of neuroblastoma. Clin Cancer Res 2016;22(4):948–60.

[177] Wood AC, et al. Dual ALK and CDK4/6 inhibition demonstrates synergy against neuroblastoma. Clin Cancer Res 2017;23(11):2856–68.

[178] Hagiwara K, et al. Combined inhibition of ALK and HDAC induces synergistic cytotoxicity in neuroblastoma cell lines. Anticancer Res 2019;39(7):3579–84.

[179] George B, et al. Dual inhibition of IGF-IR and ALK as an effective strategy to eradicate NPM-ALK. J Hematol Oncol 2019;12(1):80.

[180] Ito M, et al. Synergistic growth inhibition of anaplastic large cell lymphoma cells by combining cellular ALK gene silencing and a low dose of the kinase inhibitor U0126. Cancer Gene Ther 2010;17(9):633–44.

[181] Donato EM, et al. Brentuximab vedotin in Hodgkin lymphoma and anaplastic large-cell lymphoma: an evidence-based review. Onco Targets Ther 2018;11:4583–90.

[182] Wang R, et al. Crizotinib enhances anti-CD30-LDM induced antitumor efficacy in NPM-ALK positive anaplastic large cell lymphoma. Cancer Lett 2019;448:84–93.

[183] Laimer D, et al. PDGFR blockade is a rational and effective therapy for NPM-ALK-driven lymphomas. Nat Med 2012;18(11):1699–704.

Therapeutic strategies to overcome ALK resistance in lung cancer

Gonzalo Recondo[a] and Luc Friboulet[b]

[a]Medical Oncology, Center for Medical Education and Clinical Research (CEMIC), Buenos Aires, Argentina [b]Predictive Biomarkers and Novel Therapeutic Strategies in Oncology, INSERM U981, Gustave Roussy Cancer Campus, Paris-Saclay University, Villejuif, France

Abstract

ALK tyrosine kinase inhibitors are current standard treatments for patients with *ALK*-rearranged non-small cell lung cancer. The development of crizotinib and a newer generation of ALK TKIs like ceritinib, alectinib, brigatinib and lorlatinib have conveyed prolonged survival outcomes in patients with ALK-driven cancers. Unfortunately, tumor cells invariably acquire resistance to these drugs through a diverse spectrum of biological mechanisms. On- and off-target resistance mechanisms emerge during the evolution of the patients disease, including *ALK* kinase domain mutations or *ALK* amplification and by-pass track activation of oncogenic signaling pathways. In this chapter, we will discuss the current preclinical and clinical evidences supporting different treatment strategies aiming to prevent or overcome the biological processes that drive ALK TKI resistance.

Abbreviations

AKT	AKT serine/threonine kinase 1
ALK	anaplastic lymphoma kinase
ATP	adenosine triphosphate
CNS	central nervous system
DNA	adenosine triphosphate
EGFR	epidermal growth factor receptor
EML4	echinoderm microtubule-associated protein-like 4
EMT	epithelial-mesenchymal transition
ENU	N-ethyl-N-nitrosourea
FISH	fluorescence *in situ* hybridization
FLT3	FMS-like tyrosine kinase 3
HER2	human epidermal growth factor receptor 2
HER3	human epidermal growth factor receptor 3
IGFR1	insulin like growth factor receptor

JAK	Janus kinase 2
MAPK	mitogen-activated protein kinase
MEK	mitogen-activated protein kinase kinase
MET	hepatocyte growth factor receptor
mTOR	mammalian target of rapamycin
NCI	National Cancer Institute
NF2	neurofibromin 2
NGS	next-generation sequencing
NRG1	neuregulin 1
NSCLC	non-small cell lung cancer
NTRK	neurotrophic receptor tyrosine kinase
PD-1	programmed cell death-1
PD-L1	programmed cell death-1 ligand
PDGFR	platelet derived growth factor receptor
PDX	patient derived xenografts
PFS	progression-free survival
PI3K	phosphatidylinositol-4,5-bisphosphate 3-kinase
POTACS	proteolysis targeting chimeras
PTEN	phosphatase and tensin homolog
PTPN11	protein tyrosine phosphatase non-receptor type 11
RNA	ribonucleic acid
ROS-1	ROS proto-oncogene 1, receptor tyrosine kinase
SHP2	Src homology region 2 (SH2)-containing protein tyrosine phosphatase 2
SRC	V-Src avian sarc
ST7	suppression of tumorigenicity 7
STAT	signal transducer and activator of transcription
TKI	tyrosine kinase inhibitor
VEGFR	vascular endothelial growth factor

Conflict of interest

Gonzalo Recondo: Research grant: Amgen. Consultant/advisory role: Amgen, Bayer, BMS, Pfizer, Roche. Luc Friboulet has no conflict of interest to disclose.

Introduction

The discovery of acquired *ALK* kinase domain mutations as a common mechanism of resistance to the first generation ALK inhibitor crizotinib has fueled the development of a new generation ALK tyrosine kinase inhibitors (TKIs) [1, 2]. As previously referred in this book, second- and third-generation ALK TKIs, like ceritinib, alectinib, brigatinib and lorlatinib, were designed to bind to the ALK kinase domain in the context of specific amino acidic changes resulting from point mutations that halted crizotinib's efficacy. These inhibitors, however, have distinct ALK binding properties and, thus, the potency of these drugs to inhibit ALK is dependent on the mutation.

Several research groups have identified ALK inhibitors that can properly bind and impede ALK phosphorylation in the setting of resistance by a specific single mutation or compound mutations. This has translated into the development of a clinical strategy that relies on the sequential administration of ALK TKIs during the patient's treatment course [3]. This sequential approach starts with the first-generation inhibitor crizotinib, later moving forward to subsequent treatment with second generation ALK inhibitors like ceritinib, alectinib or brigatinib and then to the novel, third-generation inhibitor lorlatinib. Most recently, second-generation ALK inhibitors have moved to the front line setting, given the increased potency, intracranial penetration and delay in the onset of resistance.

In addition, off-target mechanisms of resistance like bypass track activation of downstream oncogenic signaling and histologic transformation like small-cell, squamous cell transformation or epithelial mesenchymal transition, also drive disease progression in patients treated with ALK TKIs [4, 5]. In this context, the molecular diagnosis can be more challenging, given that many off-target alterations are not detected by DNA or RNA sequencing, but are also dependent on the expression and activation of non-mutated nor amplified oncogenic effectors. In this context, ALK phosphorylation is properly inhibited by the ALK TKI, but cancer cell growth becomes independent of ALK, driven by other oncogenic effectors which need also to be targeted in a dual or combined fashion [6].

The current approval of ALK TKIs by drug regulatory agencies solely requires the presence of an ALK rearrangement by immunohistochemistry (clone D5F3, Ventana) but, in the setting of resistance, tissue or blood based biomarkers are not mandatory to select the specific ALK TKI that could be most beneficial based on genomic profiling. This was mainly due to the design of the clinical trials with second- and third-generation inhibitors, in which ALK inhibitors were studied in patients with clinical progression with a previous generation of ALK inhibitors, irrespective of the underlying mechanism of resistance [7–9]. However, with proper biomarker assessment using tissue and/or liquid biopsy NGS, many genomic resistance mechanisms can be identified and help tailor treatments for patients.

In this chapter, we will review the preclinical and clinical therapeutic strategies designed to overcome on-target resistance like the *ALK* kinase domain mutations and *ALK* amplification, and the wide range of resistance mechanisms that are ALK-independent also referred as "off-target" resistance.

Overcoming resistance by the **ALK** kinase domain mutations and **ALK** amplification

ALK kinase domain mutations can affect drug binding or enhance ATP affinity to the kinase domain, which hampers the inhibition of ALK phosphorylation by the ATP-competitive kinase inhibitors [2, 4, 10]. This has been studied using *in vitro* models like Ba/F3 cells infected with *EML4-ALK* fusion expressing different mutations and testing the cytotoxic potency of different ALK inhibitors. In addition, patient-derived cell lines harboring the same *ALK* mutation than the original tumor sample have also been used to test these compounds[11]. In an exploratory approach, other assays have been used to try to predict the mutations that can cause resistance to specific ALK inhibitors, per example, by *N*-ethyl-*N*-nitrosourea (ENU) mutagenesis screens using Ba/F3 [12].

Acquired *ALK* kinase domain mutations are detected in about ~20% and ALK amplification can be found in about 8% of patients progressing on treatment with crizotinib [5]. Several mutations confer resistance to crizotinib including the gatekeeper L1196M mutation which is the most common *ALK* kinase domain mutation in the setting of crizotinib resistance. Other crizotinib resistance mutations include: I1151T, L1152R, C1156Y, I1171T/N/S, F1174L/C/V, S1206C/Y, E1210K, G1269A and the solvent front G1202R, D1203N and S1206Y/C mutations

[2, 4, 13–16]. The G1269A is frequently acquired with crizotinib, this mutation is located in the ALK ATP-binding pocket [17]. In addition, the L1152P, C1156Y, F1174C/L mutations lie in N-terminal region of the kinase domain and do not affect crizotinib binding destabilizing the αC-helix mobility, enhancing ATP affinity of the kinase by stabilizing its active state [2, 18]. The G1202R mutation is relatively uncommon in the setting of crizotinib resistance, about 8% of mutations, and confers resistance to all second-generation ALK inhibitors. This mutation is located in the solvent front of the kinase and impairs binding of first- and second-generation ALK inhibitors by steric hindrance [19]. The type of EML4-ALK variant seems to be associated with the probability of acquiring G1202R mutations, as it has been reported to occur more frequently in variant 3 EML4-ALK rearrangements [20]. There is no clear explanation for this observation, however it could potentially result from enhanced stability of this shorter fusion protein associated with the loss of the TAPE domain, favoring on-target mutations.

The spectrum of activity of second generation ALK TKIs against crizotinib-resistant mutations differs (Fig. 1A). Ceritinib is an ATP competitive ALK inhibitor active compound against different crizotinib resistant mutations including: I1171X, L1196M, S1206C/Y, E1210K and G1269A [18, 21]. However, ceritinib is not active in the setting of ALK I1151X, L1152P, C1156Y and F1174L/C/V mutations. These later mutations mediate primary resistance to ceritinib if given sequentially after crizotinib, but can be targetable with other second generation inhibitors such as alectinib and brigatinib. Ceritinib has proven clinical activity in the setting of resistance to crizotinib with objective response rates reaching about 40% and median PFS of almost 6 months [22, 23].

Alectinib is a potent ALK inhibitor with high intracranial penetration and central nervous system (CNS) activity [24]. Alectinib is active against most crizotinib resistant mutations including: I1151T, L1152P, C1156Y, F1174L/C/V, L1196M, S1206C/Y, E1210K and G1269A. However, there are two mutations which cannot be targeted by this compound, and are mechanisms of innate or acquired resistance to alectinib, the I1171T/N/S and V1180L mutations, which can be selectively targeted by ceritinib or brigatinib [25] (Fig. 1A). In a biomarker analysis of two single-arm phase II clinical trials, NP28673 (NCT01801111) and NP28761 (NCT01871805) including 48 patients with detectable post-crizotinib ALK mutations treated with alectinib, median progression-free survival (PFS) was 5.6 months, and about 50% of patients with on-target resistance achieved responses [26]. In biomarker unselected population, treatment with alectinib in the setting of crizotinib resistance confers responses in about 37–51% of patients with median PFS of about 9 months [8, 27]. The ALK G1202R mutation accounts for approximately 30% of resistance mutations after treatment with a second-generation ALK TKI, being the most common acquired resistance mutation in patients treated with alectinib [28].

Brigatinib is a potent ALK and ROS-1 inhibitor and also has in vitro activity against other kinases other kinases like FLT3, EGFR and IGF1R [25]. Brigatinib can overcome resistance due to acquired kinase domain mutations that confer resistance to crizotinib, ceritinib and alectinib with the exception of the G1202R mutation (Fig. 1A). Brigatinib was initially developed to target ALK in the context of the G1202R mutation, with in vitro kinase activity against this solvent-front mutation and in vivo activity in mouse models [25]. Though plasma levels at the 180 mg dose have been predicted to achieve an IC90 for all ALK mutations, the clinical activity of brigatinib against G1202R mutant ALK-rearranged tumors seems suboptimal [29, 30]. Moreover, brigatinib has limited activity against alectinib-resistant ALK-rearranged lung cancers, with objective responses seen in about 17% of patients treated with brigatinib after

On-target Resistance

(A)

(B)

FIG. 1   (A) On target or *ALK*-dependent resistance can be caused by the acquisition of secondary kinase domain mutations that impede adequate binding of ALK inhibitors or enhance ATP affinity of the kinase. Multiple *ALK* mutations have been characterized conferring a distinct sensitivity profile to new generation ALK inhibitors as displayed. *ALK* amplification mediates resistance to crizotinib but can be targeted with more potent second- or third-generation ALK TKIs. (B) Off-target or *ALK*-independent resistance occurs by activation of by-pass track signaling in the setting of adequate ALK inhibition, and can be driven by activation of tyrosine kinase receptors (e.g., MET), intracellular kinases (MEK, SRC), or loss of inhibitory effectors of downstream signaling (e.g., merlin) alectinib progression, resulting also in a primary progression in a patient with a detectable *ALK* G1202R mutation [30]. However, clinical responses have been observed in patients with tumors harboring alectinib resistant I1171X and V1180L mutations and should be considered if these mutations are detected [30]. Besides the G1202R resistance, mechanisms involved in brigatinib progression are less known given the later development of this compound in respect to other ALK inhibitors. Compound mutations, when two ALK resistance mutations are present in the same allele or *in cis*, have been characterized to confer resistance to brigatinib, including *ALK* E1210K/S1203N and E1210K/S1206C [14]. Given the differential activity of second generation ALK inhibitors against acquired crizotinib-resistant *ALK* mutations, blood-based or tissue DNA based NGS can provide useful information that can help better select the proper second-generation TKI in a sequential treatment modality. In an approach to follow a biomarker-guided ALK TKI strategy, the NCI-NRG ALK Master protocol is enrolling patients and allocating them to treatment with specific ALK TKIs based on the results of genomic profiling at the time of resistance to a second generation ALK inhibitor. This trial is currently ongoing and will define if biomarker selection can improve objective response rates, progression-free survival and overall survival in patients with ALK-rearranged lung cancer (NCT03737994).

Lorlatinib is a third-generation ALK inhibitor, designed to overcome resistance by all known single ALK mutations including *ALK* G1202R [31] (Fig. 1A). The macrocyclic chemical design of this compound was achieved by modifying the chemical structure of the first generation inhibitor crizotinib, to improve blood-brain barrier passage, impede p-glycoprotein efflux, and allow ALK binding in the presence of single *ALK* mutations [32]. The *in vitro* potency (IC50) in Ba/F3 models against the non-mutant EML4-ALK fusion is about 2–3 nM and varies, within a range of sensitivity, against *ALK* mutations: I1151T (50 nM), C1156Y (4-15 nM), I1171T/N/S (11-41 nM), F1174L/C/V (4-8 nM), L1196M (34-43 nM), S1206C/Y (3 nM), E1210K (1.7 nM), G1269A (10 nM) [12, 14, 31]. The *in vitro* potency of lorlatinib in Ba-F3 cells harboring the G1202R mutation has been reported between 49.9 and 113 nM [12, 14, 31, 33, 34]. In mouse models, lorlatinib also inhibits tumor growth of G1202R mutant *ALK*-rearranged tumors, and has also shown significant intracranial activity [31].

Lorlatinib is currently approved for the treatment of patients with metastatic *ALK*-rearranged lung cancer who have experienced disease progression on crizotinib and a second-generation ALK inhibitor, or in patients previously treated upfront with a second-generation ALK TKI. This approval is based on the results of the phase I/II clinical trial of lorlatinib, in which patients were enrolled in different cohorts according to the type of treatment received previously (see Chapter 2) [35, 36]. A biomarker analysis of patients included in this study was done to understand the impact of genomic alterations studied in plasma and tissue NGS on lorlatinib outcomes [28]. Among the 198 patients enrolled in the clinical trial, 189 had available plasma and 164 patients had available tissue genotyping prior to lorlatinib initiation. One or more *ALK* resistance mutations were found in 24% of plasma, and 24% of tissue samples. In 28 patients with detectable *ALK* G1202R mutations at lorlatinib baseline, the objective response rate was 57%, the median duration of response with lorlatinib was 7 months and the median progression-free survival was 8.2 months [28]. Concordantly with preclinical studies, this biologically selected subgroup analysis confirms that lorlatinib is clinically active in the context of detectable *ALK* G1202R mutation and thus, is the standard treatment option in this setting [31]. In this study, the response rate with lorlatinib was higher in patients with detectable *ALK* mutations in plasma (62% vs 32%) and in tissue (69% vs 27%)

compared to patients without detectable *ALK* mutations. In addition, progression-free survival was prolonged in patients with detectable ALK resistance mutations in tumor, median PFS of 11 months compared to 5.4 months, but this benefit was not observed in patients with detectable *ALK* mutations in plasma, with a median PFS of 7.3 months *versus* 5.5 months [28].

Lorlatinib is highly selective *in vitro* to target tumor cells that harbor *ALK* resistant mutations. In preclinical studies done in ceritinib resistant patient-derived cell lines, lorlatinib did not induce cell death in cell lines that did not harbor *ALK* resistance mutations, suggesting that lorlatinib is not active in the setting of bypass mechanisms of resistance to second-generation ALK inhibitors [14]. This is different from bypass resistance in crizotinib resistant patient derived cell lines, in which tumor cells with less potent bypass activation that still have co-dependence with ALK signaling undergo apoptosis when treated with second-generation inhibitors [18]. Hence, we can hypothesis that the responses observed with lorlatinib in patients without detectable *ALK* mutations could potentially be due to false negative results of plasma NGS if ALK resistance mutations would be present at low allele frequencies, or to tumor heterogeneity in patients that underwent tissue sampling [28]. In addition, it has been documented that a significant proportion of patients may have low plasmatic effective levels of kinase inhibitors, and hypothetically, the higher potency of lorlatinib could also compensate low pharmacokinetics levels of previous ALK treatments [37]. In any case, given that some patients without detectable *ALK* mutations may respond to treatment with lorlatinib, to date lorlatinib can be prescribed to all patients who progress to a second-generation ALK inhibitor. However, if a G1202R mutation is detected upon progression to any previous generation ALK inhibitor, lorlatinib remains the treatment of choice. Given at the current dose of 100 mg daily, the lorlatinib plasma concentration exceeds the required dose to inhibit the G1202R mutation and all single ALK mutations [35].

Even when lorlatinib was recently approved, on-target resistance mutations have been known from some time by studying resistance in patients treated in the phase I/II clinical trial. The first report on resistance to lorlatinib was in a patient with that experienced progression by the sequential acquisition in *cis* of an *ALK* C1156Y mutation with crizotinib and a later *ALK* L1198F with lorlatinib exposure [38]. Clonal analysis using whole exome sequencing data from the patient tumors showed that the cancer cells containing the compound mutation at lorlatinib resistance were sub-clones derived from tumor cells that had acquired the C1156Y mutation with crizotinib. Interestingly, substitution of a leucine for a phenylalanine in position 1198 led to a steric clash with lorlatinib. Paradoxically, the L1198F mutation increased the affinity binding of crizotinib, resensitizing these cells to treatment with a previous line of ALK TKI. The patient was treated with crizotinib and experienced a significant response. This study provided two novel proof-of-concepts, the discovery of compound mutations conferring lorlatinib resistance and that compound mutations, like single *ALK* resistance mutations, may also have unique features, including re-sensitization to other ALK inhibitors. Compound mutations acquired with first and second generation ALK inhibitors given sequentially have also been reported to drive resistance to brigatinib, like the D1203N/E1210K, but this compound mutations remain sensitive to lorlatinib inhibition [14].

Several compound mutations have been later discovered to convey resistance to lorlatinib. Some have been detected in patients experiencing acquired or primary resistance to lorlatinib and others have been revealed using *in vitro* mutagenesis assays with *N*-ethyl-*N*-nitrosourea (ENU) screens with Ba/F3 cells. Some of the compound kinase domain mutations found in patients include G1202R/G1269A, I1171M/L1198F, I1171M/D1203N, L1196M/G1202R, F1174L/G1202R, D1203N/L1196M [12, 33, 34]. The ALK kinase domain mutation L1256F is predicted *in vitro* to confer resistance to lorlatinib as a single mutation but is targetable with alectinib, however this mutation has not been reported clinically [12]. With the exception of compound mutations that include L1198F mutation, which can be treated with crizotinib, there are no new novel ALK inhibitors with the capacity to overcome lorlatinib resistance by compound mutations, especially those that also involve the G1202R mutation. Therefore, chemotherapy remains the standard treatment option to treat patients with lorlatinib resistant ALK-rearranged lung cancer.

Resistance to ALK inhibitors is a dynamic and heterogeneous process, as multiple on-target and off-target genomic alterations can be found in the same patient. Two biomarker studies using plasma NGS have reported that about 7–19% of patients previously treated with first and second generation ALK inhibitors had more than one *ALK* resistance mutation detected [29, 39]. In patients progressing on lorlatinib treatment, *ALK* mutations were found in 76%, of which two or more *ALK* mutations were found in 48% of cases. The capacity of plasma or tissue NGS to detect compound mutations is limited by the size of the amplicons, as these mutations need to be found in the same read. Only mutations that are close enough to be read in the same amplicon can be called compound mutations.

Moving out of standard approaches to overcome resistance, several research groups are exploring innovative preclinical strategies to target oncogenic ALK activation, including protein degraders and EML4 dimerization blockade. Bifunctional proteolysis targeting chimeras (PROTACs) are novel compounds designed to induce targeted protein ubiquitination and proteasomal degradation by the cereblon E3 ligase complex [40]. ALK PROTACs are composed of an ALK inhibitor connected by a linker to the cereblon ligand pomalidomide. Pomalidomide recruits the E3 ubiquitin ligase complex, which proceeds to ALK ubiquitination and proteasomal degradation. Several ALK PROTACs have been designed using ceritinib and TAE684 to degrade full ALK receptor with common activating mutations in neuroblastoma (F1174L/R1275Q) and EML4-ALK *in vitro* [41, 42]. However, these PROTACs will not be able to target ALK in the context of lorlatinib-resistant compound mutations, which also confer resistance to ceritinib, thus new ALK binding compounds need to be designed in this context to induce protein degradation in lorlatinib-resistant cells.

Monomerization of EML4-ALK fusion proteins is another strategy to prevent ALK signaling. *In vitro* inhibition of spontaneous homodimerization by the EML4 coiled-coil domain is an attractive strategy to impede ALK kinase domain activation, even in the setting of resistance mutations as most do not confer constitutive kinase activation. A preclinical study has recently shown that endogenous expression and exogenous administration of mimicking peptides of the EML4 coiled-coil domain suppressed ALK-driven tumor growth *in vitro* and *in vivo* by inhibiting EML4 dimerization [43]. There are multiple challenges to develop this kind of compounds for treatment in humans, however, in the setting of complex ALK mutations that will difficult the development of novel ALK kinase inhibitors, disrupting other mechanisms of ALK activation will be necessary in the near future.

ALK amplification is another resistance mechanism reported in about 8% of patients treated with crizotinib [4, 14]. *ALK* amplification does not overlap with crizotinib resistance kinase domain mutations. *ALK* amplification can be effectively targeted with more potent second generational ALK inhibitors and has not been described to confer resistance to new generation ALK TKIs.

Overcoming resistance by off-target mechanisms

In the absence of *ALK* mutations or amplification and in the setting of adequate ALK inhibition, resistance is probably commanded by other oncogenic effectors like tyrosine kinase receptors or downstream activation of oncogenic pathways (Fig. 1B) [6]. There is a wide range of oncogenic bypass pathways that have been implied in resistance to ALK inhibitors. In any case, in order to effectively overcome resistance in this scenario, effective drug combinations targeting ALK and the acquired activated effector are needed to induce apoptosis in cancer cells. Many of these bypass activation pathways can be diagnosed using tumor or plasma NGS, as some are direct consequence of molecular alterations including activating mutations or rearrangements in oncogenes, amplification of known oncogenes and truncating mutations in tumor suppressor genes that codify for proteins that regulate oncogenic signaling. However, many bypass mechanisms of resistance are due to aberrant oncogenic activation of a non-mutated or amplified driver [6].

MET amplification is a common resistance mechanism to first, second and third generation EGFR inhibitors, however the role of MET activation in resistance to ALK inhibitors was less known until recently [44–47] (Fig. 1B). In 207 tissue/plasma samples from patients treated with next-generation ALK inhibitors, *MET* amplification was detected by NGS and fluorescence *in situ* hybridization (FISH) in about 12% of tumor biopsies at resistance to second-generation ALK inhibitors and in 22% of lorlatinib-resistant tumors [48]. In addition, *ST7-MET* fusions were also reported to be acquired at the time of resistance. *MET* rearrangements are rarely found in cancer, but have been reported in NSCLC tumors, in patients with infantile or adult glioblastomas and in infantile spindle sarcomas [49–53]. In this study, the combination of ALK-MET inhibition led to clinical responses in two patients: one treated with single agent crizotinib, and in second case with lorlatinib-crizotinib combination [48]. *MET* amplification can be assessed by using NGS or FISH and it can be rapidly adopted into patients clinical care, opening a new opportunity for combinatorial strategies, as it has been the case for patients with *EGFR*-mutant and *MET*-amplified cancers [54].

Other genomic alterations that can be associated with resistance to ALK inhibitors have been found using plasma or tissue NGS including *KRAS* mutations, activating mutations in *PI3K*, inactivating *PTEN* mutations, *MEK* mutations that can suggest that bypass pathways may be activated in the setting of clinical progression [39, 55].

When bypass track activation cannot be determined by genomic alterations, the study of off-target resistance mechanisms relies in the development of patient derived xenografts (PDX) or cell lines. By using different *in vitro* assays like drug screens, kinase assays and CRISPR-CAS9 gene editing, several non-genomic bypass mechanisms have been found and targeted *in vitro* and *in vivo* with combination therapies. Several bypass mechanisms have been discovered using patient derive models involving EGFR, IGFR1, HER2, HER3, MEK, NF2, PI3K, SHP2 and SRC [6, 14, 34, 56] (Fig. 1B).

By developing patient-derived cell lines, early preclinical studies revealed that these bypass track mechanisms of resistance occur frequently. In a patient derived cell line, an acquired *MEK* K57N mutation was found to confer resistance to ceritinib. This mutation is one of the hot spot mutations in *MEK*, leading to constitutive activation of the MAPK pathway, downstream of RAF and RAS [6]. *In vitro* and *in vivo*, the combination of ALK and MEK inhibition led to significant apoptosis, which was selective for this specific cell line and not reproduced in other models that did not harbor *MEK* mutations. In addition, this cytotoxic effect was not observed with single MEK inhibition, abrogating for the need to suppress oncogenic signaling pathways downstream of ALK other than the MAPK pathway like the PI3K/AKT/mTOR pathway. There are ongoing clinical trials combining ALK and MEK inhibitors like ceritinib with trametinib (NCT03087448), alectinib with cobimetinib (NCT03202940) and brigatinib with binimetinib (NCT04005144), translating the observed interaction *in vitro* into the clinical practice.

SRC activation has also been recurrently implied in resistance to crizotinib and most recently, in the onset of lorlatinib resistance [6, 34]. In several patient-derived models of crizotinib resistant cell lines, exposure to crizotinib induced SRC activation and signaling. SRC can be activated by a number of tyrosine kinase receptors including EGFR, IGFR-1, PDGFR, VEGFR and ALK, directly activating downstream signaling pathways. SRC is also involved in cell-to-cell adhesion through focal adhesion kinases, and in the development of epithelial mesenchymal transition [57]. Combining SRC inhibition using saracatinib (AZD0530) and ALK TKIs induced apoptosis in patients derived cell lines. In a recent pre-clinical study of lorlatinib resistance, combining lorlatinib and saracatinib effectively targeted cells undergoing SRC induced epithelial-mesenchymal transition (EMT) by a direct cytotoxic effect of this combination [34]. In a second patient with bone oligo-progression in which partial EMT was found by high levels of vimentin expression in the biopsy sample, the patients-derived cell line with EMT phenotype was also susceptible to combined ALK/SRC inhibition. EMT may be more common than suspected clinically, it has been previously reported in 40% of tissue samples from 12 ceritinib resistant tumors with E-cadherin loss and vimentin expression [14]. EMT has also been recently reported in a patient progressing on several lines of ALK inhibitors like crizotinib, ceritinib and alectinib in the setting of ALK-refractory disease [58]. The combination of the SRC inhibitor dasatinib and crizotinib was studied in a phase I trial that did not include patients with ALK-rearranged lung cancers and had a high rate of adverse events, which does not support this combination [59]. Repotrectinib (TPX-0005) is a potent ROS-1 inhibitor, and can also target ALK and NTRK and SRC [60]. It was originally designed as a dual ALK/SRC inhibitor, however *the in vitro* potency of this inhibitor against ALK was not translated in the clinical setting. Currently, there are no clinical available combination strategies against ALK/SRC co-dependence.

As mentioned earlier in this chapter, several phosphokinases can be activated in ALK TKI resistance models. Therefore, preclinical research has been conducted aiming to identify common signaling activators of intracellular signaling pathways. Src homology region 2 (SH2)-containing protein tyrosine phosphatase 2 (SHP2) encoded by *PTPN11* is a regulator of the multinetwork of signaling cascades including the MAPK, PI3K/AKT/mTOR and JAK/STAT [61]. SHP2 inhibition was identified as a hit in ceritinib resistant cell lines using a shRNA screen [56]. In this study, ERK reactivation was found to be re-induced rapidly after exposure of cancer cells to ceritinib, and combined ALK and SHP2 inhibition and was reverted by using the allosteric SHP2 inhibitor SHP099, which led to enhanced apoptosis in these models in combination with the ALK inhibitor, proving durable inhibition of ERK1/2 and p90RSK phosphorylation and inhibition of GTP loading of KRAS isoforms. This study may lead to the development of SHP2 and ALK combinations, as novel SHP2 inhibitors like TNO155 are being tested in phase I trials (NCT03114319).

Activation of the ErbB family of receptors like EGFR, HER2 and HER3 are also related to ALK TKI resistance *in vitro* due to receptor bypass signaling (217,229,230). EGFR can amplify downstream ALK signaling through adaptor proteins like GRB (229). *In vitro*, the combination of ALK and EGFR inhibitors can restore sensitivity to ALK TKIs (231), and neuregulin expression can be induced by ALK inhibition. *In vitro*, ALK dependent cell lines can acquire resistance to ALK TKIs by exposure to NRG1 and this can be reverted by combining ALK TKI with lapatinib, a HER2 tyrosine kinase inhibitor (232). Interestingly, *de novo* co-occurrence of ALK rearrangements and activating EGFR mutations have been described in case reports [62, 63]. Outside of anecdotal cases, the interactions between ALK and the ERBB family of tyrosine kinase receptors at the time of resistance to ALK TKIs have not been assessed in clinical trials.

Bypass track activation that conveys resistance to crizotinib can be potentially overcome by second generation ALK TKIs, as it has been well documented in the preclinical models and in the clinical setting, that tumors without detectable on-target resistance can be targeted by switching to new generation ALK inhibitors [18]. This suggests that mild bypass signaling activation can be sufficient to surpass crizotinib inhibition of ALK, however complete and potent ALK inhibition by new-generation ALK inhibitors can hamper the proliferation of cancer cells that still maintain a certain degree of co-dependence in ALK signaling to survive. Also, second generation ALK inhibitors may provoke off-target inhibition in other effectors of bypass signaling, like ceritinib with IGFR1 or brigatinib with EGFR and FLT3, which can also play a role in the efficacy of these agents in crizotinib-resistant models [25].

In the setting of lorlatinib resistance, there is scarce data on potential bypass mechanisms of resistance to this compound. In a preclinical study using H3122 and H2228 cell lines, incremental doses of lorlatinib induced resistance by EGFR activation *in vitro*, which has already been described with other ALK inhibitors in the H3122 cell line [64]. In a neuroblastoma cell line (CLB-GA) with full length ALK harboring the R1275Q mutation, exposure to lorlatinib favored the emergence of resistance clones that contained a truncating mutation in *NF1*, which was targetable with combination of trametinib and lorlatinib *in vitro* [64].

More recently, using patient-derived xenografts and cell lines, a bypass mechanism due to *NF2* inactivation was characterized *in vitro* in a patient progression on lorlatinib with the sequential acquisition of deleterious *NF2* mutations in different temporo-spatial metastasis [34]. *In vitro* validation of the effect of *NF2* loss in lorlatinib resistance was conducted by CRISPR-CAS9 *NF2* knockout in H3122 cells, showing high levels of pS6 even with adequate ALK inhibition, secondary to downstream activation of the PI3K/AKT/mTOR pathway due to loss of mTOR inhibition by merlin. Combining lorlatinib and the dual mTORC1–2 inhibitor vistusertib (AZD2014) reverted resistance *in vitro* and *in vivo*. More research is needed to better understand how to treat patients with off-target resistance mechanisms driving lorlatinib resistance, as more patients receive this drug in the clinical setting.

Preventing the emergence of resistance: New generation ALK inhibitors upfront

Given the complexity of on-target resistance with sequential ALK inhibitor treatments, more potent second- and third-generation ALK inhibitors have been tested upfront in clinical trials with improved efficacy over crizotinib [3]. In treatment-naïve mice models harboring

ALK-rearranged tumors, upfront treatment with alectinib or brigatinib conveyed prolonged disease control by delaying the emergence of resistance [25, 31, 65]. As referred in Chapter 2, ceritinib, alectinib and brigatinib given in the first line results in prolonged progression-free survival compared to crizotinib and delays the onset of brain metastasis in patients with metastatic *ALK*-rearranged NSCLC [66–68]. There is scarce data regarding the mechanisms of resistance to first-line treatment with a second-generation ALK TKI, however the most common resistance mutations found in patients receiving front line alectinib are the G1202R and I1171X mutations, all targetable with lorlatinib [29]. More data will be needed to understand resistance mechanisms to first line treatment with brigatinib.

Lorlatinib is also predicted to improve outcomes of patients in the first-line setting compared with crizotinib. In a subset of 30 patients treated in the first line with lorlatinib in the phase I/II trial, 90% achieved a response and intracranial responses were observed in 67% of patients, with sustained PFS benefit [36]. The phase III CROWN trial is currently ongoing, comparing front line treatment with lorlatinib and crizotinib and the results of this trial showed improved progression-free survival and intracranial activity of lorlarinib over crizotinib, however longer follow-up is required to see the impact of this approach in overall survival [69]. Upfront treatment with lorlatinib may favor the delay of on target resistance, supported by preclinical evidence with ENU mutagenesis screens in which there is absence of resistant mutations arising in BaF3 cells treated with this drug. This further suggests that, in the first line setting, a predominance of off-target resistance could be expected [33].

Drug combinations to overcome or delay resistance in ALK-rearranged NSCLC

Other synergic strategies have been explored in the treatment of patients with EGFR-mutant NSCLC, like the combination of EGFR inhibitors with platinum-pemetrexed chemotherapy and with antiangiogenic agents, prolonging progression-free survival and overall survival (only with chemotherapy combinations) in patients treated in the front line [70–72]. In EGFR-mutant lung cancer there is evidence that resistant clones may be pre-existing at the time of diagnosis, and emerge as dominant clones after treatment selection [73]. This has been documented for T790M mutations and in the setting of small-cell lung cancer transformation [74]. Combining TKIs with chemotherapy upfront could target cells that are susceptible to kinase inhibition and clones who are primarily resistant to these compounds, also potentially diminishing the pool of drug-tolerant or "persister" cells. For the moment, there is scarce data on combination of chemotherapy with ALK TKIs and it has not been explored in randomized trials [75]. There is an ongoing clinical trial combining alectinib with bevacizumab, a soluble VEGF monoclonal antibody, in patients with *ALK*-rearranged NSCLC and untreated and stable brain metastasis (NCT02521051). This phase I/II trial will assess the safety of this combination, and additionally, will study the intracranial and systemic objective response rates. If this combination is safe, it will be important to determine whether this can further delay resistance.

In addition to chemotherapy and antiangiogenic agents, programmed cell death-1 (PD-1) and its ligand (PD-L1) immune checkpoint inhibitors are current standard treatments in patients with advanced NSCLC. However, they are not effective treatments for patients with ALK-rearranged NSCLC given as monotherapy, possible due to the low immunogenicity of these tumors [76]. In an attempt to improve outcomes in patients with ALK-rearranged lung cancer, crizotinib was combined with nivolumab in a phase I trial, however this combination resulted in significant liver toxicities and was discontinued [77]. Alectinib was also combined with atezolizumab and lorlatinib with avelumab, and though these combinations showed acceptable toxicity profiles in patients, they did not seem to convey superior clinical benefit over single agent ALK inhibitors [78, 79]. In patient with *ALK*- or *EGFR*-driven tumors, who have progressed on targeted therapies, the combination of the anti PD-L1 monoclonal antibody atezolizumab with bevacizumab and chemotherapy showed improved overall survival over the chemotherapy and atezolizumab combination, in a subgroup analysis of the Impower150 trial [80, 81]. This regimen has been approved in Europe based on a subset analysis that involved a small group of patients, and though it's an available option in patients in which there are no further options of ALK inhibition, prospective studies are needed to validate the use of immunotherapy combinations in this setting. Other research groups are aiming to improve immune surveillance against ALK-rearranged lung cancer cells, by developing anti-ALK vaccines, however this approach remains experimental and clinical trials are awaited [82].

Conclusion

In summary, in this chapter we have reviewed clinical and preclinical strategies to overcome resistance to ALK inhibitors in patients with lung cancer. The road to the discovery and characterization of on-target ALK resistance mutations have successfully led to the development of effective next-generation ALK inhibitors which have been strategically sequenced to prolong patients' survival. On the other hand, treatment strategies to overcome off-target resistance have not translated fast into clinical trial designs, remaining as an unmet need for patients that require more tailored treatments in this setting. As we approach the frontier in the development of ALK inhibitors with the approval of lorlatinib, research efforts are needed to design new drugs and regimens to overcome or prevent novel forms of biological resistance. With the guidance of translational preclinical research and new drug development techniques, innovative strategies are highly awaited to further care for patients with ALK-driven cancers.

References

[1] Katayama R, Sakashita T, Yanagitani N, et al. P-glycoprotein mediates ceritinib resistance in anaplastic lymphoma kinase-rearranged non-small cell lung cancer. EBioMedicine 2016;3:54–66.
[2] Sasaki T, Okuda K, Zheng W, et al. The neuroblastoma-associated F1174L ALK mutation causes resistance to an ALK kinase inhibitor in ALK-translocated cancers. Cancer Res 2010;70(24):10038–43.
[3] Recondo G, Facchinetti F, Olaussen KA, Besse B, Friboulet L. Making the first move in EGFR-driven or ALK-driven NSCLC: first-generation or next-generation TKI? Nat Rev Clin Oncol 2018;15(11):694–708.
[4] Katayama R, Shaw AT, Khan TM, et al. Mechanisms of acquired crizotinib resistance in ALK-rearranged lung cancers. Sci Transl Med 2012;4(120):120ra17. [Internet]. Available from: www.sciencetranslationalmedicine.org/cgi/content/full/4/120/120ra17/DC1.
[5] Rotow J, Bivona TG. Understanding and targeting resistance mechanisms in NSCLC. Nat Rev Cancer 2017; 17(11):637–58.

[6] Crystal AS, Shaw AT, Sequist LV, et al. Patient-derived models of acquired resistance can identify effective drug combinations for cancer. Science 2014;346(6216):1480–6.

[7] Kim D-W, Mehra R, Tan DSW, et al. Activity and safety of ceritinib in patients with ALK-rearranged non-small-cell lung cancer (ASCEND-1): updated results from the multicentre, open-label, phase 1 trial. Lancet Oncol 2016;17(4):452–63.

[8] Yang JC-H, Ou S-HI, De Petris L, et al. Pooled systemic efficacy and safety data from the pivotal phase II studies (NP28673 and NP28761) of alectinib in ALK-positive non-small cell lung cancer. J Thorac Oncol 2017;12 (10):1552–60.

[9] Camidge DR, Kim D-W, Tiseo M, et al. Exploratory analysis of brigatinib activity in patients with anaplastic lymphoma kinase-positive non-small-cell lung cancer and brain metastases in two clinical trials. J Clin Oncol 2018;36(26):2693–701.

[10] Awad MM, Engelman JA, Shaw AT. Acquired resistance to crizotinib from a mutation in CD74-ROS1. N Engl J Med 2013;369(12):1173.

[11] Warmuth M, Kim S, Gu X, Xia G, Adrian F. Ba/F3 cells and their use in kinase drug discovery. Curr Opin Oncol 2007;19(1):55–60.

[12] Okada K, Araki M, Sakashita T, et al. Prediction of ALK mutations mediating ALK-TKIs resistance and drug re-purposing to overcome the resistance. EBioMedicine 2019;41:105–19.

[13] Choi YL, Soda M, Yamashita Y, et al. EML4-ALK mutations in lung cancer that confer resistance to ALK inhib-itors. N Engl J Med 2010;363(18):1734–9.

[14] Gainor JF, Dardaei L, Yoda S, et al. Molecular mechanisms of resistance to first- and second-generation ALK inhibitors in ALK-rearranged lung cancer. Cancer Discov 2016;6(10):1118–33.

[15] Lovly CM, Pao W. Escaping ALK inhibition: mechanisms of and strategies to overcome resistance. Sci Transl Med 2012;4(120). 120ps2.

[16] Heuckmann JM, Holzel M, Sos ML, et al. ALK mutations conferring differential resistance to structurally diverse ALK inhibitors. Clin Cancer Res 2011;17(23):7394–401.

[17] Doebele RC, Pilling AB, Aisner DL, et al. Mechanisms of resistance to crizotinib in patients with ALK gene rearranged non-small cell lung cancer. Clin Cancer Res 2012;18(5):1472–82.

[18] Friboulet L, Li N, Katayama R, et al. The ALK inhibitor ceritinib overcomes crizotinib resistance in non-small cell lung cancer. Cancer Discov 2014;4(6):662–73.

[19] Chuang YC, Huang BY, Chang HW, Yang CN. Molecular modeling of ALK L1198F and/or G1202R mutations to determine differential crizotinib sensitivity. Sci Rep 2019;9(1):11390.

[20] Lin JJ, Zhu VW, Yoda S, et al. Impact of EML4-ALK variant on resistance mechanisms and clinical outcomes in ALK-positive lung cancer. J Clin Oncol 2018;36(12):1199–206.

[21] Marsilje TH, Pei W, Chen B, et al. Synthesis, structure-activity relationships, and in vivo efficacy of the novel potent and selective anaplastic lymphoma kinase (ALK) inhibitor 5-chloro-N2-(2-isopropoxy-5-methyl-4-(piperidin-4-yl)phenyl)-N4-(2-(isopropylsulfonyl)phenyl)pyrimidine-2,4-dia. J Med Chem 2013;56(14):5675–90.

[22] Crinò L, Ahn M-J, De Marinis F, et al. Multicenter phase II study of whole-body and intracranial activity with ceritinib in patients with ALK-rearranged non–small-cell lung cancer previously treated with chemotherapy and crizotinib: results from ASCEND-2. J Clin Oncol 2016;34(24):2866–73. https://doi.org/10.1200/JCO.2015.65. 5936 [Internet].

[23] Shaw AT, Kim TM, Crino L, et al. Ceritinib versus chemotherapy in patients with ALK-rearranged non-small-cell lung cancer previously given chemotherapy and crizotinib (ASCEND-5): a randomised, controlled, open-label, phase 3 trial. Lancet Oncol 2017;18(7):874–86.

[24] Gadgeel SM, Gandhi L, Riely GJ, et al. Safety and activity of alectinib against systemic disease and brain metas-tases in patients with crizotinib-resistant ALK-rearranged non-small-cell lung cancer (AF-002JG): results from the dose-finding portion of a phase 1/2 study. Lancet Oncol 2014;15(10):1119–28.

[25] Zhang S, Anjum R, Squillace R, et al. The potent ALK inhibitor brigatinib (AP26113) overcomes mechanisms of resistance to first- and second-generation ALK inhibitors in preclinical models. Clin Cancer Res 2016; 22(22):5527–38.

[26] Noé J, Lovejoy A, Ou S-HI, et al. ALK mutation status before and after alectinib treatment in locally advanced or metastatic ALK-positive NSCLC: pooled analysis of two prospective trials. J Thorac Oncol Off Publ Int Assoc Study Lung Cancer 2020;15(4):601–8.

[27] Novello S, Mazieres J, Oh I-J, et al. Alectinib versus chemotherapy in crizotinib-pretreated anaplastic lymphoma kinase (ALK)-positive non-small-cell lung cancer: results from the phase III ALUR study. Ann Oncol Off J Eur Soc Med Oncol 2018;29(6):1409–16.

[28] Shaw AT, Solomon BJ, Besse B, et al. ALK resistance mutations and efficacy of lorlatinib in advanced anaplastic lymphoma kinase-positive non-small-cell lung cancer. J Clin Oncol 2019;37(16):1370–9.

[29] Dagogo-Jack I, Rooney MM, Lin JJ, et al. Treatment with next-generation ALK inhibitors fuels plasma ALK mutation diversity. Clin Cancer Res 2019;25(22):6662–70.

[30] Lin JJ, Zhu VW, Schoenfeld AJ, et al. Brigatinib in patients with alectinib-refractory ALK-positive NSCLC. J Thorac Oncol 2018;13(10):1530–8.

[31] Zou HY, Friboulet L, Kodack DP, et al. PF-06463922, an ALK/ROS1 inhibitor, overcomes resistance to first and second generation ALK inhibitors in preclinical models. Cancer Cell 2015;28(1):70–81.

[32] Johnson TW, Richardson PF, Bailey S, et al. Discovery of (10R)-7-amino-12-fluoro-2,10,16-trimethyl-15-oxo-10,15,16,17-tetrahydro-2H-8,4-(m etheno)pyrazolo[4,3-h][2,5,11]-benzoxadiazacyclotetradecine-3-carbonitrile (PF-06463922), a macrocyclic inhibitor of anaplastic lymphoma kinase (ALK) and c-ros. J Med Chem 2014;57 (11):4720–44.

[33] Yoda S, Lin JJ, Lawrence MS, et al. Sequential ALK inhibitors can select for lorlatinib-resistant compound ALK mutations in ALK-positive lung cancer. Cancer Discov 2018;8(6):714–29.

[34] Recondo G, Mezquita L, Facchinetti F, et al. Diverse resistance mechanisms to the third-generation ALK inhibitor lorlatinib in ALK-rearranged lung cancer. Clin Cancer Res Off J Am Assoc Cancer Res 2020;26(1):242–55.

[35] Shaw AT, Felip E, Bauer TM, et al. Lorlatinib in non-small-cell lung cancer with ALK or ROS1 rearrangement: an international, multicentre, open-label, single-arm first-in-man phase 1 trial. Lancet Oncol 2017;18(12):1590–9.

[36] Solomon BJ, Besse B, Bauer TM, et al. Lorlatinib in patients with ALK-positive non-small-cell lung cancer: results from a global phase 2 study. Lancet Oncol 2018;19(12):1654–67.

[37] Geraud A, Mezquita L, Auclin E, et al. MA21.09 tyrosine kinase inhibitors' plasma concentration and oncogene-addicted advanced non-small lung cancer (aNSCLC) resistance. J Thorac Oncol 2019;14(10):S337–8. https://doi.org/10.1016/j.jtho.2019.08.679 [Internet].

[38] Shaw AT, Friboulet L, Leshchiner I, et al. Resensitization to crizotinib by the lorlatinib ALK resistance mutation L1198F. N Engl J Med 2015;374(1):54–61. [Internet]. Available from: http://www.ncbi.nlm.nih.gov/pubmed/ 26698910.

[39] Mezquita L, Swalduz A, Jovelet C, et al. Clinical relevance of an amplicon-based liquid biopsy for detecting ALK and ROS1 fusion and resistance mutations in patients with non–small-cell lung cancer. JCO Precis Oncol 2020;4:272–82. https://doi.org/10.1200/PO.19.00281 [Internet].

[40] Sun X, Gao H, Yang Y, et al. PROTACs: great opportunities for academia and industry. Signal Transduct Target Ther 2019;4:64.

[41] Powell CE, Gao Y, Tan L, et al. Chemically induced degradation of anaplastic lymphoma kinase (ALK). J Med Chem 2018;61(9):4249–55.

[42] Zhang C, Han X-R, Yang X, et al. Proteolysis targeting chimeras (PROTACs) of anaplastic lymphoma kinase (ALK). Eur J Med Chem 2018;151:304–14.

[43] Hirai N, Sasaki T, Okumura S, Minami Y, Chiba S, Ohsaki Y. Monomerization of ALK fusion proteins as a therapeutic strategy in ALK-rearranged non-small cell lung cancers. Front Oncol 2020;10:419.

[44] Engelman JA, Zejnullahu K, Mitsudomi T, et al. MET amplification leads to gefitinib resistance in lung cancer by activating ERBB3 signaling. Science 2007;316(5827):1039–43.

[45] Sequist LV, Han J-Y, Ahn M-J, et al. Osimertinib plus savolitinib in patients with EGFR mutation-positive, MET-amplified, non-small-cell lung cancer after progression on EGFR tyrosine kinase inhibitors: interim results from a multicentre, open-label, phase 1b study. Lancet Oncol 2020;21(3):373–86.

[46] Gouji T, Takashi S, Mitsuhiro T, Yukito I. Crizotinib can overcome acquired resistance to CH5424802: is amplification of the MET gene a key factor? J. Thorac Oncol 2014;9(3):e27–8.

[47] Yamada T, Takeuchi S, Nakade J, et al. Paracrine receptor activation by microenvironment triggers bypass survival signals and ALK inhibitor resistance in EML4-ALK lung cancer cells. Clin Cancer Res 2012;18(13):3592–602.

[48] Dagogo-Jack I, Yoda S, Lennerz JK, et al. MET alterations are a recurring and actionable resistance mechanism in ALK-positive lung cancer. Clin Cancer Res 2020;26(11):2535–45.

[49] Plenker D, Bertrand M, de Langen AJ, et al. Structural alterations of MET trigger response to MET kinase inhibition in lung adenocarcinoma patients. Clin Cancer Res 2018;24(6):1337–43.

[50] Bender D, Gronych J, Warnatz HJ, et al. Recurrent MET fusion genes represent a drug target in pediatric glioblastoma. Nat Med 2016;22(11):1314–20.

[51] Ferguson SD, Zhou S, Huse JT, et al. Targetable gene fusions associate with the IDH wild-type astrocytic lineage in adult gliomas. J Neuropathol Exp Neurol 2018;77(6):437–42.

[52] Rooper LM, Karantanos T, Ning Y, Bishop JA, Gordon SW, Kang H. Salivary secretory carcinoma with a novel ETV6-MET fusion: expanding the molecular Spectrum of a recently described entity. Am J Surg Pathol 2018; 42(8):1121–6.

[53] Flucke U, van Noesel MM, Wijnen M, et al. TFG-MET fusion in an infantile spindle cell sarcoma with neural features. Genes Chromosomes Cancer 2017;56(9):663–7.

[54] Oxnard GR, Yang JC-H, Yu H, et al. TATTON: a multi-arm, phase Ib trial of osimertinib combined with selumetinib, savolitinib, or durvalumab in EGFR-mutant lung cancer. Ann Oncol Off J Eur Soc Med Oncol 2020;31(4):507–16.

[55] Horn L, Whisenant JG, Wakelee H, et al. Monitoring therapeutic response and resistance: analysis of circulating tumor DNA in patients with ALK+ lung cancer. J Thorac Oncol Off Publ Int Assoc Study Lung Cancer 2019; 14(11):1901–11.

[56] Dardaei L, Wang HQ, Singh M, et al. SHP2 inhibition restores sensitivity in ALK-rearranged non-small-cell lung cancer resistant to ALK inhibitors. Nat Med 2018;24(4):512–7.

[57] Frankson R, Yu Z-H, Bai Y, Li Q, Zhang R-Y, Zhang Z-Y. Therapeutic targeting of oncogenic tyrosine phosphatases. Cancer Res 2017;77(21):5701–5.

[58] Urbanska EM, Sørensen JB, Melchior LC, Costa JC, Santoni-Rugiu E. Changing ALK-TKI-resistance mechanisms in rebiopsies of ALK-rearranged NSCLC: ALK- and BRAF-mutations followed by epithelial-mesenchymal transition. Int J Mol Sci 2020;21(8):2847.

[59] Kato S, Jardim DL, Johnson FM, et al. Phase I study of the combination of crizotinib (as a MET inhibitor) and dasatinib (as a c-SRC inhibitor) in patients with advanced cancer. Invest New Drugs 2018;36(3):416–23.

[60] Drilon A, Ou S-HI, Cho BC, et al. Repotrectinib (TPX-0005) is a next-generation ROS1/TRK/ALK inhibitor that potently inhibits ROS1/TRK/ALK solvent-front mutations. Cancer Discov 2018;8(10):1227–36.

[61] Rehman AU, Rahman MU, Khan MT, et al. The landscape of protein tyrosine phosphatase (Shp2) and cancer. Curr Pharm Des 2018;24(32):3767–77.

[62] Won JK, Keam B, Koh J, et al. Concomitant ALK translocation and EGFR mutation in lung cancer: a comparison of direct sequencing and sensitive assays and the impact on responsiveness to tyrosine kinase inhibitor. Ann Oncol Off J Eur Soc Med Oncol 2015;26(2):348–54.

[63] Lo Russo G, Imbimbo M, Corrao G, et al. Concomitant EML4-ALK rearrangement and EGFR mutation in non-small cell lung cancer patients: a literature review of 100 cases. Oncotarget 2017;8(35):59889–900.

[64] Redaelli S, Ceccon M, Zappa M, et al. Lorlatinib treatment elicits multiple on- and off-target mechanisms of resistance in ALK-driven cancer. Cancer Res 2018;78(24):6866–80.

[65] Sakamoto H, Tsukaguchi T, Hiroshima S, et al. CH5424802, a selective ALK inhibitor capable of blocking the resistant gatekeeper mutant. Cancer Cell 2011;19(5):679–90.

[66] Soria J-C, Tan DSW, Chiari R, et al. First-line ceritinib versus platinum-based chemotherapy in advanced ALK-rearranged non-small-cell lung cancer (ASCEND-4): a randomised, open-label, phase 3 study. Lancet (London, England) 2017;389(10072):917–29.

[67] Peters S, Camidge DR, Shaw AT, et al. Alectinib versus crizotinib in untreated ALK-positive non-small-cell lung cancer. N Engl J Med 2017;377(9):829–38.

[68] Camidge DR, Kim HR, Ahn M-J, et al. Brigatinib versus crizotinib in ALK-positive non-small-cell lung cancer. N Engl J Med 2018;379(21):2027–39.

[69] Solomon B, Bauer TM, De Marinis F, et al. LBA2 Lorlatinib vs crizotinib in the first-line treatment of patients (pts) with advanced ALK-positive non-small cell lung cancer (NSCLC): results of the phase III CROWN study. Ann Oncol 2020;31:S1180–1.

[70] Noronha V, Patil VM, Joshi A, et al. Gefitinib versus gefitinib plus pemetrexed and carboplatin chemotherapy in EGFR-mutated lung cancer. J Clin Oncol Off J Am Soc Clin Oncol 2020;38(2):124–36.

[71] Hosomi Y, Morita S, Sugawara S, et al. Gefitinib alone versus gefitinib plus chemotherapy for non-small-cell lung cancer with mutated epidermal growth factor receptor: NEJ009 study. J Clin Oncol Off J Am Soc Clin Oncol 2020;38(2):115–23.

[72] Nakagawa K, Garon EB, Seto T, et al. Ramucirumab plus erlotinib in patients with untreated, EGFR-mutated, advanced non-small-cell lung cancer (RELAY): a randomised, double-blind, placebo-controlled, phase 3 trial. Lancet Oncol 2019;20(12):1655–69.

[73] Hata AN, Niederst MJ, Archibald HL, et al. Tumor cells can follow distinct evolutionary paths to become resistant to epidermal growth factor receptor inhibition. Nat Med 2016;22(3):262–9. [Internet]. Available from: http://www.ncbi.nlm.nih.gov/pmc/articles/PMC4900892/.

[74] Lee J-K, Lee J, Kim S, et al. Clonal history and genetic predictors of transformation into small-cell carcinomas from lung adenocarcinomas. J Clin Oncol 2017;35(26):3065–74.

[75] Lin JJ, Schoenfeld AJ, Zhu VW, et al. Efficacy of platinum/pemetrexed combination chemotherapy in ALK-positive NSCLC refractory to second-generation ALK inhibitors. J Thorac Oncol Off Publ Int Assoc Study Lung Cancer 2020;15(2):258–65.

[76] Mazieres J, Drilon A, Lusque A, et al. Immune checkpoint inhibitors for patients with advanced lung cancer and oncogenic driver alterations: results from the IMMUNOTARGET registry. Ann Oncol Off J Eur Soc Med Oncol 2019;30(8):1321–8.

[77] Spigel DR, Reynolds C, Waterhouse D, et al. Phase 1/2 study of the safety and tolerability of nivolumab plus crizotinib for the first-line treatment of ALK translocation-positive advanced non-small cell lung cancer (Check-Mate 370). J Thorac Oncol 2018;13(5):682–8.

[78] Kim D-W, Gadgeel SM, Gettinger SN, et al. Safety and clinical activity results from a phase Ib study of alectinib plus atezolizumab in ALK+ advanced NSCLC (aNSCLC). J Clin Oncol 2018;36(15_suppl):9009. https://doi.org/10.1200/JCO.2018.36.15_suppl.9009 [Internet].

[79] Shaw AT, et al. Avelumab (anti-PD-L1) in combination with crizotinib or lorlatinib in patients with previously treated advanced NSCLC: phase 1b results from JAVELIN Lung 101. J Clin Oncol 2018;36(suppl; abstr 9008):9008.

[80] Reck M, Mok TSK, Nishio M, et al. Atezolizumab plus bevacizumab and chemotherapy in non-small-cell lung cancer (IMpower150): key subgroup analyses of patients with EGFR mutations or baseline liver metastases in a randomised, open-label phase 3 trial. Lancet Respir Med 2019;7(5):387–401.

[81] Socinski MA, Jotte RM, Cappuzzo F, et al. Atezolizumab for first-line treatment of metastatic nonsquamous NSCLC. N Engl J Med 2018;378(24):2288–301.

[82] Awad MM, Mastini C, Blasco RB, et al. Epitope mapping of spontaneous autoantibodies to anaplastic lymphoma kinase (ALK) in non-small cell lung cancer. Oncotarget 2017;8(54):92265–74.

CHAPTER

8

Therapeutic strategies to enhance crizotinib anti-tumor efficacy in ALK + ALCL

Robert E. Hutchison

State University of New York, Upstate Medical University, Syracuse, NY, United States

Abstract

In 1985, a new type of peripheral T-cell lymphoma was discovered using a monoclonal antibody, Ki-1 (anti-CD30), raised against a Hodgkin lymphoma (HL) cell line. It, ultimately named anaplastic large cell lymphoma (ALCL), was characterized and found to usually contain translocation t(2;5)(p23;q35) fusing nucleophosmin (*NPM*) and a "new" gene, anaplastic lymphoma kinase (*ALK*). The *NPM-ALK* fusion gene results in a constitutively active ALK tyrosine kinase that drives neoplastic cell growth, in large part through activation of STAT3 and downstream signaling. ALK positive ALCL is predominantly a disease of children, adolescents, and young adults, while similar tumors in adults are ALK negative. The discovery of ALK + ALCL occurred during a time of progress in the treatment of pediatric non-Hodgkin lymphomas (NHL), and it was initially considered to have a favorable prognosis, with 60–70% event free survival compared to other pediatric NHLs for which treatment results were less favorable. Since that time, however, progress has occurred in other pediatric NHLs, but has stalled in ALCL. ALK translocations and other abnormalities have subsequently been found in a number of neoplasms including a portion of non-small cell lung cancers, and ALK tyrosine kinase inhibitors (TKIs) have been successfully used in their therapy. Tumors often develop resistance to TKIs, limiting their long-term success. TKI therapy is gradually being introduced into the treatment of ALK + ALCL in efforts to reduce toxicity of chemotherapy while preserving and increasing cure rates. Other targeted modalities are also available to aid in this quest, including the anti-CD30 antibody drug conjugate brentuximab vedotin, other immune therapies including bi-specific antibodies and immune checkpoint inhibitors, and chimeric antigen receptor T-cells (CAR-T).

Abbreviations

ABL	Abelson murine leukemia viral oncogene homolog
ADC	antibody-drug conjugate
ALCL	anaplastic large cell lymphoma
ALK	anaplastic lymphoma kinase

ATIC	5-aminoimidazole-4-carboxamide ribonucleotide formyltransferase/IMP cyclohydrolase
ATP	adenosine tri-phosphate
BAG6/BAT3	BAG chaperone 6
BCL	B cell lymphoma protein
BCNU	carmustine
BCR	breakpoint cluster region
BFM	Berlin-Frankfurt-Munich cooperative treatment group
BimAb	bispecific monoclonal antibody
BRAF	B-Raf proto-oncogene serine/threonine-protein kinase
BV	brentuximab vedotin
CAR-T	chimeric antigen receptor T-cells
CCSG	Children's Cancer Study Group
CD	cluster designation
CDK	cyclin D kinase
CEBP	CCAAT enhancer protein
CHOP	cyclophosphamide, hydroxydaunorubicin, Oncovin, prednisone
COG	Children's Oncology Group
COMP	myocet (doxorubicin)-cyclophosphamide-vincristine, prednisone
ctDNA	circulating tumor deoxyribonucleic acid
CTLA-4	cytotoxic T-lymphocyte-associated protein 4
DLBCL	diffuse large B-cell lymphoma
DLCL	diffuse large cell lymphoma
DM1	ermtansine/mertansine
DM4	ravtansine/soravtansine
DNA	deoxyribonucleic acid
DUSP	dual specificity phosphatase
EBV	Epstein-Barr virus
EFS	event-free survival
EML4	echinoderm microtubule-associated protein-like 4
ERK	extracellular-signal-regulated kinase
FRAP	synonym for MTOR
GLI1	glioma-associated oncogene
HER-2	human epidermal growth factor receptor 2
HiDAC	high dose cytosine arabinoside
HL	Hodgkin lymphoma
HM	series of treatment protocols of the French Society for pediatric Oncology
ICH	immune checkpoint inhibitor
IGF	insulin-like growth factor
IL	interleukin
IMT	inflammatory myofibroblastic tumor
IRF4	interferon regulatory factor 4
JAK	Janus kinase
kDa	kilodalton
KIT	stem cell factor receptor
KRAS	*GTPase from* Kirsten sarcoma virus
LDM	lidamycin
LSA2-L2	early treatment protocol for lymphoid neoplasms
MAPK	mitogen-activated protein kinase
MCL1	myeloid cell leukemia 1
MDD	minimal disseminated disease
MET	Mesenchymal-epithelial transition factor-1
MMAE	monomethyl auristatin E
MMAF	monomethyl auristatin F

MRD	minimal residual disease
MSKCC	Memorial Sloan Kettering Cancer Center
MTOR	mammalian target of rapamycin
MTX	methotrexate
MYC	MYC proto-oncogene
NfkB	nuclear factor kappa B
NHL	non-Hodgkin lymphoma
NPM	nucleophosmin
OS	overall survival
PD1	programmed death 1
PDGF	platelet derived growth factor
PDL-1/2	programmed death ligand 1/2
PI3K	phosphatidylinositol-3-kinase
POG	Pediatric Oncology Group
PPAR	peroxisome proliferator-activated receptor
PTCL	peripheral T-cell lymphoma
PTEN	phosphatase and tensin homolog
PTPN1	protein tyrosine phosphatase non-receptor type 1
RAS	GTPase from Rous sarcoma virus
RET	RET proto-oncogene
RNA	ribonucleic acid
ROS1	ROS Proto-Oncogene 1, Receptor Tyrosine Kinase
RT	radiation therapy
SD	stable disease
SERPINA1	serine protease inhibitor alpha-1
SGN-30/35	anti-CD30 and anti-CD30-drug conjugate
SHH	Sonic Hedgehog Signaling Molecule
Shp	Src homology 2 domain-containing protein tyrosine phosphatase
Src	sarcoma
STAT	signal transducer and activator of transcription
SUDHL-1	Stanford Unifersity diffuse histiocytic lymphoma cell line
TCR	T-cell receptor
TKI	tyrosine kinase inhibitor
TNFR	tumor necrosis factor receptor
TP	tumor protein
TPM	tropomysin
VBL	vinblastine

Conflict of interest

No potential conflicts of interest were disclosed.

Introduction

In 1985, Stein et al., reported on a monoclonal antibody, Ki-1, raised against a Hodgkin lymphoma (HL) cell line [1]. It reacted with cells of HL, lymphomatoid papulosis, some T cell lymphomas, activated T and B cells, and a group of tumors with anaplastic morphology previously diagnosed as histiocytic neoplasms. A paraffin-reactive analogue to Ki-1 (BER-H2) was developed and the antigen was assigned cluster of differentiation (CD) 30 [2]. The tumors previously considered histiocytic became recognized as a new entity, Ki-1 positive anaplastic large cell lymphoma (ALCL) [3, 4]. Soon after the discovery of this "new" tumor, cases of histiocytic and large cell lymphomas, mostly in children, were noted to have translocations t (2;5)(p23;q35). These included about 1/3 of pediatric large cell lymphomas, or approximately 10–15% of pediatric non-Hodgkin lymphomas (NHLs) [4–6].

In 1994, Morris et al. characterized the t(2;5)(p23;q35) translocation as a fusion between a novel tyrosine kinase gene, anaplastic lymphoma kinase (*ALK*) at 2p23, with the nucleophosmin (*NPM1*) nucleolar phosphoprotein gene at 5q35 [7]. *ALK* was described as a normal gene of the insulin receptor kinase subfamily, weakly present in a number of tissues, particularly fetal brain, but not present in normal hematopoietic cells. The *ALK* portion of the neoplastic fusion gene is truncated with a constitutively activated catalytic domain due to a promoter in the NPM portion. *NPM* was subsequently found not to be rearranged [8]. Shiota et al., also described the *NPM-ALK* fusion product, designated as p80, and focused on pathologic aspects [9].

Histologic studies of ALCL were facilitated by the development of anti-ALK antibodies effective in paraffin sections, including ALK1 and p80$^{NPM/ALK}$. These were tested against relatively large numbers of cases by several cooperative treatment groups, with similar findings across the studies [10–18]. The majority of cases were found to occur in children and young adults.

Pathology of ALK + ALCL

The pathology of ALK + ALCL was well-described from clinical studies and is now included as an entity in the WHO Classification of Haematopoietic Neoplasms [19]. Histology shows heterogeneous morphology including a common histologic type with the presence of large cells with indented, reniform, or horseshoe shaped nuclei and abundant cytoplasm. These are often called "hallmark cells." Lobated nuclei and multinucleated cells are sometimes seen, and there is frequent involvement of lymph node sinuses (Fig. 1). The immunophenotype is that of activated T-cells, with expression of CD30 and CD25 (IL2-R), although T-cell markers are usually incomplete or aberrant. CD45, CD43, CD45RO, CD2 and CD4 are frequent, CD8 less often, HLA-DR typically positive, CD13 often aberrantly positive, with CD3, CD5, and CD7 variable [20, 21]. The epithelial membrane antigen is frequent, and histiocyte associated markers including CD68 are sometimes seen [22]. CD30 expression shows a membrane and Golgi pattern, and both cytoplasmic and nuclear ALK staining are prototypical, while cytogenetic variants show only cytoplasmic labeling (Fig. 1).

T cell markers are absent in up to 30%. Most cases exhibit the T-cell receptor (TCR) alpha gene rearrangement but not the TCR beta rearrangement. This latter finding may relate to the origin from an early thymic precursor [23]. Clinically, clonal TCR gamma rearrangement is often present by PCR. Histologic variants include monomorphic, small cell, lymphohistiocytic, Hodgkin-like, and composites. All are characterized by the expression of CD30 and ALK. Only the small cell variant (in which mostly perivascular large cells show ALK staining), and possibly the overlapping lymphohistiocytic variant, are noted to have worse prognosis [19].

Similar ALK+ anaplastic lymphomas also occur, with fusion of *ALK* to other partner genes [24, 25]. These partners include tropomysin 3 and tropomysin 4 (*TPM3-ALK* and *TPM4-ALK*), 5′aminoamidozole-4-carboxyamide ribonucleotide formyltransferase/IMP cyclohydrolase (*ATIC-ALK*), TRK-fused gene (*TFG-ALK*), myosin heavy chain (*MYH9-ALK*), moesin (*MSN-ALK*), and clatherin-like heavy chain (*CLTC-ALK*). All but *CLTC-ALK* have immunophenotypes similar to those with *NPM-ALK* except that ALK antibody staining is

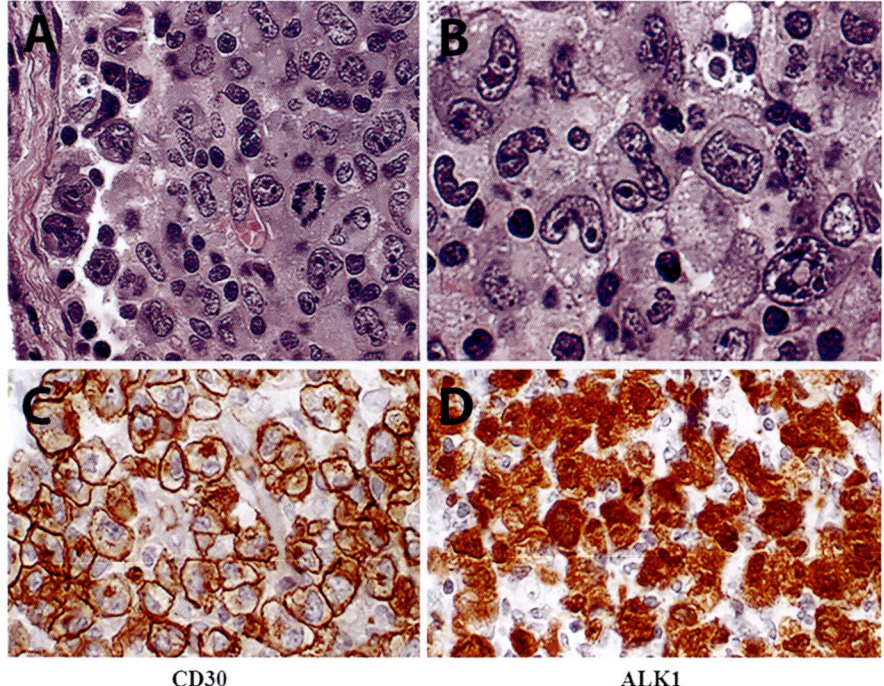

CD30   **ALK1**

FIG. 1   (A) H&E stained slide showing lymph node sinus involvement by ALK+ ALCL (500×). (B) Neoplastic cells showing indented or lobated nuclei with prominent nucleoli and abundant cytoplasm (1000×). (C) Anti-CD30 immunohistochemical stain showing Golgi and cytoplasmic positivity. (D) ALK1 immunohistochemical stain showing typical nuclear and cytoplasmic labeling.

only cytoplasmic in variants, rather than nuclear and cytoplasmic. The *CLTC-ALK* lymphoma is associated with B-lineage, similar to some early mouse models with knock-in *NPM-ALK*, and, peculiarly, with the absence of both CD30 and CD20 [26, 27].

ALK negative ALCL also deserves mention to limit confusion between it and ALK+ disease, and there are several distinct types. Systemic ALK − ALCL histologically resembles ALK+ disease but typically occurs in middle age or later. Clinical behavior may be influenced by genetic findings. Cases with DUSP22 rearrangement are considered favorable, similar to ALK + ALCL, while those with TP63 rearrangements (less common) are considered adverse. A majority of cases are "triple negative" for ALK, DUSP22, and TP63 abnormalities, and have intermediate prognosis [28, 29].

Cutaneous ALCL is a subgroup of primary cutaneous CD30+ lymphoproliferative disorder, characterized by localized sheets of CD30+ anaplastic large cells in the dermis, without involvement of the epithelium [19]. This finding is sometimes mimicked by cutaneous involvement by systemic ALK+ or ALK − ALCL, and thus clinical staging is important. Primary cutaneous cases are typically relapsing/remitting diseases and frequently indolent, cautioning against aggressive treatment in absence of systemic disease. Some have DUSP22 rearrangements.

Breast implant ALK − ALCL is a distinct entity involving the region of prior breast implants. It is usually triple negative but has constitutive activation of STAT3 and expression of PD1/PD-L1 [30]. Essentially all ALK+ and ALK − ALCLs show STAT3 activation, but in the ALK negative cases STAT3 is activated by different factors.

To assign targeted therapy for ALK+ ALC, it is important to consider how *NPM-ALK* acts as an oncogene. This has been studied and reviewed previously [31, 32]. Principle among the activities of the NPM-ALK fusion protein as a tyrosine kinase is phosphorylation and activation of STAT3, both directly and indirectly [32, 33]. The interaction of IL-2, IL-9, IL-21, and related cytokines with JAK3 plays a role, along with Src kinases, SHP1, JAK2, and STAT5.

The PI3K/Akt signaling pathway is constitutively active in ALK + ALCL, and this may be related to phosphorylation of the negative regulatory tail of PTEN [34]. CD30 expression is associated with MAPK/JunB signaling, and activation of NFkB [35]. Myc is activated, and *CEBPB*, *BCL6*, *PTPN1*, *MCL1* and *SERPINA1* are upregulated [31].

A simplified view of the functions of NPM-ALK fusion protein would be activation of the JAK3/STAT3, AKT/PI3K, RAS/ERK/MAPK, and sonic hedgehog (SHH/GLI1) pathways, with coordinated downstream involvement by numerous gene and protein modulations, increase in proliferation, and decrease in apoptosis [36]. Non-coding RNAs, particularly micro RNAs (miRNAs) are involved, interact with proliferative and apoptotic modulators, and are involved in differential methylation patterns [37, 38]. NPM may play a role as well. Although the *NPM-ALK* fusion gene is not rearranged, it is likely deregulated. NPM normally associates with cyclin D kinases (CDK) and p53 and is important in cell cycle regulation [39]. With crizotinib-induced ALK inhibition in ALCL cell lines, *CDK* expressions become downregulated, along with many of the other genes annotated to ALK + ALCL [40] (Fig. 2).

Comparative proteomic analysis by mass spectrometry, utilizing cell lines with and without *NPM/ALK* translocation, has shown analogous findings [41]. Differentially expressed proteins included those involved in survival, cell cycle regulation, proliferation, growth factor/cytokine signaling, antiapoptosis, neoangiogenesis, adhesion and migration. Signaling pathways represented included those of PI3K/AKT, JAK/STAT, IGF1, IL4, IL6, NFkB, PPAR, p38/MAPK, ERK/MAPK, FRAP/MTOR, g-protein, PDGF, Wnt/B-catenin, and integrin.

Evolution of standard therapy of ALK + ALCL

Discussion of targeted therapy for pediatric neoplasms also requires discussion of standard therapies. The first successful treatment protocol for pediatric NHL was LSA2-L2, initiated in 1971 at Memorial Sloan-Kettering Cancer Center (MSKCC), for NHL of all histologies (including "reticulum cell sarcoma" or large cell types) [42], and all stages. It utilized nine chemotherapy agents plus radiation therapy (RT), with induction, consolidation, and maintenance phases [43]. Drugs included cyclophosphamide, vincristine, methotrexate, prednisone, daunomycin, arabinosylcytosine, thioguanine, ʟ-asparaginase, and BCNU. The first treatment group included six cases of "diffuse histiocytic" lymphoma, likely including some of ALCL, and overall actuarial survival in follow-up was 80% [44]. A prior MSKCC protocol utilizing cyclophosphamide and RT had shown survival of only about 33%. Modifications of LSA2-L2 have continued to be utilized for lymphoblastic lymphoma [45].

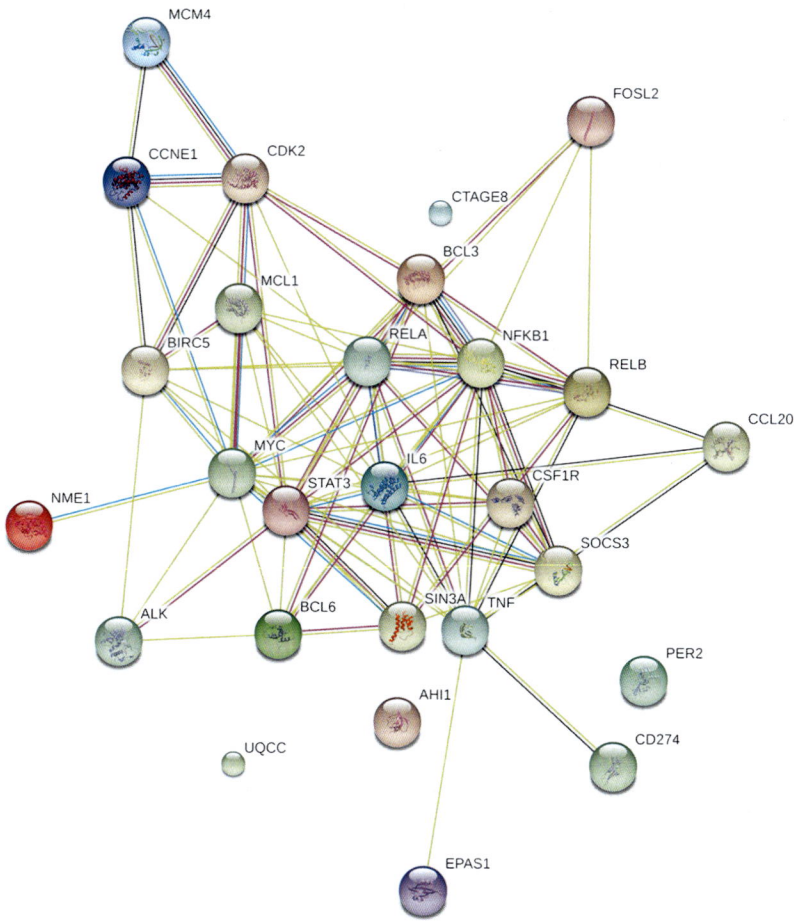

FIG. 2 Protein network of genes differentially expressed and associated with apoptosis in ALK+ ALCL cell lines treated with crizotinib.

Around the times that ALCL and NPM-ALK were described, pediatric lymphomas were being treated in various ways per institutional or regional cooperative treatment group strategies. Initially during that period, patients were separated into lymphoblastic versus non-lymphoblastic groups [46]. The Pediatric Oncology group found importance in separating small noncleaved cell (Burkitt) lymphoma from diffuse large cell lymphoma (DLCL), which included both B and T-cell types, although immunophenotypes were not determined [47]. The Nebraska Lymphoma Study separated peripheral T-cell lymphoma (PTCL), including ALCL, and found benefit in aggressive treatment [48]. Investigators of Bologna University treated cases of ALCL, then recently included in the Updated Kiel Classification, with modified LSA2-L2 and found somewhat favorable 4 years event-free survival of 62.9% [49–51]. The Children's Cancer Study Group compared LSA2-L2 with COMP therapy, finding that large cell types (both B and T) fared equally with the two therapies, although 5 years EFS were only 52% and 43%, respectively [52].

The Berlin-Frankfurt-Munster (BFM) cooperative group reported results from 62 patients with ALCL on three studies with good results (up to 9 years EFS of 83%) using short pulse chemotherapy. Drugs included prednisone, cyclophosphamide, dexamethasone, methotrexate (MTX), intrathecal chemotherapy, ifosfamide (or cyclophosphamide and doxorubicin), cytarabine (Ara-C), and etoposide (VP-16) [53]. A subsequent study, BFM-90, further tested the concepts of short pulse therapy for ALCL in a large prospective multicenter trial, stratified for stage, and found that of 89 patients, the overall 5-year event free survival to be 76%, with increased risk in patients with B symptoms [54].

The French Society for Pediatric Oncology reported on two protocols for ALCL, HM89 and HM91. These utilized cyclophosphamide, vincristine, prednisone and doxorubicin as first line therapy, with heterogenous treatments for relapse. Combined 3 years EFS was 66%, varying with stage [14]. A subsequent report from that group showed that relapsed ALCL remains sensitive to vinblastine-based chemotherapy [55].

The Pediatric Oncology Group treated 180 patients of diffuse large cell lymphoma (DLCL), including 86 ALCL, 75 B-cell (DLBCL), 10 PTCL of other types, and 9 unclassified. Treatment was APO (doxorubicin, vincristine, prednisone, 6-mercaptopurine, and methotrexate) versus APO with intermediate dose methotrexate and high dose cytarabine (HiDAC). Overall 4 years EFS was 67.4%, with no significant differences between arms or between DLBCL versus ALCL [56]. APO was also the backbone of a subsequent Children's Oncology Group (COG) trial for ALCL which showed that adding daily vinblastine did not increase efficacy, but did increase toxicity [57].

A large multi-national multi-institutional trial, ALCL-99, based on the previously mentioned BFM-90 protocol showed 2-year event free survival of 352 patients of approximately 74%. The initial report of this study showed that methotrexate schedule could be safely reduced from the prior studies, and the results were found particularly favorable because of relatively good survival combined with lower cumulative toxicity than other regimens [58]. Lymphohistiocytic and small cell histologic variants were found to have inferior prognosis [59].

There have been no randomized trials of ALK + ALCL in adult patients, although the disease represents about 3% of adult NHL, mostly patients less than 35 years, and most adult treatments have been with anthracycline based CHOP-like therapies [16, 60].

Thus, treatment results from clinical trials are based on pediatric studies, and the majority of patients with ALK + ALCL are from pediatric, adolescent, and young adult populations. Most trials have shown long-term survivals between 65% and 75%, and ALCL99 has emerged as the most used front line therapy. Identified risk factors include mediastinal or visceral involvement with skin lesions; minimal disseminated disease (MDD) in blood, bone marrow, and/or CNS; molecular minimal residual disease (MRD) after 4 weeks of treatment; lymphohistiocytic or small cell histology; and low titers of anti-ALK antibodies [61]. Relapsed patients may be treated by salvage therapy (chemo or RT) followed by autologous bone marrow transplantation. Other relapse options include praletrexate (anti-folate agent), romidepsin (HDAC1 inhibitor), or vinblastine, the latter reported to have immunomodulatory properties as well as being an anti-tubulin chemotherapy [60, 61].

Results of treatment of ALK + ALCL have not consistently exceeded 70% EFS or 90% OS (including salvage therapy), and although toxicities have decreased, they remain significant. Further decrease in toxicity might be achieved by single agent vinblastine as a first line agent, but this is yet to be tested and would likely be long-term therapy. Targeted therapy with ALK inhibition (i.e., crizotinib or other ALK inhibitors) and/or anti-CD30 antibody-drug conjugates (i.e., brentuximab vedotin) also offers promise, but the use such agents is limited by the relative success of chemotherapy, to some extent by cost of the drugs, and by the lack of interest by pharmaceutical companies in pursuing this course for a rare disease [62].

ALK tyrosine kinase inhibitors

The first tyrosine kinase inhibitors were developed in the 1980s and 1990s and found to inhibit products of oncogenes such as *v-SRC*, *v-ABL*, and *BCR-ABL*. One, signal transduction inhibitor 571 (STI571, imatinib), inhibited PDGFRa, c-KIT, and BCR-ABL [63]. It was found to be active against cultured chronic myeloid leukemia cells and became a revolutionary treatment for that and other neoplasms as well, including gastrointestinal stromal tumors (GIST), myeloproliferative neoplasm with eosinophilia, mast cell neoplasm, and others [63, 64]. Many other TKI's have also been successful in treating cancer, however, treatment is usually long-term and emergence of resistance is a generalized function, related in part to ongoing mutations that effect ATP-binding sites with which TKIs interact [63, 65].

Crizotinib, originally known as PF-2341066, was described as an inhibitor of c-MET and of ALK in 2007 [66, 67]. It inhibits proliferation with G1-S–phase arrest and induces apoptosis in commonly used ALCL cell lines. Signal transduction effects attributed to NPM-ALK in lymphoma, which are mostly up-regulations, are negated by crizotinib at physiologic doses (Fig. 2). Crizotinib is effective and FDA-approved for treatment of non-small cell lung cancer (NSCLC) with ALK translocation and for NSCLC with ROS1 rearrangement. Acquired resistance to crizotinib is common in lung cancer treatment due to a variety of mechanism, and second and third generation ALK inhibitors are often effective in relapse/progression [65].

Crizotinib in pediatric neoplasms

ALK translocation, mutation, or expression has been found in several neoplasms of young people. These include, in addition to ALCL, familial neuroblastoma due to germline mutations at ALK tyrosine kinase domains G1128A, R1192P, and R1275Q, neuroblastoma without ALK mutation but with expression on the cell surface, which is nearly ubiquitous, and rhabdomyosarcoma [68, 69]. ALK translocations also occur in half of pediatric inflammatory myofibroblastic tumors (IMT). Activity of Crizotinib in refractory/relapsed ALCL and in myofibroblastic tumors was tested by the Children's Oncology Group in a phase I/II study, with complete response in most ALCL and in 36% of IMTs [70].

So the current state of therapy for ALK + ALCL in young people is with treatment according to ALCL 99, with an overall long-term EFS of about 70%. This outcome has not changed substantially in 30 years, although toxicity is reduced. This is in contrast to results of treatment of other pediatric lymphomas, for which treatment results have improved during this time period, with long-term EFS of overall 80–90% for Burkitt and DLBCL [71]. Overall survival (OS) for ALCL in young people is indeed about 90%, with a variety of salvage therapies, principally autologous BMT. However, salvage therapies impart increased long-term toxicity, very significant for young people, and those who progress on therapy or fail salvage have dismal prognoses. Goals for ALCL, as for other pediatric cancers which have relatively high long-term survival, include preserving outcomes while reducing toxicity and increasing quality of life.

Targeted agents including crizotinib and the CD30 antibody-drug conjugate brentuximab vedotin (BV) are very slowly being introduced into relapse and front-line therapy, but results are not yet available. These include a COG protocol based on ALCL99 (NCT01979536) that introduces a randomization of crizotinib versus BV in the treatment of newly diagnosed ALCL, and a phase II treatment of relapsed refractory ALCL using crizotinib with VBL (CRISP trial, ITCC053) [62].

As ALK inhibitors such as crizotinib find potential roles in ALCL therapy, acquired resistance is expected [72]. In a study of 11 patients with relapsed/refractory ALK + ALCL, progression-free survival at 2 years was 63.7%, and all patients who progressed showed molecular evidence of mutations associated with acquired ALK resistance [73].

Crizotinib in ALK + ALCL

Single agent ALK inhibitor crizotinib has been used with moderate success in patients with relapsed/refractory ALCL. This approach is difficult to test in front-line clinical trials because of relatively good success with chemotherapy, limiting availability of patients due to ethical considerations [62]. Additional study of relapsed/refractory patients may lead to more confidence in this approach, particularly if combined with monitoring prognostic biomarkers. Single agent therapy, which is likely to be prolonged even when effective, may be difficult to accept in pediatric oncology, a field in which cure is the over-riding goal, and almost all curative progress has occurred with combination therapies.

Another approach is to combine ALK inhibitor with chemotherapy. This is currently being tested in one arm of COG protocol ANHL12P1 (NCT01979536) in a backbone of ALCL99, but data is not yet available. Accrual has been limited by unexpected toxicity on that arm.

One cell line study suggests that there is antagonism between crizotinib and doxorubicin, which is a component of ALCL99 and of many chemotherapy regimens. Expression profiling of treated cell lines showed that while crizotinib restored normal cell signaling to ALK driven cell lines, with associated apoptosis, doxorubicin disrupted cell signaling and antagonized apoptosis [40]. Neither brentuximab vedotin (BV) nor vinblastine had significant effects on cell signaling. BV synergized apoptosis, with vinblastine showing a trend toward synergy. While it may be surprising that BV and crizotinib could act synergistically since BV is an anti-CD30 drug conjugate and crizotinib downregulates CD30 [35], drugs in that study were introduced into cell cultures almost simultaneously, allowing at least an initial CD30 target, and results were scored at 24 h to preserve viable cells for expression analysis. Preliminary data of apoptosis and growth inhibition for that study had not identified antagonism between crizotinib and other drugs commonly used in pediatric NHL (unpublished data), so this antagonism may be specific to doxorubicin (and likely other anthracyclines) rather than being a general effect of chemotherapy. It has been argued that anthracycline-based therapies (such as CHOP) are ineffective for peripheral T-cell lymphomas with the exception of ALK + ALCL, which responds relatively well [74, 75]. If there is interference between ALK inhibition and anthracycline, and since ALCL99 is the favored backbone for ALK + ALCL, then regimen modifications including timing, dosages, and constituent drugs will be critical for inclusion of ALK inhibitors, as is usual in therapeutic design. In regard to other chemotherapy, a European trial (ITCC053) including use of vinblastine in combination with crizotinib in relapsed/refractory ALCL with ALK aberrations was opened but is currently halted. Other drugs suggested to enhance ALK inhibition include the mTOR inhibitor Everolimus, and a bromodomain inhibitor, OTX015/MK-8628 [76, 77].

Immune based therapies

Brentuximab vedotin, produced by Seattle Genetics, was the first antibody-drug conjugate (ADC) to be approved by the US FDA. Approval was granted in 2011 for adults with relapsed/refractory Hodgkin lymphoma (HL) and ALCL. This was extended to first line treatment in combination with chemotherapy (doxorubicin, vinblastine, dacarbazine) in 2018 and also includes other CD30+ T-cell lymphomas. BV consists of a chimeric monoclonal anti-CD30 linked to the compound monomethyl auristatin E (MMAE), a potent anti-tubulin agent. The ADC attaches to the tumor cell, is internalized by endocytosis, and its' effects are focused on the neoplastic cells with little bystander effect. ADC design revolves around three components: the antibody, the linker, and the conjugated drug [78].

Generally speaking, the antibody should be directed to an antigen differentially expressed on tumor cells to facilitate anti-tumor effect and minimize off-target toxicities, have sufficient affinity to the target, and be internalized by the cell. Linkers must be stable enough for the complex to circulate to the tumor, and allow release of the drug; both cleavable and non-cleavable linkers are in use. The drug component must be sufficiently potent, chemically able to be attached to the linker, and both soluble and stable under physiologic conditions. Auristatins such as MMAE are analogs of dolastatin-10, derived from the sea hare Dolabella auricularia, which inhibit tubulin polymerization with a potency 20–50 times that of vinblastine [78]. Monomethy auristatin F (MMAF) is related but is soluble and thus can diffuse regionally. Another class of tubulin inhibitor used in ADCs is that of maytansinoids, derived from the bark of an African shrub and which acts similarly to vinca alkaloids. These include ermtansine/mertansine (DM1), ravtansine/soravtansine (DM4), and ansamitocin [79]. A third class of drug is derived from the potent antibiotic calicheamicin, which cleaves double stranded DNA, and those of a fourth commonly used class, the duocarmycins, are DNA alkylating agents [78].

ADCs developed for the treatment of lymphoma and leukemia are predominantly targeted toward B-cells, for use in acute lymphoblastic leukemia (ALL) and B-cell lymphomas. These include naratuximab emtansine, a CD37 ADC for NHL, polatuzumab vedotin, a CD79b ADC for treatment of diffuse large B-cell lymphoma, coltuximab ravtansine, a CD19 specific antibody labeled with DM4, denintuzumab mafodotin, a CD19 antibody conjugated to MMAF, inotuzumab ozogamicin, anti-CD22 ADC useful for B ALL, and others including those specific for CD25 [80, 81]. Anti-CD25 ADCs could presumably be utilized for ALCL, since many or most ALCL are positive for CD25. Anti-CD25 ADCs have previously been tested on other lymphomas including HL and adult T-cell leukemia/lymphoma, with promising results [82, 83]. Gemtuzumab ozogamicin is an anti-CD33 ADC useful in treatment of acute myeloid leukemia [84].

CD30 is an ideal target for ADC immune therapy. It is highly expressed on tumor cells of ALCL, HL, and some other lymphomas, but is normally only expressed on activated lymphocytes with limited distribution, predominantly in lymph node perifollicular regions, and in the paracortex in infections such as Epstein Barr Virus (EBV), thus limiting potential toxicity. It is a member of the tumor necrosis factor receptor (TNFR) superfamily and is integral to the pathobiology of both ALCL and HL, however, it's cell signaling functions vary with tumor type and cell lines studied. It is also released as circulating CD30 in lymphoma and in infections [85].

A variety of unconjugated anti-CD30 monoclonal antibodies have been tested against CD30+ lymphomas, each with differing immunologic effects. One chimeric (murine/human) anti-CD30 antibody, cAC10, was tested and found safe in phase I trials, but without demonstrated subsequent efficacy. This drug, SGN-30, was modified with conjugation to MMAE, and became SGN-35 (BV) [85].

Bi-specific monoclonal antibodies (BimAbs) offer another potential option for targeted immune therapy which may be combined with other relevant treatments. These antibodies are engineered to have two binding sites, one for tumor cells, and another for T-cell costimulatory molecules. These bring activated T-cells to the tumor cells to facilitate tumor cell killing [86, 87]. Blinatumomab, an anti-CD19 bi-specific, has shown promise in treatment of B-ALL and B-cell lymphomas. Several anti-CD30 BimAbs have been tested in preclinical studies. Neither unconjugated antibodies nor BimAbs have thus far shown the efficacy of BV [85].

Chimeric antigen receptor T-cells (CAR-T) have also found a role in the treatment of B-ALL and B-cell NHL [86]. CD30-specific CAR-T have been tested in patients with ALCL and HL, found to be safe, with 5 of 7 HL patients and 1 of 2 ALCL patients showing CR of variable durations in one study [88]. Another study of 17 patients with HL and 1 with cutaneous ALCL, showed the treatment to be well tolerated, with 7 partial remissions (PR) and 6 patients with stable disease (SD) [89].

At this time, BV is the principle ADC and principle immune based therapy available for ALCL. It was tested in phase I and phase II trials of adult ALCL, both ALK positive and ALK negative, and other CD30+ lymphomas which had failed therapy, with favorable results (up to 64% 4 years OS) [72, 90]. BV was approved by the FDA in 2011 for treatment of ALCL in relapse/progression after failure of at least one multi-agent chemotherapy protocol. Multiple trials have been performed on HL in adults, with promising results, and one protocol combining BV and chemotherapy for pediatric HL is ongoing (active, not recruiting; NCT02166463). The BV arm of COG ANHL12P1 (NCT0179536), trialing BV or crizotinib in a backbone of ALCL99 has completed accrual [72]. BV has large molecular weight (~150 kDa) and likely does not easily penetrate the blood brain barrier. It has been described in a case study, however, in combination with intrathecal methotrexate, to successfully treat CNS progression of ALK + ALCL on front line therapy of ALCL99 [91]. Whether BV contributed substantially is difficult to know.

A novel anti-CD30 ADC has recently been described that is conjugated to lidamycin (enediyne containing antibiotic) and performed well on CD30 cells lines in vitro and in mouse models [92]. Another utilizing a novel anti-CD30 antibody conjugated to DMI was found active in ALCL cell lines in xenographs [93]. Yet another anti-CD30 lidamycin (LDM) conjugate was produced and found to synergize with crizotinib in xenographs of Karpas 299 and

SUDHL-1 ALCL cell lines [94]. These cell lines, incidentally, are utilized in virtually all cell line studies of ALK+ ALCL reported in this review. Another ADC utilizes an ALK antibody (CDX-0125) labeled with thienoindole (TEI) DNA alkylating agent, and found to be active against ALK+ neuroblastoma cell lines [95]. This may be helpful in those cases of ALK+ ALCL that express surface ALK, on the order of possibly 33% [21].

Immunotherapy utilizing immune checkpoint inhibitors including those targeting programmed cell death 1 (PD1), it's ligands PD-L1 and PD-L2, and CTLA-4, may also be useful to augment other therapy. ALK+ ALCL is positive for PD-L1 in approximately 76% of cases [96]. This appears to be mediated through STAT3 activation and subsequent action of IRF4 and BATF3 on the enhancer region of CD274 (PD-L1). This suggests that ALCL may be amenable to treatment with PD1/PD-L1 blockade [97]. Immune checkpoint blockade has shown some success in HL, and nivolumab (anti-PD1) is currently being evaluated in a phase II trial of pediatric and adult relapsed/refractory ALK+ ALCL [98].

Introduction of immune checkpoint inhibitors to front-line treatment of ALK+ ALCL faces the same challenges as do other novel therapies in a disease which usually responds to standard therapies and for which there are few refractory/relapsed patients to test. Pre-clinical screening for potential efficacy of mixed modality targeted therapy is more complicated than screening cell lines for drug effects, in that immune therapies often require an intact immune system. In-vitro systems to assess immune response, such as using allogeneic or autologous immune cells to measure anti-tumor effects as well as to effect therapy, are advancing [99]. ALK+ ALCL are ideal candidates for these advances since they have strong immune components, including eliciting antibody responses, and having high levels of PD-L1 expression [100].

Conclusion

In the future, we may test patient tumor cells for treatment sensitivities similarly to how bacterial diseases are tested for antibiotic sensitivity. Eventually, combinations of targeted therapies with increased efficacy and less toxicity than chemotherapy will likely emerge. In the short term, clinical trials of targeted therapy for ALK+ ALCL will proceed on relapsed/refractory patients, and as promising combinations are found, and as risk factors for treatment failure are clarified, these therapies may move to the front line. ALK inhibitors might best be used transiently, but if use is prolonged enough for development of resistance, the experience obtained from NSCLC will provide invaluable guidance.

References

[1] Stein H, Mason DY, Gerdes J, O'Connor N, Wainscoat J, Pallesen G, Gatter K, Falini B, Delsol G, Lemke H. The expression of the Hodgkin's disease associated antigen Ki-1 in reactive and neoplastic lymphoid tissue: evidence that reed-Sternberg cells and histiocytic malignancies are derived from activated lymphoid cells. Blood 1985;
66(4):848–58.
[2] McMichael AJ, Beverley PCL, Cobbled S, Crumpton MJ, Gilks W, Gotch FM, Waldmann H, editors. Leucocyte typing III: white cell differentiation antigens. Oxford: Oxford University Press; 1987.

[3] Schwarting R, Gerdes J, Durkop H, Falini B, Pileri S, Stein H. BER-H2: a new anti-Ki-1 (CD30) monoclonal antibody directed at a formol-resistant epitope. Blood 1989;74(5):1678–89.

[4] Stein H. Ki-1-anaplastic large cell lymphoma: is it a discrete entity? Leuk Lymphoma 1993;10(Suppl):81–4.

[5] Vannier JP, Bastard C, Rossi A, Hemet J, Thomine E, Tron P. Chromosomal t(2; 5) and hematological malignancies. Pediatr Hematol Oncol 1987;4(2):177–8.

[6] Mason DY, Bastard C, Rimokh R, Dastugue N, Huret JL, Kristoffersson U, Magaud JP, Nezelof C, Tilly H, Vannier JP. CD30-positive large cell lymphomas ('Ki-1 lymphoma') are associated with a chromosomal translocation involving 5q35. Br J Haematol 1990;74(2):161–8.

[7] Morris SW, Kirstein MN, Valentine MB, Dittmer KG, Shapiro DN, Saltman DL, Look AT. Fusion of a kinase gene, ALK, to a nucleolar protein gene, NPM, in non-Hodgkin's lymphoma. Science 1994;263(5151):1281–4.

[8] Bullrich F, Morris SW, Hummel M, Pileri S, Stein H, Croce CM. Nucleophosmin (NPM) gene rearrangements in Ki-1-positive lymphomas. Cancer Res 1994;54(11):2873–7.

[9] Shiota M, Fujimoto J, Takenaga M, Satoh H, Ichinohasama R, Abe M, Nakano M, Yamamoto T, Mori S. Diagnosis of t(2;5)(p23;q35)-associated Ki-1 lymphoma with immunohistochemistry. Blood 1994;84(11):3648–52.

[10] Weisenburger DD, Gordon BG, Vose JM, Bast MA, Chan WC, Greiner TC, Anderson JR, Sanger WG. Occurrence of the t(2;5)(p23;q35) in non-Hodgkin's lymphoma. Blood 1996;87(9):3860–8.

[11] Nakamura S, Shiota M, Nakagawa A, Yatabe Y, Kojima M, Motoori T, Suzuki R, Kagami Y, Ogura M, Morishima Y, Mizoguchi Y, Okamoto M, Seto M, Koshikawa T, Mori S, Suchi T. Anaplastic large cell lymphoma: a distinct molecular pathologic entity: a reappraisal with special reference to p80(NPM/ALK) expression. Am J Surg Pathol 1997;21(12):1420–32.

[12] Shiota M, Mori S. Anaplastic large cell lymphomas expressing the novel chimeric protein p80NPM/ALK: a distinct clinicopathologic entity. Leukemia 1997;11(Suppl 3):538–40.

[13] Hutchison RE, Banki K, Shuster JJ, Barrett D, Dieck C, Berard CW, Murphy SB, Link MP, Pick TE, Laver J, Schwenn M, Mathew P, Morris SW. Use of an anti-ALK antibody in the characterization of anaplastic large-cell lymphoma of childhood. Ann Oncol 1997;8(Suppl 1):37–42.

[14] Brugieres L, Deley MC, Pacquement H, Meguerian-Bedoyan Z, Terrier-Lacombe MJ, Robert A, Pondarre C, Leverger G, Devalck C, Rodary C, Delsol G, Hartmann O. CD30(+) anaplastic large-cell lymphoma in children: analysis of 82 patients enrolled in two consecutive studies of the French Society of Pediatric Oncology. Blood 1998;92(10):3591–8.

[15] Benharroch D, Meguerian-Bedoyan Z, Lamant L, Amin C, Brugieres L, Terrier-Lacombe MJ, Haralambieva E, Pulford K, Pileri S, Morris SW, Mason DY, Delsol G. ALK-positive lymphoma: a single disease with a broad spectrum of morphology. Blood 1998;91(6):2076–84.

[16] Falini B, Pileri S, Zinzani PL, Carbone A, Zagonel V, Wolf-Peeters C, Verhoef G, Menestrina F, Todeschini G, Paulli M, Lazzarino M, Giardini R, Aiello A, Foss HD, Araujo I, Fizzotti M, Pelicci PG, Flenghi L, Martelli MF, Santucci A. ALK+ lymphoma: clinico-pathological findings and outcome. Blood 1999;93(8):2697–706.

[17] Cataldo KA, Jalal SM, Law ME, Ansell SM, Inwards DJ, Fine M, Arber DA, Pulford KA, Strickler JG. Detection of t(2;5) in anaplastic large cell lymphoma: comparison of immunohistochemical studies, FISH, and RT-PCR in paraffin-embedded tissue. Am J Surg Pathol 1999;23(11):1386–92.

[18] Perkins SL, Pickering D, Lowe EJ, Zwick D, Abromowitch M, Davenport G, Cairo MS, Sanger WG. Childhood anaplastic large cell lymphoma has a high incidence of ALK gene rearrangement as determined by immunohistochemical staining and fluorescent in situ hybridisation: a genetic and pathological correlation. Br J Haematol 2005;131(5):624–7.

[19] Swerdlow S, Campo E, Harris NL, Jaffe ES, Pileri AS. WHO classification of tumours of haematopoietic and lymphoid tissues. 4th ed. Lyon: IARC; 2017.

[20] Carbone A, Gloghini A, Volpe R. Paraffin section immunohistochemistry in the diagnosis of Hodgkin's disease and anaplastic large cell (CD30+) lymphomas. Virchows Arch A Pathol Anat Histopathol 1992;420 (6):527–32.

[21] Juco J, Holden JT, Mann KP, Kelley LG, Li S. Immunophenotypic analysis of anaplastic large cell lymphoma by flow cytometry. Am J Clin Pathol 2003;119(2):205–12.

[22] Carbone A, Gloghini A, De Re V, Tamaro P, Boiocchi M, Volpe R. Histopathologic, immunophenotypic, and genotypic analysis of Ki-1 anaplastic large cell lymphomas that express histiocyte-associated antigens. Cancer 1990;66(12):2547–56.

[23] Malcolm TIM, Villarese P, Fairbairn CJ, Lamant L, Trinquand A, Hook CE, Burke GA, Brugieres L, Hughes K, Payet D, Merkel O, Schiefer AI, Ashankyty I, Mian S, Wasik M, Turner M, Kenner L, Asnafi V, Macintyre E, Turner SD. Anaplastic large cell lymphoma arises in thymocytes and requires transient TCR expression for thymic egress. Nat Commun 2016;7:10087.

[24] Colleoni GW, Bridge JA, Garicochea B, Liu J, Filippa DA, Ladanyi M. ATIC-ALK: a novel variant ALK gene fusion in anaplastic large cell lymphoma resulting from the recurrent cryptic chromosomal inversion, inv(2) (p23q35). Am J Pathol 2000;156(3):781–9.

[25] Damm-Welk C, Klapper W, Oschlies I, Gesk S, Rottgers S, Bradtke J, Siebert R, Reiter A, Woessmann W. Distribution of NPM1-ALK and X-ALK fusion transcripts in paediatric anaplastic large cell lymphoma: a molecular-histological correlation. Br J Haematol 2009;146(3):306–9.

[26] Gascoyne RD, Lamant L, Martin-Subero JI, Lestou VS, Harris NL, Muller-Hermelink HK, Seymour JF, Campbell LJ, Horsman DE, Auvigne I, Espinos E, Siebert R, Delsol G. ALK-positive diffuse large B-cell lymphoma is associated with Clathrin-ALK rearrangements: report of 6 cases. Blood 2003;102(7):2568–73.

[27] Giuriato S, Turner SD. Twenty years of modelling NPM-ALK-induced lymphomagenesis. Front Biosci 2015;7:236–47.

[28] Xing X, Feldman AL. Anaplastic large cell lymphomas: ALK positive, ALK negative, and primary cutaneous. Adv Anat Pathol 2015;22(1):29–49.

[29] Hapgood G, Ben-Neriah S, Mottok A, Lee DG, Robert K, Villa D, Sehn LH, Connors JM, Gascoyne RD, Feldman AL, Farinha P, Steidl C, Scott DW, Slack GW, Savage KJ. Identification of high-risk DUSP22-rearranged ALK-negative anaplastic large cell lymphoma. Br J Haematol 2019;186(3):e28–31.

[30] Gerbe A, Alame M, Dereure O, Gonzalez S, Durand L, Tempier A, De Oliveira L, Tourneret A, Costes-Martineau V, Cacheux V, Szablewski V. Systemic, primary cutaneous, and breast implant-associated ALK-negative anaplastic large-cell lymphomas present similar biologic features despite distinct clinical behavior. Virchows Arch 2019;475(2):163–74.

[31] Amin HM, Lai R. Pathobiology of ALK+ anaplastic large-cell lymphoma. Blood 2007;110(7):2259–67.

[32] Lai R, Ingham RJ. The pathobiology of the oncogenic tyrosine kinase NPM-ALK: a brief update. Ther Adv Hematol 2013;4(2):119–31.

[33] Chiarle R, Simmons WJ, Cai H, Dhall G, Zamo A, Raz R, Karras JG, Levy DE, Inghirami G. Stat3 is required for ALK-mediated lymphomagenesis and provides a possible therapeutic target. Nat Med 2005;11(6):623–9.

[34] Thakral C, Hutchison RE, Shrimpton A, Barrett D, Laver J, Link M, Halleran DR, Hudson S. ALK+ anaplastic large cell lymphoma exhibits phosphatidylinositol-3 kinase/Akt activity with retained but inactivated PTEN—a report from the Children's oncology group. Pediatr Blood Cancer 2012;59(3):440–7.

[35] Hsu FYY, Johnston PB, Burke KA, Zhao Y. The expression of CD30 in anaplastic large cell lymphoma is regulated by nucleophosmin-anaplastic lymphoma kinase-mediated JunB level in a cell type-specific manner. Cancer Res 2006;66(18):9002–8.

[36] Mosse YP, Wood A, Maris JM. Inhibition of ALK signaling for cancer therapy. Clin Cancer Res 2009; 15(18):5609–14.

[37] Hoareau-Aveilla C, Meggetto F. Crosstalk between microRNA and DNA methylation offers potential biomarkers and targeted therapies in ALK-positive lymphomas. Cancer 2017;9(8). https://doi.org/10.3390/cancers9080100.

[38] Fuchs S, Naderi J, Meggetto F. Non-coding RNA networks in ALK-positive anaplastic-large cell lymphoma. Int J Mol Sci 2019;20(9). https://doi.org/10.3390/ijms20092150.

[39] Naoe T, Suzuki T, Kiyoi H, Urano T. Nucleophosmin: a versatile molecule associated with hematological malignancies. Cancer Sci 2006;97(10):963–9.

[40] Hudson S, Wang D, Middleton F, Nevaldine BH, Naous R, Hutchison RE. Crizotinib induces apoptosis and gene expression changes in ALK+ anaplastic large cell lymphoma cell lines; brentuximab synergizes and doxorubicin antagonizes. Pediatr Blood Cancer 2018;65(8), e27094.

[41] Lim MS, Carlson ML, Crockett DK, Fillmore GC, Abbott DR, Elenitoba-Johnson OF, Tripp SR, Rassidakis GZ, Medeiros LJ, Szankasi P, Elenitoba-Johnson KS. The proteomic signature of NPM/ALK reveals deregulation of multiple cellular pathways. Blood 2009;114(8):1585–95.

[42] Jaffe ES, Harris NL, Stein H, Isaacson PG. Classification of lymphoid neoplasms: the microscope as a tool for disease discovery. Blood 2008;112(12):4384–99.

[43] Wollner N, Lieberman P, Exelby P, D'angio G, Burchenal J, Fang S, Murphy ML. Non-Hodgkin's lymphoma in children: results of treatment with LSA2-L2 protocol. Br J Cancer Suppl 1975;2:337–42.

[44] Wollner N, Exelby PR, Lieberman PH. Non-Hodgkin's lymphoma in children: a progress report on the original patients treated with the LSA2-L2 protocol. Cancer 1979;44(6):1990–9.

[45] Pillon M, Arico M, Mussolin L, Carraro E, Conter V, Sala A, Buffardi S, Garaventa A, D'Angelo P, Lo Nigro L, Santoro N, Piglione M, Lombardi A, Porta F, Cesaro S, Moleti ML, Casale F, Mura R, d'Amore ES, Basso G, Rosolen A. Long-term results of the AIEOP LNH-97 protocol for childhood lymphoblastic lymphoma. Pediatr Blood Cancer 2015;62(8):1388–94.

[46] Nachman J. Therapy for childhood non-Hodgkin's lymphomas, nonlymphoblastic type. Review of recent studies and current recommendations. Am J Pediatr Hematol Oncol 1990;12(3):359–66.

[47] Hvizdala EV, Berard C, Callihan T, Falletta J, Sabio H, Shuster JJ, Sullivan M, Wharam MD. Nonlymphoblastic lymphoma in children—histology and stage-related response to therapy: a Pediatric Oncology Group Study. J Clin Oncol 1991;9(7):1189–95.

[48] Gordon BG, Weisenburger DD, Warkentin PI, Anderson J, Sanger WG, Bast M, Gnarra D, Vose JM, Bierman PJ, Armitage JO. Peripheral T-cell lymphoma in childhood and adolescence. A clinicopathologic study of 22 patients. Cancer 1993;71(1):257–63.

[49] Vecchi V, Burnelli R, Pileri S, Rosito P, Sabattini E, Civino A, Pericoli R, Paolucci G. Anaplastic large cell lymphoma (Ki-1+/CD30+) in childhood. Med Pediatr Oncol 1993;21(6):402–10.

[50] Stansfeld AG, Diebold J, Noel H, Kapanci Y, Rilke F, Kelenyi G, Sundstrom C, Lennert K, van Unnik JA, Mioduszewska O. Updated Kiel classification for lymphomas. Lancet 1988;1(8580):292–3.

[51] Hastrup N, Hamilton-Dutoit S, Ralfkiaer E, Pallesen G. Peripheral T-cell lymphomas: an evaluation of reproducibility of the updated Kiel classification. Histopathology 1991;18(2):99–105.

[52] Anderson JR, Jenkin RD, Wilson JF, Kjeldsberg CR, Sposto R, Chilcote RR, Coccia PF, Exelby PR, Siegel S, Meadows AT. Long-term follow-up of patients treated with COMP or LSA2L2 therapy for childhood non-Hodgkin's lymphoma: a report of CCG-551 from the Childrens Cancer Group. J Clin Oncol 1993;11(6):1024–32.

[53] Reiter A, Schrappe M, Tiemann M, Parwaresch R, Zimmermann M, Yakisan E, Dopfer R, Bucsky P, Mann G, Gadner H. Successful treatment strategy for Ki-1 anaplastic large-cell lymphoma of childhood: a prospective analysis of 62 patients enrolled in three consecutive Berlin-Frankfurt-Munster group studies. J Clin Oncol 1994;12(5):899–908.

[54] Seidemann K, Tiemann M, Schrappe M, Yakisan E, Simonitsch I, Janka-Schaub G, Dorffel W, Zimmermann M, Mann G, Gadner H, Parwaresch R, Riehm H, Reiter A. Short-pulse B-non-Hodgkin lymphoma-type chemotherapy is efficacious treatment for pediatric anaplastic large cell lymphoma: a report of the Berlin-Frankfurt-Munster group trial NHL-BFM 90. Blood 2001;97(12):3699–706.

[55] Brugieres L, Quartier P, Le Deley MC, Pacquement H, Perel Y, Bergeron C, Schmitt C, Landmann J, Patte C, Terrier-Lacombe MJ, Delsol G, Hartmann O. Relapses of childhood anaplastic large-cell lymphoma: treatment results in a series of 41 children—a report from the French Society of Pediatric Oncology. Ann Oncol 2000;11 (1):53–8.

[56] Laver JH, Kraveka JM, Hutchison RE, Chang M, Kepner J, Schwenn M, Tarbell N, Desai S, Weitzman S, Weinstein HJ, Murphy SB. Advanced-stage large-cell lymphoma in children and adolescents: results of a randomized trial incorporating intermediate-dose methotrexate and high-dose cytarabine in the maintenance phase of the APO regimen: a Pediatric Oncology Group phase III trial. J Clin Oncol 2005;23(3):541–7.

[57] Alexander S, Kraveka JM, Weitzman S, Lowe E, Smith L, Lynch JC, Chang M, Kinney MC, Perkins SL, Laver J, Gross TG, Weinstein H. Advanced stage anaplastic large cell lymphoma in children and adolescents: results of ANHL0131, a randomized phase III trial of APO versus a modified regimen with vinblastine: a report from the children's oncology group. Pediatr Blood Cancer 2014;61(12):2236–42.

[58] Brugieres L, Le Deley MC, Rosolen A, Williams D, Horibe K, Wrobel G, Mann G, Zsiros J, Uyttebroeck A, Marky I, Lamant L, Reiter A. Impact of the methotrexate administration dose on the need for intrathecal treatment in children and adolescents with anaplastic large-cell lymphoma: results of a randomized trial of the EICNHL group. J Clin Oncol 2009;27(6):897–903.

[59] Lamant L, McCarthy K, d'Amore E, Klapper W, Nakagawa A, Fraga M, Maldyk J, Simonitsch-Klupp I, Oschlies I, Delsol G, Mauguen A, Brugieres L, Le Deley MC. Prognostic impact of morphologic and phenotypic features of childhood ALK-positive anaplastic large-cell lymphoma: results of the ALCL99 study. J Clin Oncol 2011; 29(35):4669–76.

[60] Eyre TA, Khan D, Hall GW, Collins GP. Anaplastic lymphoma kinase-positive anaplastic large cell lymphoma: current and future perspectives in adult and paediatric disease. Eur J Haematol 2014;93(6):455–68.

[61] Minard-Colin V, Brugieres L, Reiter A, Cairo MS, Gross TG, Woessmann W, Burkhardt B, Sandlund JT, Williams D, Pillon M, Horibe K, Auperin A, Le Deley MC, Zimmerman M, Perkins SL, Raphael M, Lamant L, Klapper W, Mussolin L, Poirel HA, Macintyre E, Damm-Welk C, Rosolen A, Patte C. Non-Hodgkin lymphoma in children and adolescents: progress through effective collaboration, current knowledge, and challenges ahead. J Clin Oncol 2015;33(27):2963–74.

[62] Larose H, Burke GAA, Lowe EJ, Turner SD. From bench to bedside: the past, present and future of therapy for systemic paediatric ALCL, ALK. Br J Haematol 2019;185(6):1043–54.

[63] Hunter T. Treatment for chronic myelogenous leukemia: the long road to imatinib. J Clin Invest 2007;117(8):2036–43.

[64] Druker BJ, Tamura S, Buchdunger E, Ohno S, Segal GM, Fanning S, Zimmermann J, Lydon NB. Effects of a selective inhibitor of the Abl tyrosine kinase on the growth of Bcr-Abl positive cells. Nat Med 1996;2(5):561–6.

[65] Jiao Q, Bi L, Ren Y, Song S, Wang Q, Wang YS. Advances in studies of tyrosine kinase inhibitors and their acquired resistance. Mol Cancer 2018;17(1):36.

[66] Zou HY, Li Q, Lee JH, Arango ME, McDonnell SR, Yamazaki S, Koudriakova TB, Alton G, Cui JJ, Kung PP, Nambu MD, Los G, Bender SL, Mroczkowski B, Christensen JG. An orally available small-molecule inhibitor of c-Met, PF-2341066, exhibits cytoreductive antitumor efficacy through antiproliferative and antiangiogenic mechanisms. Cancer Res 2007;67(9):4408–17.

[67] Christensen JG, Zou HY, Arango ME, Li Q, Lee JH, McDonnell SR, Yamazaki S, Alton GR, Mroczkowski B, Los G. Cytoreductive antitumor activity of PF-2341066, a novel inhibitor of anaplastic lymphoma kinase and c-Met, in experimental models of anaplastic large-cell lymphoma. Mol Cancer Ther 2007;6(12 Pt 1):3314–22.

[68] Mosse YP, Laudenslager M, Longo L, Cole KA, Wood A, Attiyeh EF, Laquaglia MJ, Sennett R, Lynch JE, Perri P, Laureys G, Speleman F, Kim C, Hou C, Hakonarson H, Torkamani A, Schork NJ, Brodeur GM, Tonini GP, Rappaport E, Devoto M, Maris JM. Identification of ALK as a major familial neuroblastoma predisposition gene. Nature 2008;455(7215):930–5.

[69] Mosse YP. Anaplastic lymphoma kinase as a cancer target in pediatric malignancies. Clin Cancer Res 2016;22(3):546–52.

[70] Mosse YP, Voss SD, Lim MS, Rolland D, Minard CG, Fox E, Adamson P, Wilner K, Blaney SM, Weigel BJ. Targeting ALK with crizotinib in pediatric anaplastic large cell lymphoma and inflammatory myofibroblastic tumor: a Children's Oncology Group Study. J Clin Oncol 2017;35(28):3215–21.

[71] Sandlund JT. Non-Hodgkin lymphoma in children. Curr Hematol Malig Rep 2015;10(3):237–43.

[72] Prokoph N, Larose H, Lim MS, Burke GAA, Turner SD. Treatment options for paediatric anaplastic large cell lymphoma (ALCL): current standard and beyond. Cancer 2018;10(4). https://doi.org/10.3390/cancers10040099.

[73] Gambacorti Passerini G, Farina F, Stasia A, Redaelli S, Ceccon M, Mologni L, Messa C, Guerra L, Giudici G, Sala E, Mussolin L, Deeren D, King MH, Steurer M, Ordemann R, Cohen AM, Grube M, Bernard L, Chiriano G, Antolini L, Piazza R. Crizotinib in advanced, chemoresistant anaplastic lymphoma kinase-positive lymphoma patients. J Natl Cancer Inst 2014;106(2), djt378.

[74] Savage KJ. Therapies for peripheral T-cell lymphomas. Hematology Am Soc Hematol Educ Program 2011;2011:515–24.

[75] Hapgood G, Savage KJ. The biology and management of systemic anaplastic large cell lymphoma. Blood 2015;126(1):17–25.

[76] Xu W, Kim JW, Jung WJ, Koh Y, Yoon SS. Crizotinib in combination with Everolimus synergistically inhibits proliferation of anaplastic lymphoma kinase positive anaplastic large cell lymphoma. Cancer Res Treat 2018;50(2):599–613.

[77] Boi M, Todaro M, Vurchio V, Yang SN, Moon J, Kwee I, Rinaldi A, Pan H, Crescenzo R, Cheng M, Cerchietti L, Elemento O, Riveiro ME, Cvitkovic E, Bertoni F, Inghirami G. AIRC 5xMille consortium 'Genetics-Driven Targeted management of lymphoid malignancies'. Therapeutic efficacy of the bromodomain inhibitor OTX015/MK-8628 in ALK-positive anaplastic large cell lymphoma: an alternative modality to overcome resistant phenotypes. Oncotarget 2016;7(48):79637–53.

[78] Birrer MJ, Moore KN, Betella I, Bates RC. Antibody-drug conjugate-based therapeutics: state of the science. J Natl Cancer Inst 2019;111(6):538–49.

[79] Cassady JM, Chan KK, Floss HG, Leistner E. Recent developments in the maytansinoid antitumor agents. Chem Pharm Bull (Tokyo) 2004;52(1):1–26.

[80] Herrera AF, Molina A. Investigational antibody-drug conjugates for treatment of B-lineage malignancies. Clin Lymphoma Myeloma Leuk 2018;18(7):452–68. e4.

[81] Rossi C, Chretien ML, Casasnovas RO. Antibody-drug conjugates for the treatment of hematological malignancies: a comprehensive review. Target Oncol 2018;13(3):287–308.

[82] Kreitman RJ, Stetler-Stevenson M, Jaffe ES, Conlon KC, Steinberg SM, Wilson W, Waldmann TA, Pastan I. Complete remissions of adult T-cell leukemia with anti-CD25 recombinant immunotoxin LMB-2 and chemotherapy to block immunogenicity. Clin Cancer Res 2016;22(2):310–8.

[83] Janik JE, Morris JC, O'Mahony D, Pittaluga S, Jaffe ES, Redon CE, Bonner WM, Brechbiel MW, Paik CH, Whatley M, Chen C, Lee JH, Fleisher TA, Brown M, White JD, Stewart DM, Fioravanti S, Lee CC, Goldman CK, Bryant BR, Junghans RP, Carrasquillo JA, Worthy T, Corcoran E, Conlon KC, Waldmann TA. 90Y-daclizumab, an anti-CD25 monoclonal antibody, provided responses in 50% of patients with relapsed Hodgkin's lymphoma. Proc Natl Acad Sci U S A 2015;112(42):13045–50.

[84] Godwin CD, Gale RP, Walter RB. Gemtuzumab ozogamicin in acute myeloid leukemia. Leukemia 2017;31 (9):1855–68.

[85] Schirrmann T, Steinwand M, Wezler X, Ten Haaf A, Tur MK, Barth S. CD30 as a therapeutic target for lymphoma. BioDrugs 2014;28(2):181–209.

[86] Crisci S, Di Francia R, Mele S, Vitale P, Ronga G, De Filippi R, Berretta M, Rossi P, Pinto A. Overview of targeted drugs for mature B-cell non-hodgkin lymphomas. Front Oncol 2019;9:443.

[87] Velasquez MP, Bonifant CL, Gottschalk S. Redirecting T cells to hematological malignancies with bispecific antibodies. Blood 2018;131(1):30–8.

[88] Ramos CA, Ballard B, Zhang H, Dakhova O, Gee AP, Mei Z, Bilgi M, Wu MF, Liu H, Grilley B, Bollard CM, Chang BH, Rooney CM, Brenner MK, Heslop HE, Dotti G, Savoldo B. Clinical and immunological responses after CD30-specific chimeric antigen receptor-redirected lymphocytes. J Clin Invest 2017;127(9):3462–71.

[89] Grover NS, Savoldo B. Challenges of driving CD30-directed CAR-T cells to the clinic. BMC Cancer 2019;19(1):203–9.

[90] Pro B, Advani R, Brice P, Bartlett NL, Rosenblatt JD, Illidge T, Matous J, Ramchandren R, Fanale M, Connors JM, Fenton K, Huebner D, Pinelli JM, Kennedy DA, Shustov A. Five-year results of brentuximab vedotin in patients with relapsed or refractory systemic anaplastic large cell lymphoma. Blood 2017;130(25):2709–17.

[91] Mitsunobu T, Nishikawa T, Kusuda M, Nakagawa S, Kodama Y, Okamoto Y, Kawano Y. Brentuximab Vedotin and high-dose methotrexate administrated alternately for refractory anaplastic large-cell lymphoma with central nervous system disease. J Pediatr Hematol Oncol 2019;3.

[92] Gong J, Guo F, Cheng W, Fan H, Miao Q, Yang J. Preliminary biological evaluation of 123I-labelled anti-CD30-LDM in CD30-positive lymphomas murine models. Artif Cells Nanomed Biotechnol 2020;48(1):408–14.

[93] Shen Y, Yang T, Cao X, Zhang Y, Zhao L, Li H, Zhao T, Xu J, Zhang H, Guo Q, Cai J, Gao B, Yu H, Yin S, Song R, Wu J, Guan L, Wu G, Jin L, Su Y, Liu Y. Conjugation of DM1 to anti-CD30 antibody has potential antitumor activity in CD30-positive hematological malignancies with lower systemic toxicity. MAbs 2019;11(6):1149–61.

[94] Wang R, Li L, Duan A, Li Y, Liu X, Miao Q, Gong J, Zhen Y. Crizotinib enhances anti-CD30-LDM induced antitumor efficacy in NPM-ALK positive anaplastic large cell lymphoma. Cancer Lett 2019;448:84–93.

[95] Sano R, Krytska K, Larmour CE, Raman P, Martinez D, Ligon GF, Lillquist JS, Cucchi U, Orsini P, Rizzi S, Pawel BR, Alvarado D, Mosse YP. An antibody-drug conjugate directed to the ALK receptor demonstrates efficacy in preclinical models of neuroblastoma. Sci Transl Med 2019;11(483). https://doi.org/10.1126/scitranslmed.aau9732.

[96] Shen J, Li S, Medeiros LJ, Lin P, Wang SA, Tang G, Yin CC, You MJ, Khoury JD, Iyer SP, Miranda RN, Xu J. PD-L1 expression is associated with ALK positivity and STAT3 activation, but not outcome in patients with systemic anaplastic large cell lymphoma. Mod Pathol 2020;33(3):324–33.

[97] Zhang JP, Song Z, Wang HB, Lang L, Yang YZ, Xiao W, Webster DE, Wei W, Barta SK, Kadin ME, Staudt LM, Nakagawa M, Yang Y. A novel model of controlling PD-L1 expression in ALK(+) anaplastic large cell lymphoma revealed by CRISPR screening. Blood 2019;134(2):171–85.

[98] Pearson ADJ, Rossig C, Lesa G, Diede SJ, Weiner S, Anderson J, Gray J, Geoerger B, Minard-Colin V, Marshall LV, Smith M, Sondel P, Bajars M, Baldazzi C, Barry E, Blackman S, Blanc P, Capdeville R, Caron H, Cole PD, Jimenez JC, Demolis P, Donoghue M, Elgadi M, Gajewski T, Galluzzo S, Ilaria Jr R, Jenkner A, Karres D, Kieran M, Ligas F, Lowy I, Meyers M, Oprea C, Peddareddigari VGR, Sterba J, Stockman PK, Suenaert P, Tabori U, van Tilburg C, Yancey T, Weigel B, Norga K, Reaman G, Vassal G. Accelerate and European Medicines Agency Paediatric Strategy Forum for medicinal product development of checkpoint inhibitors for use in combination therapy in paediatric patients. Eur J Cancer 2020;127:52–66.

[99] Gao X, Mi Y, Guo N, Xu H, Xu L, Gou X, Jin W. Cytokine-induced killer cells as pharmacological tools for cancer immunotherapy. Front Immunol 2017;8:774.

[100] Wang L, Lui VWY. Emerging roles of ALK in immunity and insights for immunotherapy. Cancer 2020;12(2). https://doi.org/10.3390/cancers12020426.

Drug combinations: A strategy to enhance anti-tumor activity and overcome drug resistance to ALK inhibitors in neuroblastoma

Libo Zhang[a,b] and Sylvain Baruchel[c,d]

[a]Department of Molecular Medicine, The Hospital for Sick Children Research Institute, Toronto, ON, Canada [b]Department of Anesthesia, The Hospital for Sick Children Research Institute, Toronto, ON, Canada [c]Department of Pediatrics, The Hospital for Sick Children Research Institute, Toronto, ON, Canada [d]Institute of Medical Sciences, University of Toronto, Toronto, ON, Canada

Abstract

Alterations in the *ALK* gene are found in 8% of familial and sporadic neuroblastoma cases. The frequency of ALK aberrations in high-risk neuroblastomas is 14% (10% point mutation and 4% amplification) at the time of diagnosis, with increasing number of aberrations at tumor relapse (Bresler et al., 2011). Despite recent advancements in ALK-targeted therapies, ALK inhibitor potency eventually diminishes, and drug resistance emerges after initial successful treatment. To improve drug efficacy and overcome resistance to ALK inhibitors, combined regimens with chemotherapy or other targeted therapies showed superior anti-tumor effects in preclinical and early clinical studies, making it a promising strategy for neuroblastoma treatment.

Abbreviations

ALCL	anaplastic large-cell lymphoma
ALK	anaplastic lymphoma kinase
ATRX	alpha thalassemia/mental retardation syndrome X-linked
CDKN2A	cyclin dependent kinase inhibitor 2A
CR	complete response
CTX	cyclophosphamide
DDR	discoidin domain receptor

EFS	event free survival
ERK	extracellular signal-regulated kinase
FOXM1	forkhead box M1
IGF1R	insulin-like growth factor-1 receptor
INRG	International Neuroblastoma Risk Group
INSR	insulin receptor
JAK	Janus Activated Kinase
MAPK	mitogen-activated protein kinase
MDM2	mouse double minute 2 homolog
MET	mesenchymal epithelial transition
MS	mass spectrometry
MTC	medullary thyroid cancer
MTD	maximum tolerated dose
NSCLC	non-small cell lung cancer
PI3K	phosphoinositide 3-kinase
PK	pharmacokinetics
RET	Ret proto-oncogene
ROS1	ROS proto-oncogene 1
RP2D	recommended phase 2 dose
RPTOR	regulatory associated protein of MTOR complex 1
SD	stable disease
STAT	signal transducer and activator of transcription
TKI	tyrosine kinase inhibitor
TOPO	topotecan
TRK	tropomyosin receptor kinase

Conflict of interest

No potential conflicts of interest were disclosed.

Introduction

Neuroblastoma

Neuroblastoma is the most common extracranial pediatric solid tumor originating from the sympathoadrenal lineage of the neural crest. Primary tumors usually initiate in the adrenal glands, but can also develop in the neck, chest, abdomen, or spine. The International Neuroblastoma Risk Group (INRG) Task Force has established a classification system to stratify neuroblastoma patients into four categories: very low-risk, low-risk, intermediate-risk, or high-risk. Classification is based on different prognostic factors: the stage of the disease according to the INRG staging system, age at time of diagnosis, tumor histology/differentiation, MYCN amplification, Chromosome 11q status, and tumor cell ploidy. Low- and intermediate-risk forms of neuroblastoma are highly curable, while the survival rate of high-risk disease is only 40% despite aggressive multimodal therapy. Survivors of high-risk disease also often suffer from severe side-effects associated with drug toxicities [1, 2]. With a greater understanding of tumor biomarkers and molecular aberrations in neuroblastoma, targeted therapies are being developed to explore novel, less toxic, and potentially more efficacious treatments for this malicious disease [3]. However, relapse and fatality may occur due to de novo or acquired drug resistance, especially in high-risk neuroblastoma patients [4, 5].

Recent genome-wide sequencing analyses in large neuroblastoma patient cohorts contribute to spotlighting the genetic basis of neuroblastoma. Next-generation sequencing-based genomic profiling identified the most frequent alterations including MYCN (26.5%), ALK (17.8%), ATRX (6.5%), CDKN2A (4.8%), and RPTOR (4.8%) in 230 neuroblastoma patient samples. [6–10].

ALK variants in neuroblastoma

In 2008, several groups discovered ALK mutations, including germline missense mutations and somatically acquired mutations, in high-risk neuroblastoma patients [11–13]. Mutated kinases were autophosphorylated and displayed increased kinase activity compared with the wild-type ALK. So far, more than 35 ALK variants, predominantly point mutations, have been detected in neuroblastoma [14], with fewer cases involving a truncated extracellular domain [15, 16] and BEND5-ALK fusion protein [10]. Almost all cases of familial neuroblastoma (<2% of all neuroblastoma) bear ALK mutations [17]. Three common pathogenic variants, p.Arg1275Gln (R1275Q), p.Gly1128Ala (G1128A) and p.Phe1174Leu (F1174L), were identified in familial neuroblastoma patients, among which R1275Q accounts for 45% of ALK germline mutations [7, 11]. ALK mutations have also been reported in 6–10% of sporadic neuroblastoma cases [18]. Twelve different ALK active mutations have been identified to associate with sporadic neuroblastoma, including two of the most common ALK variants, F1174L and R1275Q [19]. ALK amplifications have also been reported in 2% of primary neuroblastoma tumors [7]. More interestingly, ALK mutations appear to be further enriched in relapsed patients, either through subclonal mutation selection or additional mutations acquired during therapy [20–22].

In neuroblastoma, most of ALK pathogenic variants are found within the tyrosine kinase domain of ALK and cause constitutive autophosphorylation and activation of the ALK protein and downstream cellular pathways, including the MAPK and RAS-related protein 1 signal pathways [23, 24], the PI3K (phosphatidylinositol 3-kinase)/Akt [25], and the JAK/STAT (Janus activated kinase/signal transducer and activator of transcription) [26] pathways.

ALK inhibitors in neuroblastoma

Crizotinib is the most extensively studied ALK inhibitor in neuroblastoma. It is an oral small-molecule tyrosine kinase inhibitor, originally developed as a c-MET inhibitor, and later found to be an inhibitor for ALK phosphorylation [27, 28]. In vitro studies demonstrated that crizotinib was potent in neuroblastoma cell lines with ALK amplification or the R1275Q mutation, one of the most common ALK variants in neuroblastoma [29], whereas cells bearing F1174L mutation are relatively crizotinib-resistant [11, 12, 30, 31]. In vivo, crizotinib treatment causes complete and sustained regression of xenografts with R1275Q mutation, but it has limited effects on the growth of F1174L-positive tumors [14]. Single agent crizotinib has been tested for the treatment of pediatric solid tumors through a COG Phase 1/2 trial which determined the RP2D to be $280 \, mg/m^2$/dose twice daily [32]. From the 11 enrolled patients with ALK mutations and rearrangements, only one patient had a CR and two had SD. Drug resistance becomes a real challenge with crizotinib.

In order to address secondary resistance to crizotinib, new generations of ALK TKIs, including ceritinib, AP26113, alectinib, ensartinib (X-396), entrectinib (RXDX-101), belizatinib (TSR-011) and lorlatinib (PF06463922) have been developed in clinical use for adult patients [33, 34]. Lorlatinib, a selective next-generation ROS1/ALK inhibitor, has high potency across ALK variants, R1275Q, F1174L and F1245C mutations. It induces complete tumor regression in both crizotinib-resistant and crizotinib-sensitive neuroblastoma xenograft models, as well as in patient-derived xenografts [34]. A pediatric phase 1 trial of lorlatinib (NCT03107988) is underway in patients with relapsed/refractory neuroblastoma. Lorlatinib will be utilized both as a single agent and in combination with chemotherapy with cyclophosphamide and topotecan. A new ALK/IGF1R inhibitor AZD3463 is designed by AstraZeneca to overcome the acquired resistance to crizotinib. This new drug suppresses cell proliferation of neuroblastoma cell lines with wild-type ALK as well as ALK-activating mutations (F1174L and D1091N) by blocking the ALK-mediated PI3K/AKT/mTOR pathway. In addition, AZD3463 also exhibits significant therapeutic effects on the growth of the neuroblastoma tumors bearing an ALK^{F1174L} mutation in orthotopic xenograft mouse models [35]. A novel ALK inhibitor alectinib (5-chloro-2,4-diaminophenylpyrimidine) has been tested in neuroblastoma preclinical models and shows substantial inhibitory effects against tumors with ALK mutations, including ALK^{L1152R}, ALK^{F1174L}, and ALK^{D1091N} [36, 37]. Most recently, the next-generation inhibitor repotrectinib (TPX-0005), which targets the active kinase conformations of ALK, ROS1 and TRK receptors, is capable of inhibiting signaling activity of a range of ALK mutant variants and reducing tumor growth in a neuroblastoma xenograft model [38].

Combined therapy with ALK inhibitors

It is becoming increasingly clear that monotherapy simply inhibiting a single upstream target, in many cases, is not enough to hinder cancer cell growth. In a Children's Oncology Group phase 1 consortium study, 11 patients with neuroblastoma with known ALK mutations were treated at doses ranging from 100 to 365 mg/m^2 per dose. Out of 11 patients, only one had a complete response (9%), and two had stable disease [39], which suggests that crizotinib monotherapy is not efficient to treat neuroblastoma. Combined therapy with other antitumor agents may help to maximize treatment benefit and overcome the limitations of single-agent ALK inhibitors for neuroblastoma.

In addition, similar to other small-molecule tyrosine kinase inhibitors (TKI), crizotinib invariably loses its potency, and drug resistance emerges after initial successful crizotinib treatment in neuroblastoma patients, which has also occurred in other types of cancer [40, 41]. Crizotinib resistance may develop as a consequence of secondary mutations in the ALK tyrosine kinase domain, or amplification of the ALK locus, which is also called ALK-dependent resistance [42, 43]. In other cases, crizotinib resistance may also arise from pre-existing minority ALK variant populations with drug resistance due to intratumor heterogeneity in neuroblastoma tumors. Another mechanism of ALK inhibitor resistance is ALK-independent, by which tumor cells switch to alternative signaling pathways to evade ALK inhibitor-induced apoptosis.

To overcome drug resistance to ALK inhibitors, alternative treatment approaches would be required, such as increased dose of ALK inhibitors, transition to newer generations of ALK inhibitors, or addition of inhibitor(s) to the alternative pathways into existing ALK inhibition therapy. Unfortunately, with increased dose of crizotinib or newer generations of ALK inhibitors, despite exhibiting more potent initial response to ALK inhibition, many patients eventually develop resistance with secondary mutations. Therefore, targeting these bypass mechanisms becomes crucial to enhancing treatment efficacy and reversing drug resistance, by which a superior anti-tumor activity and reduction of tumor relapse could be achieved in neuroblastomas with ALK mutations. Several alternative signaling pathways and oncogenic drivers have been identified in neuroblastoma, including cyclin-dependent kinase, p53, PI3K-mTOR, MEK/ERK, ALK-ETV5-RET axis, insulin-like growth factor-1 receptor (IGF1R)/INSR, etc. [22].

Crizotinib and chemotherapies

Topotecan alone and in combination with chemotherapies such as cyclophosphamide, has been studied in several single and multi-institutional phase I and II trials and proved tolerable and effective in relapsed and newly diagnosed neuroblastoma [44–46]. In the Children's Oncology Group (COG) study, 57 patients were treated with topotecan and cyclophosphamide (TOPO/CTX), and 62 patients were treated with topotecan alone for up to 1 year. TOPO/CTX proved superior to TOPO in PFS, with a trend toward superior response rate, but no improvement in OS. Currently, TOPO/CTX regimens are also being incorporated into several phase I and II trials as backbones to combine with targeted therapies for the treatment of high-risk or refractory neuroblastoma.

ALK inhibitor in combination with TOPO/CTX was also studied in neuroblastoma. Krytska et al. [47] combined crizotinib with TOPO/CTX in human neuroblastoma-derived cell lines and patient-derived xenograft models. In their study, combined regimen showed enhanced cytotoxic effects compared to crizotinib or chemotherapy alone in vitro. In vivo, combined therapy not only induced rapid and sustained tumor regression with improved EFS, but also restored sensitivity in preclinical models harboring ALK aberrations (both ALK mutation and amplification). Complete responses were maintained even 24 weeks after treatment was completed. Another preclinical study tested crizotinib combined with low-dose metronomic topotecan in neuroblastoma and demonstrated that single-agent crizotinib showed limited anti-tumor activity in ALK^{F1174L}-mutated neuroblastoma xenograft models. However, when combined with low-dose metronomic administration of topotecan, significantly delayed tumor development was observed. In addition, relapsed tumors remained responsive to combined therapy [48]. This enhanced anti-tumor activity was probably achieved through targeting hypoxia-related pathways. Hypoxia regulates tumor cell proliferation, migration, and invasiveness through the expression of a group of transcription factors called hypoxia-inducible factors (HIFs) [49, 50]. Topotecan, especially daily metronomic topotecan, induces oxidative stress and down-regulates HIF-1 alpha expression in cancer cells [51–53]. It was also demonstrated that ALK specifically regulates HIF-1α expression under hypoxia conditions in both ALCL and NSCLC [54]. A COG phase 1 trial of crizotinib in combination with dose intensity TOPO/CTX has been conducted in children with relapsed and refractory solid tumors. The RP2D of crizotinib was 215 mg/m^2/dose twice daily when combined with chemotherapy (NCT01606878) [41]. A phase 3 clinical trial is underway to determine whether the addition of crizotinib to standard therapy improves the survival of patients with newly-diagnosed high-risk neuroblastoma (NCT03126916).

ALK inhibitor ceritinib and CDK4/6 inhibitor ribociclib

Ribociclib (LEE011) is an orally bioavailable, small molecule inhibitor of both CDK4 and CDK6. In vitro, ribociclib caused cell-cycle arrest and cellular senescence with reduced phosphorylation of the retinoblastoma (*Rb*) protein and the transcription factor FOXM1. In vitro screening of a panel of more than 500 cell lines constituting the Novartis Cancer Cell Line Encyclopedia identified neuroblastoma to be among the most sensitive cell lines to ribociclib treatment [55]. Ribociclib significantly reduced cell proliferation in 12 out of 17 human neuroblastoma-derived cell lines, and caused tumor growth delay in vivo [56]. From a Phase 1 clinical trial [57], ribociclib demonstrated acceptable safety and pharmacokinetics (PK) in pediatric patients with malignant rhabdoid tumors, neuroblastoma, and other solid tumors. MTD ($470 \, mg/m^2$) and RP2D ($350 \, mg/m^2$) were equivalent to those in adults.

Wood and colleagues [58] investigated the combination of ceritinib and ribociclib against 17 comprehensively characterized human neuroblastoma-derived cell lines. In their study, the combination of ribociclib and the ALK inhibitor ceritinib demonstrated synergistic antitumor effects in cell lines with ALK mutations. Compared to single-agent treatment, combined therapy significantly reduced cell proliferation, enhanced cell cycle arrest, and caspase-independent cell death. In vivo, combination therapy induced complete regression in xenograft models with ALK-F1174L and F1245C mutations. For ceritinib-resistant SH-SY5Y xenografts, combined ceritinib and ribociclib treatment also achieved complete and sustained regressions, and prolonged event-free survival (EFS) compared to either of the single agents. Their synergistic antitumor activity was achieved by targeting both the receptor tyrosine kinase ALK and cell cycle oncogenic network. Based on this study, The Children's Hospital of Philadelphia recently initiated Next Generation Personalized Neuroblastoma Therapy (NEPENTHE) trial (NCT02780128), a Phase 1 clinical trial, to study the combination therapy of ceritinib and ribociclib in relapsed or refractory neuroblastoma patients. Their qualified participants with ALK mutations are selected based on genetic sequencing results from tumor biopsy.

ALK inhibitor ceritinib and p53 activation

Another drug combination study targeting both ALK and p53 pathways also showed prominent anti-tumoral potential in p53 wild-type neuroblastoma. In p53 wild-type tumors, p53 function is frequently impaired by two major negative regulators, MDM2 and its homolog MDMX. MDM2 induces p53 degradation via its E3 ligase activity. It can also bind to the N-terminal transactivation domain of p53, blocking its transcriptional activity [59]. Small molecules targeting MDM2, such as NVP-CGM097, block the p53-binding site of MDM2 and prevents its interaction with p53. Thus, NVP-CGM097 exerts its antitumor activity by stabilizing p53 and activating the p53 pathway.

Wang et al. [60] examined the anti-proliferative effect of the MDM2 inhibitor NVP-CGM097 in a panel of neuroblastoma cell lines, including five TP53 mutant and six TP53 wild-type cell lines. Consistent with the molecular mechanisms of MDM2 inhibitors, the TP53 wild-type neuroblastoma cell lines were significantly more responsive to NVP-CGM097 treatment than the TP53 mutant cell lines.

In their in vivo study, NVP-CGM097 alone at 50 mg/kg/day demonstrated no inhibition of tumor growth, while the combination of ceritinib with NVP-CGM097 promoted apoptosis in the ALK mutant/TP53 wild-type neuroblastoma. Combination therapy also induced complete and durable tumor regression and prolonged animal survival in neuroblastoma xenograft models. More interestingly, combined therapy overcomed acquired ceritinib resistance caused by MYCN upregulation in an ALK-driven neuroblastoma model. This group also investigated the molecular mechanism of synergy between the ALK and MDM2 inhibitor. Ceritinib inhibited phospho-ALK, phospho-AKT and phospho-ERK signaling, which demonstrated effective ALK pathway inhibition. NVP-CGM097 treatment resulted in an increase in p53 expression and induction of its target genes, including p21, MDM2 and PUMA, in TP53 wild-type neuroblastoma cells, while p53 protein or its downstream effectors were not affected in TP53-mutant cells. Thus, ceritinib and NVP-CGM097 target two complementary signal pathways, both anti-proliferative and apoptosis-stimulating signals, which causes synergistic antitumor effects and overcomes ceritinib drug resistance [60].

Targeting MYCN signals

ALK^{F1174L} is the most aggressive ALK mutations in neuroblastoma. It possesses high transforming potential and is often accompanied by the *MYCN* oncogene amplification. Both *ALK* and *MYCN* genes are located in chromosome 2p, a chromosomal alteration identified as a statistically significant prognostic factor [61]. It has been shown that ALK and MYCN drive tumor malignancy cooperatively. Activation of ALK increases the expression of MYCN by enhancing the activity of the MYCN promoter and stabilizing MYCN protein likely via activation of the AKT and ERK pathways [24, 25, 62]. In vivo, compared to ALK^{F1174L} and MYCN alone, co-expression of these two oncogenes leads to the development of neuroblastoma tumors with earlier onset, higher penetrance and enhanced lethality [25, 63, 64]. In our study, neuroblastoma cells harboring both ALK^{F1174L} mutation and MYCN amplification showed less response to an ALK inhibitor crizotinib than to other variants [48].

Directly targeting MYCN has proven to be technically challenging due to a lack of appropriate surfaces on its DNA binding domain for small molecule binding [65]. Also, since Myc is predominantly located in the cell nucleus, targeting Myc with monoclonal antibodies is technically impractical. Alternative approaches to indirectly abrogate Myc functions are being investigated. The PI3K/AKT/mTOR pathway appears to play an important role in MYCN stabilization, therefore, targeting the PI3K/AKT/mTOR pathway becomes a good option to attenuate MYCN signals.

Moore et al. [66] showed that in ALK^{F1174L}/MYCN-amplified neuroblastoma cells, crizotinib alone did not affect mTORC1 activity as evidenced by persistent RPS6 phosphorylation. Combined treatment with crizotinib and an ATP-competitive mTOR inhibitor, Torin2, inhibited RPS6 phosphorylation, enhanced antitumor activity and prolonged survival in ALK^{F1174L}/MYCN-amplified neuroblastoma models compared to single-agent treatment. However, in MYCN wild-type tumors, this combination induced mTORC1 downregulation, but at the same time caused upregulation of PI3K activity. Therefore, no synergistic or additive cytotoxicity was observed in MYCN wild-type neuroblastoma. To overcome the limitation of this selective mTOR inhibitor in MYCN wild-type tumors, Moore et al. tried to block both mTOR and PI3K signals with PF-05212384, a dual inhibitor against both mTOR and PI3K, and combined PF-05212384 with crizotinib treatment. As expected, synergistic activity was achieved in neuroblastoma cells with non-amplified *MYCN*. As indicated in this study, understanding the molecular basis of drug resistance is crucial for designing new combination strategies to improve antitumor activity and overcome ALK-TKI drug resistance.

Targeting both ALK and its downstream signals

Here is another strategy to enhance the treatment efficacy and avoid ALK-TKI resistance by blocking both ALK activity and its downstream signals together. In ALK-positive tumors, several downstream signal pathways have been reported in ALK activation, including ERK5, RET, IGF1R/INSR, PI3K-AKT-mTOR, RAS-MAPK, etc. [66–70].

Extracellular signal-regulated kinase 5 (ERK5) serving as a downstream target in the ALK pathway, is known as big mitogen-activated protein kinase 1 (MAPK1), a member of the MAPK family. ERK5 has autophosphorylation activity, which phosphorylates MEK5 and activates downstream transcription factors. ERK5 is required for epidermal growth factor (EGF)-induced cell proliferation, cell cycle promotion, and cellular transformation [71]. In neuroblastoma, ERK5 activation was also required for ALK-induced transcription of the oncogene *MYCN* and the stimulation of cell proliferation [69]. Umapathy and colleagues showed that treatment with either crizotinib or PI3K pathway inhibitors diminished the levels of phosphorylated ERK5 in the nucleus and reduced MYCN expression, which indicates that ALK-driven ERK5 activation is mediated by the PI3K-AKT pathway [72]. With ALK-positive, MYCN-amplified neuroblastoma cell lines, in vitro studies showed inhibition of cell proliferation by either the ERK5 inhibitor XMD8-92 or ERK5 siRNA. Combined treatment with crizotinib and XMD8-92 synergistically suppressed neuroblastoma cell proliferation. Further in vivo, dual inhibition of ALK and ERK5 abrogated tumor growth synergistically and resulted in reduced expression of MYCN in tumor cells. These findings suggest that concomitant ALK and ERK5 inhibition may provide another effective regimen to suppress oncogenic MYCN signaling, and potentially overcome crizotinib resistance in ALK-positive neuroblastoma.

The oncogene *RET* has also been reported as a molecular target of ALK activation in both human neuroblastoma cell lines and primary tumors [73, 74]. Cazes et al. developed a neuroblastoma mouse model with endogenous expression of mutated ALK in a *MYCN* transgenic context (MYCN/ALK^{mut}). They compared transcriptomic profiling between mouse MYCN/ALK^{mut} and MYCN/ALK wild-type tumors and found that the *RET* oncogene was upregulated in ALK-mutated tumors [73, 74]. They further demonstrated a strong correlation between RET and ETV5 expressions and showed that ALK activation induced ETV5 protein upregulation in a MEK/ERK-dependent manner. Inhibition ETV5 with siRNA decreased RET expression both at the protein and mRNA levels. Therefore, RET upregulation by activated ALK was achieved through an ALK-ETV5-RET axis. With ChIP-seq analysis, Cazes and colleagues further confirmed that ETV5 took part in the transcriptional regulation of the *RET* gene by binding on the RET promoter and acted as an upstream enhancer. In mouse models, tumor growth of MYCN/ALK^{mut} tumors was inhibited by a RET inhibitor, vandetanib, indicating RET as a potential therapeutic target in ALK-mutated neuroblastoma.

They also investigated the effect of the crizotinib/vandetanib combination therapy after allograft of a MYCN/Alk^{F1178L} tumor in nude mice. Tumor-bearing mice were treated with vehicle, crizotinib (100 mg/kg/day) or a combination of crizotinib (100 mg/kg/day) and RET inhibitor vandetanib (30 mg/kg/day). A strong reduction in tumor growth was observed in mice treated with the combination agents compared to the vehicle control, and this effect was more potent than each single agent. A prolonged event-free survival was also achieved in mice treated with the combination therapy. There is an ongoing multicenter phase 1/2 trial with LOXO-292, a highly selective RET inhibitor, in patients with advanced solid tumors, including RET fusion-positive solid tumors, medullary thyroid cancer (MTC), and other tumors with RET activation (NCT03157128). Although pediatric neuroblastoma is not included in this study. Once the drug safety profiles, pharmacokinetic properties and initial treatment responses are determined in RET activated tumors, it would provide another treatment option for ALK-resistant neuroblastoma.

Combined therapy directed by proteomics or genomic profiling

The rapid advances in proteomics-based technologies, especially quantitative proteomic research, provide us great opportunities to identify drug targets and their downstream signals. To identify downstream signaling components regulated by ALK network in neuroblastoma cells, Eynden et al. [75] performed phosphoproteomics and RNA-sequencing (RNA-seq) analysis after exposing neuroblastoma cells to the first- and third-generation ALK TKIs, crizotinib and lorlatinib, and generated both phosphoproteomic and gene expression signatures reflecting ALK signaling events in different neuroblastoma cell lines. From Mass Spectrometry (MS)-based phosphoproteomic analysis, with ALK positive cell lines, crizotinib treatment resulted in a specific inhibition of tyrosine phosphorylation in ALK, whereas lorlatinib treatment caused decreased phosphorylation of discoidin domain receptor 1 (DDR1), DDR2, and insulin-like growth factor-1 receptor/insulin receptor (IGF1R/INSR) in addition to ALK inhibition. Neither the selective DDR1 inhibitor DDR1-IN-1 nor the selective EGFR inhibitor afatinib affected the growth of neuroblastoma cells, which left IGF1R/INSR as a potential drug target. Emdal and colleagues selected neuroblastoma cell lines with different ALK profiles: CLB-BAR (gain of function, Δexon4–11 truncated ALK), CLB-GE (gain of function, ALK^{Phe1174Val} mutation), IMR-32 (wild-type ALK, ligand-dependent activation), and SK-N-AS (wild-type ALK) for further drug evaluation. All cell lines were treated with either ALK inhibitor (crizotinib or lorlatinib), or an IGF1R/INSR inhibitor (linsitinib), or the combination of both. It was shown that linsitinib treatment reduced growth of all neuroblastoma cell lines, whereas crizotinib and lorlatinib were active only in ALK positive CLB-BAR and CLB-GE cells. Furthermore, linsitinib had an additive effect on the cell proliferation inhibition comparing to the single-agent crizotinib or lorlatinib treatment. Although in vitro results showed promising effects by targeting both ALK and IGF1R/INSR in ALK positive neuroblastoma cells, further in vivo studies are required to validate this regimen.

Genome-wide CRISPR activation (CRISPRa) screening was also used to identify bypass mechanisms of resistance to ALK inhibitors in ALK-positive neuroblastomas. Trigg and colleagues [76] conducted genome-wide CRISPRa screens in two ALK mutant NB cell lines SH-SY5Y (ALKF1174L) and CHLA-20 (ALKR1275Q) exposed to brigatinib or ceritinib for

14 days. They identified *PIM1* as a putative resistance gene which is often associated with high-risk disease and poor survival outcomes in neuroblastoma. They further confirmed that overexpression or knockdown of PIM1 induced resistance or sensitization to ALK inhibitors, respectively. These findings suggested that combining ALK inhibitors with AZD1208, a small-molecule pan-PIM inhibitor, may overcome ALK drug resistance. To evaluate the efficacy of combined ALK and PIM inhibition in vivo, both MYCN wild-type and MYCN-amplified neuroblastoma cells were tested in their xenograft models. Superior antitumor activity was achieved in ceritinib and AZD1208 combination therapy compared to single-agent ceritinib or AZD1208 treatment. A significant increase in event-free survival was also observed in animals with combined treatment relative to single-agent therapies. In addition, combined PIM and ALK inhibition is effective independent of MYCN status in neuroblastoma cells.

Conclusion

Among all identified oncogenic mutations in neuroblastoma, *ALK* is the second most frequent alterations after *MYCN*. It is also one of the most well-studied druggable molecular targets. During the last decades, considerable advances have been made to understand the association between ALK genetic aberrations and disease prognosis in neuroblastoma. Incorporation of ALK alteration into neuroblastoma stratification may help predict the risk of recurrence or progression, guide standard therapeutic strategies for neuroblastoma and provide greater clinical benefit for patients.

For neuroblastoma treatment, ALK inhibitors and their combination with chemotherapeutic drugs are being evaluated at the different stages of clinical trials. Those clinical data provide more and more evidence about the development of ALK mutations during the treatment with ALK inhibitors. Molecular profiling of tumor tissues before and after a tyrosine kinase inhibitors (TKIs) exposure have been a common method to monitor therapy response and identify drug resistance mechanisms. Therefore, repeated biopsy of neuroblastoma tumors at recurrence is critical to understand the underlying molecular changes causing acquired resistance to ALK-TKI. Also, given the growing number of genetic alterations detected, small sequencing panels that focus on a limited number of genes may not be sufficient, especially in highly heterogeneous neuroblastoma tumors. Indeed, molecular profiling with next-generation sequencing (NGS), RNA-seq or Mass spectrometry-based proteomics could be used as a standard approach to discover new oncogenic drivers and identify novel drug targets in neuroblastoma. Of course, tumor molecular profiling for each individual patient at different stages of treatment will assist physicians choosing more-efficient therapeutic interventions or adding a second-line treatment to overcome ALK-TKI resistance.

References

[1] Maris JM. Recent advances in neuroblastoma. N Engl J Med 2010;362:2202–11.
[2] Smith MA, Altekruse SF, Adamson PC, Reaman GH, Seibel NL. Declining childhood and adolescent cancer mortality. Cancer 2014;120:2497–506.
[3] Greengard EG. Molecularly targeted therapy for neuroblastoma. Children (Basel) 2018;5.

[4] Matthay KK, Villablanca JG, Seeger RC, Stram DO, Harris RE, Ramsay NK, Swift P, Shimada H, Black CT, Brodeur GM, Gerbing RB, Reynolds CP. Treatment of high-risk neuroblastoma with intensive chemotherapy, radiotherapy, autologous bone marrow transplantation, and 13-cis-retinoic acid. Children's Cancer Group. N Engl J Med 1999;341:1165–73.

[5] Pearson AD, Pinkerton CR, Lewis IJ, Imeson J, Ellershaw C, Machin D, European Neuroblastoma Study Group, Children's Cancer and Leukaemia Group (CCLG formerly United Kingdom Children's Cancer Study Group). High-dose rapid and standard induction chemotherapy for patients aged over 1 year with stage 4 neuroblastoma: a randomised trial. Lancet Oncol 2008;9:247–56.

[6] Trochet D, Bourdeaut F, Janoueix-Lerosey I, Deville A, de Pontual L, Schleiermacher G, Coze C, Philip N, Frebourg T, Munnich A, Lyonnet S, Delattre O, Amiel J. Germline mutations of the paired-like homeobox 2B (PHOX2B) gene in neuroblastoma. Am J Hum Genet 2004;74:761–4.

[7] Janoueix-Lerosey I, Lequin D, Brugieres L, Ribeiro A, de Pontual L, Combaret V, Raynal V, Puisieux A, Schleiermacher G, Pierron G, Valteau-Couanet D, Frebourg T, Michon J, Lyonnet S, Amiel J, Delattre O. Somatic and germline activating mutations of the ALK kinase receptor in neuroblastoma. Nature 2008;455:967–70.

[8] Schleiermacher G, Janoueix-Lerosey I, Delattre O. Recent insights into the biology of neuroblastoma. Int J Cancer 2014;135:2249–61.

[9] Campbell K, Gastier-Foster JM, Mann M, Naranjo AH, Van Ryn C, Bagatell R, Matthay KK, London WB, Irwin MS, Shimada H, Granger MM, Hogarty MD, Park JR, DuBois SG. Association of MYCN copy number with clinical features, tumor biology, and outcomes in neuroblastoma: a report from the Children's Oncology Group. Cancer 2017;123:4224–35.

[10] Chmielecki J, Bailey M, He J, Elvin J, Vergilio JA, Ramkissoon S, Suh J, Frampton GM, Sun JX, Morley S, Spritz D, Ali S, Gay L, Erlich RL, Ross JS, Buxhaku J, Davies H, Faso V, Germain A, Glanville B, Miller VA, Stephens PJ, Janeway KA, Maris JM, Meshinchi S, Pugh TJ, Shern JF, Lipson D. Genomic profiling of a large set of diverse pediatric cancers identifies known and novel mutations across tumor spectra. Cancer Res 2017;77:509–19.

[11] Mosse YP, Laudenslager M, Longo L, Cole KA, Wood A, Attiyeh EF, Laquaglia MJ, Sennett R, Lynch JE, Perri P, Laureys G, Speleman F, Kim C, Hou C, Hakonarson H, Torkamani A, Schork NJ, Brodeur GM, Tonini GP, Rappaport E, Devoto M, Maris JM. Identification of ALK as a major familial neuroblastoma predisposition gene. Nature 2008;455:930–5.

[12] Chen Y, Takita J, Choi YL, Kato M, Ohira M, Sanada M, Wang L, Soda M, Kikuchi A, Igarashi T, Nakagawara A, Hayashi Y, Mano H, Ogawa S. Oncogenic mutations of ALK kinase in neuroblastoma. Nature 2008;455:971–4.

[13] George RE, Sanda T, Hanna M, Frohling S, Luther W, 2nd JZ, Ahn Y, Zhou W, London WB, McGrady P, Xue L, Zozulya S, Gregor VE, Webb TR, Gray NS, Gilliland DG, Diller L, Greulich H, Morris SW, Meyerson M, Look AT. Activating mutations in ALK provide a therapeutic target in neuroblastoma. Nature 2008;455:975–8.

[14] Bresler SC, Weiser DA, Huwe PJ, Park JH, Krytska K, Ryles H, Laudenslager M, Rappaport EF, Wood AC, McGrady PW, Hogarty MD, London WB, Radhakrishnan R, Lemmon MA, Mosse YP. ALK mutations confer differential oncogenic activation and sensitivity to ALK inhibition therapy in neuroblastoma. Cancer Cell 2014;26:682–94.

[15] Cazes A, Louis-Brennetot C, Mazot P, Dingli F, Lombard B, Boeva V, Daveau R, Cappo J, Combaret V, Schleiermacher G, Jouannet S, Ferrand S, Pierron G, Barillot E, Loew D, Vigny M, Delattre O, Janoueix-Lerosey I. Characterization of rearrangements involving the ALK gene reveals a novel truncated form associated with tumor aggressiveness in neuroblastoma. Cancer Res 2013;73:195–204.

[16] Okubo J, Takita J, Chen Y, Oki K, Nishimura R, Kato M, Sanada M, Hiwatari M, Hayashi Y, Igarashi T, Ogawa S. Aberrant activation of ALK kinase by a novel truncated form ALK protein in neuroblastoma. Oncogene 2012;31:4667–76.

[17] Maris JM, Weiss MJ, Mosse Y, Hii G, Guo C, White PS, Hogarty MD, Mirensky T, Brodeur GM, Rebbeck TR, Urbanek M, Shusterman S. Evidence for a hereditary neuroblastoma predisposition locus at chromosome 16p12-13. Cancer Res 2002;62:6651–8.

[18] Pugh TJ, Morozova O, Attiyeh EF, Asgharzadeh S, Wei JS, Auclair D, Carter SL, Cibulskis K, Hanna M, Kiezun A, Kim J, Lawrence MS, Lichenstein L, McKenna A, Pedamallu CS, Ramos AH, Shefler E, Sivachenko A, Sougnez C, Stewart C, Ally A, Birol I, Chiu R, Corbett RD, Hirst M, Jackman SD, Kamoh B, Khodabakshi AH, Krzywinski M, Lo A, Moore RA, Mungall KL, Qian J, Tam A, Thiessen N, Zhao Y, Cole KA, Diamond M, Diskin SJ, Mosse YP, Wood AC, Ji L, Sposto R, Badgett T, London WB, Moyer Y, Gastier-Foster JM, Smith MA, Guidry Auvil JM, Gerhard DS, Hogarty MD, Jones SJ, Lander ES, Gabriel SB, Getz G, Seeger RC, Khan J, Marra MA, Meyerson M, Maris JM. The genetic landscape of high-risk neuroblastoma. Nat Genet 2013;45:279–84.

[19] De Brouwer S, De Preter K, Kumps C, Zabrocki P, Porcu M, Westerhout EM, Lakeman A, Vandesompele J, Hoebeeck J, Van Maerken T, De Paepe A, Laureys G, Schulte JH, Schramm A, Van Den Broecke C, Vermeulen J, Van Roy N, Beiske K, Renard M, Noguera R, Delattre O, Janoueix-Lerosey I, Kogner P, Martinsson T, Nakagawara A, Ohira M, Caron H, Eggert A, Cools J, Versteeg R, Speleman F. Meta-analysis of neuroblastomas reveals a skewed ALK mutation spectrum in tumors with MYCN amplification. Clin Cancer Res 2010;16: 4353–62.

[20] Schleiermacher G, Javanmardi N, Bernard V, Leroy Q, Cappo J, Rio Frio T, Pierron G, Lapouble E, Combaret V, Speleman F, de Wilde B, Djos A, Ora I, Hedborg F, Trager C, Holmqvist BM, Abrahamsson J, Peuchmaur M, Michon J, Janoueix-Lerosey I, Kogner P, Delattre O, Martinsson T. Emergence of new ALK mutations at relapse of neuroblastoma. J Clin Oncol 2014;32:2727–34.

[21] Eleveld TF, Oldridge DA, Bernard V, Koster J, Colmet Daage L, Diskin SJ, Schild L, Bentahar NB, Bellini A, Chicard M, Lapouble E, Combaret V, Legoix-Ne P, Michon J, Pugh TJ, Hart LS, Rader J, Attiyeh EF, Wei JS, Zhang S, Naranjo A, Gastier-Foster JM, Hogarty MD, Asgharzadeh S, Smith MA, Guidry Auvil JM, Watkins TB, Zwijnenburg DA, Ebus ME, van Sluis P, Hakkert A, van Wezel E, van der Schoot CE, Westerhout EM, Schulte JH, Tytgat GA, Dolman ME, Janoueix-Lerosey I, Gerhard DS, Caron HN, Delattre O, Khan J, Versteeg R, Schleiermacher G, Molenaar JJ, Maris JM. Relapsed neuroblastomas show frequent RAS-MAPK pathway mutations. Nat Genet 2015;47:864–71.

[22] Martinsson T, Eriksson T, Abrahamsson J, Caren H, Hansson M, Kogner P, Kamaraj S, Schonherr C, Weinmar J, Ruuth K, Palmer RH, Hallberg B. Appearance of the novel activating F1174S ALK mutation in neuroblastoma correlates with aggressive tumor progression and unresponsiveness to therapy. Cancer Res 2011;71:98–105.

[23] Souttou B, Carvalho NB, Raulais D, Vigny M. Activation of anaplastic lymphoma kinase receptor tyrosine kinase induces neuronal differentiation through the mitogen-activated protein kinase pathway. J Biol Chem 2001;276:9526–31.

[24] Schonherr C, Ruuth K, Kamaraj S, Wang CL, Yang HL, Combaret V, Djos A, Martinsson T, Christensen JG, Palmer RH, Hallberg B. Anaplastic lymphoma kinase (ALK) regulates initiation of transcription of MYCN in neuroblastoma cells. Oncogene 2012;31:5193–200.

[25] Berry T, Luther W, Bhatnagar N, Jamin Y, Poon E, Sanda T, Pei D, Sharma B, Vetharoy WR, Hallsworth A, Ahmad Z, Barker K, Moreau L, Webber H, Wang W, Liu Q, Perez-Atayde A, Rodig S, Cheung NK, Raynaud F, Hallberg B, Robinson SP, Gray NS, Pearson AD, Eccles SA, Chesler L, George RE. The ALK(F1174L) mutation potentiates the oncogenic activity of MYCN in neuroblastoma. Cancer Cell 2012;22:117–30.

[26] Zamo A, Chiarle R, Piva R, Howes J, Fan Y, Chilosi M, Levy DE, Inghirami G. Anaplastic lymphoma kinase (ALK) activates Stat3 and protects hematopoietic cells from cell death. Oncogene 2002;21:1038–47.

[27] Rodig SJ, Shapiro GI. Crizotinib, a small-molecule dual inhibitor of the c-Met and ALK receptor tyrosine kinases. Curr Opin Investig Drugs 2010;11:1477–90.

[28] Cui JJ, Tran-Dube M, Shen H, Nambu M, Kung PP, Pairish M, Jia L, Meng J, Funk L, Botrous I, McTigue M, Grodsky N, Ryan K, Padrique E, Alton G, Timofeevski S, Yamazaki S, Li Q, Zou H, Christensen J, Mroczkowski B, Bender S, Kania RS, Edwards MP. Structure based drug design of crizotinib (PF-02341066), a potent and selective dual inhibitor of mesenchymal-epithelial transition factor (c-MET) kinase and anaplastic lymphoma kinase (ALK). J Med Chem 2011;54:6342–63.

[29] Azarova AM, Gautam G, George RE. Emerging importance of ALK in neuroblastoma. Semin Cancer Biol 2011;21:267–75.

[30] Bresler SC, Wood AC, Haglund EA, Courtright J, Belcastro LT, Plegaria JS, Cole K, Toporovskaya Y, Zhao H, Carpenter EL, Christensen JG, Maris JM, Lemmon MA, Mosse YP. Differential inhibitor sensitivity of anaplastic lymphoma kinase variants found in neuroblastoma. Sci Transl Med 2011;3, 108ra114.

[31] Mosse YP, Lim MS, Voss SD, Wilner K, Ruffner K, Laliberte J, Rolland D, Balis FM, Maris JM, Weigel BJ, Ingle AM, Ahern C, Adamson PC, Blaney SM. Safety and activity of crizotinib for paediatric patients with refractory solid tumours or anaplastic large-cell lymphoma: a Children's Oncology Group phase 1 consortium study. Lancet Oncol 2013;14:472–80.

[32] Mosse YP, Deyell RJ, Berthold F, Nagakawara A, Ambros PF, Monclair T, Cohn SL, Pearson AD, London WB, Matthay KK. Neuroblastoma in older children, adolescents and young adults: a report from the International Neuroblastoma Risk Group project. Pediatr Blood Cancer 2014;61:627–35.

[33] Tucker ER, Danielson LS, Innocenti P, Chesler L. Tackling crizotinib resistance: the pathway from drug discovery to the pediatric clinic. Cancer Res 2015;75:2770–4.

[34] Infarinato NR, Park JH, Krytska K, Ryles HT, Sano R, Szigety KM, Li Y, Zou HY, Lee NV, Smeal T, Lemmon MA, Mosse YP. The ALK/ROS1 inhibitor PF-06463922 overcomes primary resistance to crizotinib in ALK-driven neuroblastoma. Cancer Discov 2016;6:96–107.

[35] Wang Y, Wang L, Guan S, Cao W, Wang H, Chen Z, Zhao Y, Yu Y, Zhang H, Pang JC, Huang SL, Akiyama Y, Yang Y, Sun W, Xu X, Shi Y, Zhang H, Kim ES, Muscal JA, Lu F, Yang J. Novel ALK inhibitor AZD3463 inhibits neuroblastoma growth by overcoming crizotinib resistance and inducing apoptosis. Sci Rep 2016;6:19423.

[36] Tchekmedyian N, Ali SM, Miller VA, Haura EB. Acquired ALK L1152R mutation confers resistance to ceritinib and predicts response to alectinib. J Thorac Oncol 2016;11:e87–8.

[37] Lu J, Guan S, Zhao Y, Yu Y, Woodfield SE, Zhang H, Yang KL, Bieerkehazhi S, Qi L, Li X, Gu J, Xu X, Jin J, Muscal JA, Yang T, Xu GT, Yang J. The second-generation ALK inhibitor alectinib effectively induces apoptosis in human neuroblastoma cells and inhibits tumor growth in a TH-MYCN transgenic neuroblastoma mouse model. Cancer Lett 2017;400:61–8.

[38] Cervantes-Madrid D, Szydzik J, Lind DE, Borenas M, Bemark M, Cui J, Palmer RH, Hallberg B. Repotrectinib (TPX-0005), effectively reduces growth of ALK driven neuroblastoma cells. Sci Rep 2019;9:19353.

[39] Mosse YP, Voss SD, Lim MS, Rolland D, Minard CG, Fox E, Adamson P, Wilner K, Blaney SM, Weigel BJ. Targeting ALK with crizotinib in pediatric anaplastic large cell lymphoma and inflammatory myofibroblastic tumor: a Children's Oncology Group Study. J Clin Oncol 2017;35:3215–21.

[40] Katayama R. Therapeutic strategies and mechanisms of drug resistance in anaplastic lymphoma kinase (ALK)-rearranged lung cancer. Pharmacol Ther 2017;177:1–8.

[41] Fukuda K, Takeuchi S, Arai S, Katayama R, Nanjo S, Tanimoto A, Nishiyama A, Nakagawa T, Taniguchi H, Suzuki T, Yamada T, Nishihara H, Ninomiya H, Ishikawa Y, Baba S, Takeuchi K, Horiike A, Yanagitani N, Nishio M, Yano S. Epithelial-to-mesenchymal transition is a mechanism of ALK inhibitor resistance in lung cancer independent of ALK mutation status. Cancer Res 2019;79:1658–70.

[42] Katayama R, Shaw AT, Khan TM, Mino-Kenudson M, Solomon BJ, Halmos B, Jessop NA, Wain JC, Yeo AT, Benes C, Drew L, Saeh JC, Crosby K, Sequist LV, Iafrate AJ, Engelman JA. Mechanisms of acquired crizotinib resistance in ALK-rearranged lung cancers. Sci Transl Med 2012;4, 120ra117.

[43] Tanizaki J, Okamoto I, Okabe T, Sakai K, Tanaka K, Hayashi H, Kaneda H, Takezawa K, Kuwata K, Yamaguchi H, Hatashita E, Nishio K, Nakagawa K. Activation of HER family signaling as a mechanism of acquired resistance to ALK inhibitors in EML4-ALK-positive non-small cell lung cancer. Clin Cancer Res 2012;18:6219–26.

[44] Ashraf K, Shaikh F, Gibson P, Baruchel S, Irwin MS. Treatment with topotecan plus cyclophosphamide in children with first relapse of neuroblastoma. Pediatr Blood Cancer 2013;60:1636–41.

[45] London WB, Frantz CN, Campbell LA, Seeger RC, Brumback BA, Cohn SL, Matthay KK, Castleberry RP, Diller L. Phase II randomized comparison of topotecan plus cyclophosphamide versus topotecan alone in children with recurrent or refractory neuroblastoma: a Children's Oncology Group Study. J Clin Oncol 2010;28:3808–15.

[46] Park JR, Scott JR, Stewart CF, London WB, Naranjo A, Santana VM, Shaw PJ, Cohn SL, Matthay KK. Pilot induction regimen incorporating pharmacokinetically guided topotecan for treatment of newly diagnosed high-risk neuroblastoma: a Children's Oncology Group Study. J Clin Oncol 2011;29:4351–7.

[47] Krytska K, Ryles HT, Sano R, Raman P, Infarinato NR, Hansel TD, Makena MR, Song MM, Reynolds CP, Mosse YP. Crizotinib synergizes with chemotherapy in preclinical models of neuroblastoma. Clin Cancer Res 2016;22:948–60.

[48] Zhang L, Wu B, Baruchel S. Oral metronomic topotecan sensitizes crizotinib antitumor activity in ALKF1174L drug-resistant neuroblastoma preclinical models. Transl Oncol 2017;10:604–11.

[49] Acker T, Plate KH. Hypoxia and hypoxia inducible factors (HIF) as important regulators of tumor physiology. Cancer Treat Res 2004;117:219–48.

[50] Semenza GL. HIF-1: mediator of physiological and pathophysiological responses to hypoxia. J Appl Physiol 2000;88:1474–80.

[51] Timur M, Akbas SH, Ozben T. The effect of Topotecan on oxidative stress in MCF-7 human breast cancer cell line. Acta Biochim Pol 2005;52:897–902.

[52] Beppu K, Nakamura K, Linehan WM, Rapisarda A, Thiele CJ. Topotecan blocks hypoxia-inducible factor-1alpha and vascular endothelial growth factor expression induced by insulin-like growth factor-I in neuroblastoma cells. Cancer Res 2005;65:4775–81.

[53] Rapisarda A, Zalek J, Hollingshead M, Braunschweig T, Uranchimeg B, Bonomi CA, Borgel SD, Carter JP, Hewitt SM, Shoemaker RH, Melillo G. Schedule-dependent inhibition of hypoxia-inducible factor-1alpha protein

accumulation, angiogenesis, and tumor growth by topotecan in U251-HRE glioblastoma xenografts. Cancer Res 2004;64:6845–8.

[54] Martinengo C, Poggio T, Menotti M, Scalzo MS, Mastini C, Ambrogio C, Pellegrino E, Riera L, Piva R, Ribatti D, Pastorino F, Perri P, Ponzoni M, Wang Q, Voena C, Chiarle R. ALK-dependent control of hypoxia-inducible factors mediates tumor growth and metastasis. Cancer Res 2014;74:6094–106.

[55] Kim S, Loo A, Chopra R, Caponigro G, Huang A, Vora S, Parasuraman S, Howard S, Keen N, Sellers W, Brain C. Abstract PR02: LEE011: an orally bioavailable, selective small molecule inhibitor of CDK4/6-reactivating Rb in cancer. Mol Cancer Ther 2013;12:PR02.

[56] Rader J, Russell MR, Hart LS, Nakazawa MS, Belcastro LT, Martinez D, Li Y, Carpenter EL, Attiyeh EF, Diskin SJ, Kim S, Parasuraman S, Caponigro G, Schnepp RW, Wood AC, Pawel B, Cole KA, Maris JM. Dual CDK4/CDK6 inhibition induces cell-cycle arrest and senescence in neuroblastoma. Clin Cancer Res 2013;19:6173–82.

[57] Geoerger B, Bourdeaut F, DuBois SG, Fischer M, Geller JI, Gottardo NG, Marabelle A, Pearson ADJ, Modak S, Cash T, Robinson GW, Motta M, Matano A, Bhansali SG, Dobson JR, Parasuraman S, Chi SN. A phase I study of the CDK4/6 inhibitor Ribociclib (LEE011) in pediatric patients with malignant rhabdoid tumors, neuroblastoma, and other solid tumors. Clin Cancer Res 2017;23:2433–41.

[58] Wood AC, Krytska K, Ryles HT, Infarinato NR, Sano R, Hansel TD, Hart LS, King FJ, Smith TR, Ainscow E, Grandinetti KB, Tuntland T, Kim S, Caponigro G, He YQ, Krupa S, Li N, Harris JL, Mosse YP. Dual ALK and CDK4/6 inhibition demonstrates synergy against neuroblastoma. Clin Cancer Res 2017;23:2856–68.

[59] Li Q, Lozano G. Molecular pathways: targeting Mdm2 and Mdm4 in cancer therapy. Clin Cancer Res 2013;19:34–41.

[60] Wang HQ, Halilovic E, Li X, Liang J, Cao Y, Rakiec DP, Ruddy DA, Jeay S, Wuerthner JU, Timple N, Kasibhatla S, Li N, Williams JA, Sellers WR, Huang A, Li F. Combined ALK and MDM2 inhibition increases antitumor activity and overcomes resistance in human ALK mutant neuroblastoma cell lines and xenograft models. elife 2017;6.

[61] Janoueix-Lerosey I, Schleiermacher G, Michels E, Mosseri V, Ribeiro A, Lequin D, Vermeulen J, Couturier J, Peuchmaur M, Valent A, Plantaz D, Rubie H, Valteau-Couanet D, Thomas C, Combaret V, Rousseau R, Eggert A, Michon J, Speleman F, Delattre O. Overall genomic pattern is a predictor of outcome in neuroblastoma. J Clin Oncol 2009;27:1026–33.

[62] Chesler L, Schlieve C, Goldenberg DD, Kenney A, Kim G, McMillan A, Matthay KK, Rowitch D, Weiss WA. Inhibition of phosphatidylinositol 3-kinase destabilizes Mycn protein and blocks malignant progression in neuroblastoma. Cancer Res 2006;66:8139–46.

[63] Zhu S, Lee JS, Guo F, Shin J, Perez-Atayde AR, Kutok JL, Rodig SJ, Neuberg DS, Helman D, Feng H, Stewart RA, Wang W, George RE, Kanki JP, Look AT. Activated ALK collaborates with MYCN in neuroblastoma pathogenesis. Cancer Cell 2012;21:362–73.

[64] Heukamp LC, Thor T, Schramm A, De Preter K, Kumps C, De Wilde B, Odersky A, Peifer M, Lindner S, Spruessel A, Pattyn F, Mestdagh P, Menten B, Kuhfittig-Kulle S, Kunkele A, Konig K, Meder L, Chatterjee S, Ullrich RT, Schulte S, Vandesompele J, Speleman F, Buttner R, Eggert A, Schulte JH. Targeted expression of mutated ALK induces neuroblastoma in transgenic mice. Sci Transl Med 2012;4, 141ra191.

[65] Prochownik EV, Vogt PK. Therapeutic targeting of Myc. Genes Cancer 2010;1:650–9.

[66] Moore NF, Azarova AM, Bhatnagar N, Ross KN, Drake LE, Frumm S, Liu QS, Christie AL, Sanda T, Chesler L, Kung AL, Gray NS, Stegmaier K, George RE. Molecular rationale for the use of PI3K/AKT/mTOR pathway inhibitors in combination with crizotinib in ALK-mutated neuroblastoma. Oncotarget 2014;5:8737–49.

[67] Hrustanovic G, Bivona TG. RAS-MAPK in ALK targeted therapy resistance. Cell Cycle 2015;14:3661–2.

[68] Hrustanovic G, Bivona TG. RAS-MAPK signaling influences the efficacy of ALK-targeting agents in lung cancer. Mol Cell Oncol 2016;3, e1091061.

[69] Umapathy G, El Wakil A, Witek B, Chesler L, Danielson L, Deng X, Gray NS, Johansson M, Kvarnbrink S, Ruuth K, Schonherr C, Palmer RH, Hallberg B. The kinase ALK stimulates the kinase ERK5 to promote the expression of the oncogene MYCN in neuroblastoma. Sci Signal 2014;7:ra102.

[70] Yang L, Li G, Zhao L, Pan F, Qiang J, Han S. Blocking the PI3K pathway enhances the efficacy of ALK-targeted therapy in EML4-ALK-positive nonsmall-cell lung cancer. Tumour Biol 2014;35:9759–67.

[71] Wang X, Tournier C. Regulation of cellular functions by the ERK5 signalling pathway. Cell Signal 2006;18:753–60.

[72] Anon. ERK5 is a potential therapeutic target in ALK-positive neuroblastoma. Cancer Discov 2014;4:1363.

[73] Lambertz I, Kumps C, Claeys S, Lindner S, Beckers A, Janssens E, Carter DR, Cazes A, Cheung BB, De Mariano M, De Bondt A, De Brouwer S, Delattre O, Gibbons J, Janoueix-Lerosey I, Laureys G, Liang C, Marchall GM, Porcu M, Takita J, Trujillo DC, Van Den Wyngaert I, Van Roy N, Van Goethem A, Van Maerken T, Zabrocki P, Cools J, Schulte JH, Vialard J, Speleman F, De Preter K. Upregulation of MAPK negative feedback regulators and RET in mutant ALK neuroblastoma: implications for targeted treatment. Clin Cancer Res 2015;21:3327–39.

[74] Cazes A, Lopez-Delisle L, Tsarovina K, Pierre-Eugene C, De Preter K, Peuchmaur M, Nicolas A, Provost C, Louis-Brennetot C, Daveau R, Kumps C, Cascone I, Schleiermacher G, Prignon A, Speleman F, Rohrer H, Delattre O, Janoueix-Lerosey I. Activated Alk triggers prolonged neurogenesis and Ret upregulation providing a therapeutic target in ALK-mutated neuroblastoma. Oncotarget 2014;5:2688–702.

[75] Van den Eynden J, Umapathy G, Ashouri A, Cervantes-Madrid D, Szydzik J, Ruuth K, Koster J, Larsson E, Guan J, Palmer RH, Hallberg B. Phosphoproteome and gene expression profiling of ALK inhibition in neuroblastoma cell lines reveals conserved oncogenic pathways. Sci Signal 2018;11.

[76] Trigg RM, Lee LC, Prokoph N, Jahangiri L, Reynolds CP, Amos Burke GA, Probst NA, Han M, Matthews JD, Lim HK, Manners E, Martinez S, Pastor J, Blanco-Aparicio C, Merkel O, de Los Fayos Alonso IG, Kodajova P, Tangermann S, Hogler S, Luo J, Kenner L, Turner SD. The targetable kinase PIM1 drives ALK inhibitor resistance in high-risk neuroblastoma independent of MYCN status. Nat Commun 2019;10:5428.

State of the art and future perspectives

Francesco Facchinetti and Luc Friboulet

Predictive Biomarkers and Novel Therapeutic Strategies in Oncology, INSERM U981, Gustave Roussy Cancer Campus, Paris-Saclay University, Villejuif, France

Abstract

The clinical success of ALK inhibition through targeted therapies is manifestly evident in ALK-rearranged non-small cell lung cancer (NSCLC). Deciphering resistance mechanisms to targeted agents in ALK-positive NSCLC has been crucial to develop novel inhibitors and novel treatment strategies, with a continued and reciprocal exchange of evidence between clinical and preclinical research. With the advent of the third-generation ALK inhibitor lorlatinib in the clinical setting, as well as with the availability of second-generation agents (e.g., alectinib) in the upfront treatment of ALK-rearranged NSCLC, novel scenarios of resistance have recently emerged. In parallel, moving ALK inhibition in disease setting earlier than the advanced/metastatic one would likely demand new concepts and methodologies.

Whereas the evidence of resistance mechanisms to ALK inhibitors are scarce in ALK-dependent tumors other than lung cancer, biological and clinical specificities can be recognized. The relevant results obtained in NSCLC should serve as an example for other ALK-driven diseases.

Abbreviations

ALK	anaplastic lymphoma kinase
NSCLC	non-small cell lung cancer
ACLC	anaplastic large cell lymphoma
IMT	inflammatory myofibroblastic tumor
TKI	tyrosine kinase inhibitor
EGFR	epidermal growth factor receptor
PFS	progression-free survival
OS	overall survival
HR	hazard ratio
CI	confidence interval
NR	not reached
CNS	central nervous system
ctDNA	circulating tumor DNA
PD-L1	programmed death-ligand1
CT	computed tomography
PET	positron emission tomography
HSCT	hematopoietic stem cell transplantation

Conflict of interest

Francesco Facchinetti participated to editorial activities sponsored by BMS and Roche. Luc Friboulet has no conflict of interest to disclose.

Introduction

Over the last 25 years, the understanding of the biological underpinnings of the *ALK* oncogene has led to parallel major clinical improvements for patients affected by ALK-driven cancers.

Firstly, the identification of ALK rearrangements and mutations allowed to identify specific pathological-molecular entities across tumor types. This is the case for ALK-rearranged non-small cell lung cancer (NSCLC), anaplastic large cell lymphoma (ACLC), inflammatory myofibroblastic tumor (IMT) and ALK-mutant neuroblastoma. Of note, taking into account the detection of activating events involving ALK in additional malignancies at lower frequencies, virtually every solid tumor type can be characterized by a subset driven by ALK. ALK-positive diseases can be globally defined as rare molecular entities of common-to-rare cancers (NSCLC, lymphomas, pediatric malignancies and other), acquiring a relevant epidemiologic significance if taken together [1]. With the broad diffusion of next-generation sequencing techniques, hopefully no cancer harboring an ALK-driven event should be missed. Even more importantly, the identification of ALK driver events holds clinical and therapeutic repercussions in every single patient, as an eminent example of precision medicine.

Secondly indeed, the intrinsic physiological nature of ALK as a receptor tyrosine kinase, together with its constitutive activation in case of molecular oncogenic events, boosted the development of specific tyrosine kinase inhibitors (TKIs). The clinical results obtained since the introduction of these targeted agents are recapitulated by the outstanding achievements observed in patients suffering from ALK-rearranged, advanced/metastatic NSCLC. With regard to these patients, the quest for longer survival outcomes has been made possible by the rapid availability of several ALK-TKIs and by novel strategies regulating the timing of their administration (sequential versus more-potent TKI upfront) [2]. The mentioned ALK-driven tumors other than advanced NSCLC are characterized by their predominance in younger ages and by potential curability (see Chapter 4). In this setting, the integration of ALK inhibitors into the treatment armamentarium should aim to increase the rate of cure, with reduced long-term toxicities.

As approached in the previous Sections, deciphering the ways ALK-driven tumors escape to targeted agents is the pillar for achieving the best outcomes for today's and tomorrow patients, with a continuous and reciprocal interaction between preclinical research and clinical practice (Fig. 1).

In this last Chapter, we resume the current state of the art, the benchmark for further developments and applications in the field of ALK resistance mechanisms.

Dealing with resistance in ALK-rearranged NSCLC: Actualities and perspectives

After the unrevealing of EGFR activating mutations as the responsible for major responses to EGFR-TKIs [3,4], ALK rearrangements are the second targetable driver events in NSCLC,

Evolution of ALK inhibition and resistance mechanisms discovery

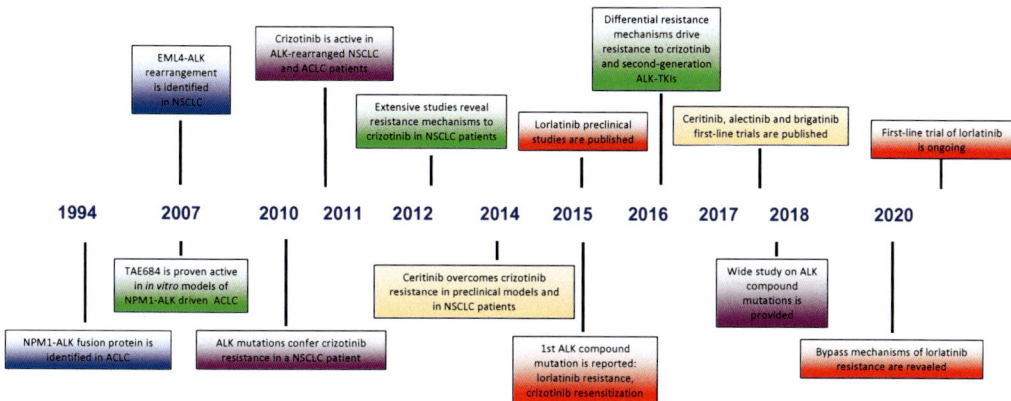

FIG. 1 Timeline of major discovery in the field of ALK inhibition and research of resistance mechanisms. *ACLC*, anaplastic large cell lymphoma; *NSCLC*, non-small cell lung cancer; *TKIs*, tyrosine kinase inhibitors.

both in terms of their frequency and the chronological availability of active ALK-TKIs [5]. Considering the impressive survival outcomes observed in ALK-positive advanced NSCLC patients, this subset of disease represents, thus far, the most successful model of targeted therapy at least in lung cancer, outdoing the initial successes of EGFR inhibition and representing a model for other molecular entities (e.g., ROS1, BRAF, MET, RET, HER2, NTRK). Since the clinical availability of crizotinib, huge scientific efforts have been in parallel addressed to identify resistance mechanisms. Sequential biopsies performed at the progression on ALK-TKIs allowed to detect genomic events putatively in cause of resistance [6]. The functional studies in engineered cellular models (e.g., Ba/F3), the establishment of primary cell cultures and in vivo experiments, validating genomic analyses insights and providing independent evidence, represent the cornerstone of ALK resistance research [7–13]. These elements are moreover the bases for the virtuous harmony between clinical and preclinical research aiming to improve patients' outcomes, making of ALK the best example in the oncology field.

The global achievements in the field of ALK-rearranged NSCLC are recapitulated by the updates of PROFILE-1014 and ALEX trials (see Chapter 2) [14,15], from which we take a cue to provide a global vision of the study of resistance in ALK-positive cancers.

State of the art in sequential treatment strategy: From clinical results to resistance mechanisms

In PROFILE-1014, the first-generation inhibitor crizotinib was compared to platinum/pemetrexed chemotherapy in the first-line setting [16]. After having demonstrated its positivity satisfying its primary objective progression-free survival (PFS), the study resulted informative providing overall survival (OS) data. Formally, a trend for longer OS was observed for crizotinib experimental arm compared to chemotherapy [hazard ratio (HR) 0.760; 95% confidence interval (CI), 0.548–1.053; $P = .0978$], with OS medians not reached (NR; 95% CI: 45.8 months-NR) and 47.5 months (95% CI: 32.2-NR), respectively [14]. This is in line with other trials of targeted agents compared to standard-of-care chemotherapy, when crossover to the targeted therapy is allowed at progression to cytotoxic treatment [17,18]. When virtually adjusting for crossover indeed (modeling the lack of cross-over to crizotinib from the chemotherapy arm and vice-versa), OS resulted significantly better in the experimental arm [HR 0.346; 95% bootstrap CI: 0.081–0.718; median OS 59.8 (46.6-NR) and 19.2 (13.6-NR) months, respectively [14]. Besides reinforcing the role of ALK-TKIs in the first-line setting (already incorporated into guidelines), the trial update was informative for other aspects of interest. As envisaged and confirmed by the data, the administration of post-study novel-generation TKIs dramatically impacted on survival, with better results observed for patients in the crizotinib arm receiving subsequent targeted agents. This represents indeed the sequential strategy of ALK-TKIs adopted in the clinical practice, made possible by the availability of several generation of drugs developed on the basis of preclinical/translational evidence on resistance mechanisms [19].

All second- (ceritinib, alectinib, brigatinib) and third-generation (lorlatinib) approved ALK inhibitors share indeed four peculiar characteristics making them more performant compared to the previous generation:

(1) Higher potency in inhibiting ALK allowing the complete abrogation of its direct signaling, useful in the case of bypass/downstream pathway activation, not sufficient per se of maintaining cell viability, but still requiring a contribution from ALK signaling.
(2) Activity against the largest spectrum of ALK resistance mutations.
(3) Pharmacokinetics properties in terms of penetration of the blood-brain-barrier, with regard to the frequent CNS (central nervous system) localizations of ALK-rearranged disease, both at diagnosis and at progression to early-generation ALK-TKIs [20,21].
(4) Increased specificity for ALK, in order to avoid off-target adverse events.

These features, progressively incorporated in the second-generation ALK-TKIs, are recapitulated at their higher degree in the third-generation inhibitor lorlatinib [9]. CNS progression represented a major issue when only first- or second-generation inhibitors were available. With its impressive blood-brain penetration, as its concentration in the cerebrospinal fluid is the 75% of that in the blood, lorlatinib allows a relevant control of brain disease and "relegates" CNS progression after the extra-CNS, systemic one [22,23]. Dealing with this latter, two major molecular responsible have been identified.

First, bypass/downstream pathway activations are responsible for both primary and acquired resistance to lorlatinib [11,13]. If their role in acquired resistance can be easily transposed from other inhibitors and targets, de novo resistance requires some peculiar understanding. If the onset of a bypass track is the responsible for resistance to a second-generation inhibitor, the higher potency against ALK exerted by lorlatinib cannot be useful, as direct ALK signaling has already been completely abrogated by the second-generation TKI itself.

Second, cancer cells find the way to acquire new ALK mutations able to elude lorlatinib inhibition, still acknowledging that the third-generation TKI is active against all known single mutants (including the recalcitrant G1202R) [11–13]. Both in vitro experiments and in clinical samples, cancer cells and patients exposed of sequential generations of ALK-TKIs can indeed develop complex mutations, named compound mutations, occurring on the same ALK allele modifying the structure of the ALK-rearranged receptor, preventing lorlatinib binding and activity. Of major interest, the occurrence of compound mutations in in vitro and in vivo models responsible for lorlatinib resistance seems to require exposure to first- and second-generation inhibitors, as a pre-existing single mutation is the prerequisite to develop additional ones [11].

Dealing with resistance to lorlatinib caused by bypass/downstream pathways activation, the putative combination of lorlatinib with an additional inhibitor is meant to be the strategy of choice. Nevertheless, despite increasing preclinical validation [13], such combinatorial treatments have not really taken place yet in the clinics thus far. Identifying specific, recurrent alterations in patient-derived samples at lorlatinib progression could hopefully boost the development of such strategies, as seen for EGFR-driven NSCLC [24–26]. The implementation of circulating tumor DNA (ctDNA) in ALK-rearranged diseases, already useful in detecting simple and compound mutations involving ALK, will ostensibly power the recognition of such molecular events [27].

With regard to compound mutations, only anecdotical reports have witnessed the potential re-sensitization to crizotinib exerted by ALK compound mutants (namely the ones containing L1198F mutation) engendering resistance lorlatinib and to second-generation ALK TKIs [28]. Developing a fourth-generation inhibitor, capable to act against variety of compound mutants potentially emerging during lorlatinib treatment, appears to be a relevant challenge.

This brings us to consider the other treatment strategy currently adopted in the clinics (novel generation ALK-TKIs upfront) and its implications in approaching ALK resistance.

State of the art in new generation-upfront strategy: Improving clinical results, waiting for evidence on resistance mechanisms

As seen for the EGFR-mutant NSCLC, moving novel-generation ALK-TKIs upfront fulfills two major goals: to make ALK-positive NSCLC patients live longer and with a reduced toxicity burden [29]. When administered as the first-line treatment, as expected, the three second-generation inhibitors ceritinib, alectinib and brigatinib achieved longer PFS compared to crizotinib [15,30,31]. Taking into account the current wide upfront administration of alectinib in the clinical practice, the careful methodology used (with particular regard to the evaluation of CNS disease at screening and systematically during the study), and the recent update, ALEX trial comparing alectinib with crizotinib is the best example of the benefits of this treatment strategy [15,20].

After having set a median PFS at 34.8 months (versus 10.9 in the crizotinib arm), the estimations on OS, exploratory and still immature, are impressive, as median OS was NR and 57.4 months (95% CI 34.6–NR) with alectinib and crizotinib, turning out in a 5-year survival rate of 62.5% and (95% CI: 54.3–70.8) and 45.5% (95% CI: 33.6–57.4), respectively [15]. Despite no cross-over to alectinib was formally allowed within the study to patients progressing on crizotinib, the majority of them had access to second-line targeted agents, recapitulating a sequential treatment strategy. Of interest, dealing with chronic treatments, the toxicity spectrum was in favor of alectinib with regard both to mild and serious adverse events. In addition to these observations, alectinib features of action against and prevention of CNS metastases vouch for its administration as first-line treatment in ALK-positive NSCLC patients [32,33].

Acknowledging these utmost results, no evidence is thus far available with regard to the resistance mechanisms to alectinib in the upfront setting. It can be envisaged that the repartition between on-target (i.e., ALK mutations) and off-target (i.e., bypass/downstream pathways activation) mechanisms may recapitulate the one observed when second-generation

TKIs are administered after crizotinib, with an enrichment in the first ones (namely with regard to G1202R mutation) [10]. This knowledge will put the bases to assess lorlatinib sensitivity and resistance in this new setting (see previous paragraph).

Another tempting strategy is to directly move the "best-in-class" of ALK-TKI, lorlatinib, directly upfront, as currently evaluated in the phase III CROWN trial (NCT03052608). Besides providing a formal comparison between the third-generation inhibitor and crizotinib, this study will allow to estimate the PFS of first-line lorlatinib. With regard to resistance mechanisms awaited, we take advantage of both clinical evidence from the pretreated setting and from preclinical work. CNS progression, relatively rare when lorlatinib is administered within a sequential treatment approach [23], appears even less probable in mouse models of brain metastases when given upfront [9]. Similarly, no compound mutation is expected to raise in the lack of a pre-lorlatinib single ALK mutant [11]. Therefore, if our expectations are correct, activation of bypass/downstream tracks will likely be identified as major players in lorlatinib resistance. Countermoves in terms of combinatorial treatments development would then be a putative solution to overcome (or to prevent) lorlatinib resistance.

Moving ALK inhibitions in NSCLC settings other than advanced disease: Implications for resistance

In the last years, targeted agents have shown their role in early stages, as EGFR-TKIs as adjuvant treatment in EGFR-driven lung cancer [34–36]. The implementation of ALK-TKIs in the adjuvant setting is still exploratory, but the positive results observed in oncogene-driven disease sustain this possibility, also considering that the administration of a brain-penetrating ALK inhibitor could prevent or at least delay the occurrence of CNS relapses, compared to chemotherapy, after the surgical removal of the tumor. Dealing with locally-advanced NSCLC potentially suitable of radical cure, several case reports have reported the potential role of ALK-TKIs as neoadjuvant treatment before surgery [37–41], and also induction treatment with such inhibitors may be envisaged in order to reduce tumor burden before (chemo)radiotherapy administration. These observations will require a systematic evaluation of the ALK status (thus far limited to the advanced/metastatic NSCLC setting) even in resectable and locally-advanced stages. Of note, these two are setting of current peculiar interest: the first, as the spread of lung cancer screening programs will hopefully move the NSCLC epidemiology towards more early stages [42], the second as the locally-advanced diseases have recently seen the improvement since the introduction of immunotherapy consolidation with durvalumab (an anti-PD-L1 agent) [43]. Given the low activity and efficacy of immune checkpoint inhibitors in the advanced setting of oncogene-addicted diseases [44], it should be discussed if locally-advanced ALK-positive NSCLC would better benefit from a treatment strategy involving ALK TKIs.

Still dealing with advanced-metastatic NSCLC, the "oligometastatic" disease entity has recently came into the spotlight [45,46], since dedicated treatment strategies have been proposed [47,48]. The variable and complex definitions of oligometastatic NSCLC are beyond the objective of this chapter, nevertheless the application of this knowledge in ALK-driven lung cancer may foster research in the field of resistance mechanisms research. Patients suffering from ALK-rearranged NSCLC experiencing isolated progression under ALK-TKIs can benefit from local treatments (surgery, radiotherapy, interventional radiology techniques), while maintaining the same TKI, in the scenario of oligoprogressive disease [49,50]. The oligometastatic disease, on the other hand, represents advanced lung cancers whose limited number or metastases allows to envisage a local treatment of all disease sites, once systemic therapy guarantees disease control. This strategy has been mainly tested with chemotherapy agents, but its potential with targeted agents, namely ALK inhibitors, is promising.

These settings of early/locally advanced/oligometastatic ALK-rearranged NSCLC may drive the concept of resistance to new directions [51]. These clinical situations indeed imply that local treatment will be combined to ALK-TKIs in order to act locally against every macroscopic site of the disease identified at CT and PET scans. Novel clinical scenarios will therefore need to be faced to assess resistance (Fig. 2):

(1) early disease undergoing surgical resection and adjuvant ALK-TKI will stimulate the study of molecular factors on the resected tumor that could affect disease relapse.
(2) early or locally advanced disease undergoing neoadjuvant ALK-TKI will allow the availability of large surgical samples of ALK-positive tumors after targeted treatment, boosting to get insights on cancer cells which still remains vital.
(3) locally advanced NSCLC patients who undergo induction ALK-TKIs before (chemo) radiotherapy could be sequentially biopsied in order to seek differential molecular features engendered by targeted agents and radiotherapy.
(4) local treatment of oligometastic ALK-positive disease, concomitantly to (in the case of surgery) or preceded by tumor biopsy (in the case of radiotherapy or interventional radiology techniques) would allow to compare the molecular status of the different disease sites after ALK-TKI exposure.

These four clinical scenarios, still hypothetical but hopefully achievable, will share some relevant aspects of research on ALK resistance mechanisms. Indeed, thus far the molecular evidence (still extremely precious) has been developed from genomic analyses performed mainly on biopsy tumor samples obtained at progression to ALK-TKIs (Fig. 2A). Moving ALK inhibition into early stages will allow to deal with larger amounts of tumor material (surgical resections) obtained at the moment of disease response. Although counterintuitive, this approach could be strategic in obtaining insights on the prediction of following disease relapse/progression through complex molecular analyses. These latter would likely encompass genomic, epigenomic, proteomic, metabolomics analyses, likely to be connected with clinicoradiological information (e.g., disease- and progression-free survivals, risk of brain disease localizations) into deep-learning algorithms [52]. In parallel, focusing into these novel scenarios will require particular attention towards the cells of origin of potential relapse or progression ("drug-tolerant cells") [53–56], and the concept of minimal residual disease. This latter, with the refinement of ctDNA detecting techniques, will be ostensibly monitored with liquid biopsy analyses, potentially informative of the presence of micro-metastatic disease before its clinical-radiological manifestations [57,58].

State of the art and perspective on ALK-dependent malignancies other that lung cancer

Tumors other than NSCLC that are often driven by ALK (namely ACLC, IMT and neuroblastoma) share some characteristics that impact on their treatment and on the study on ALK

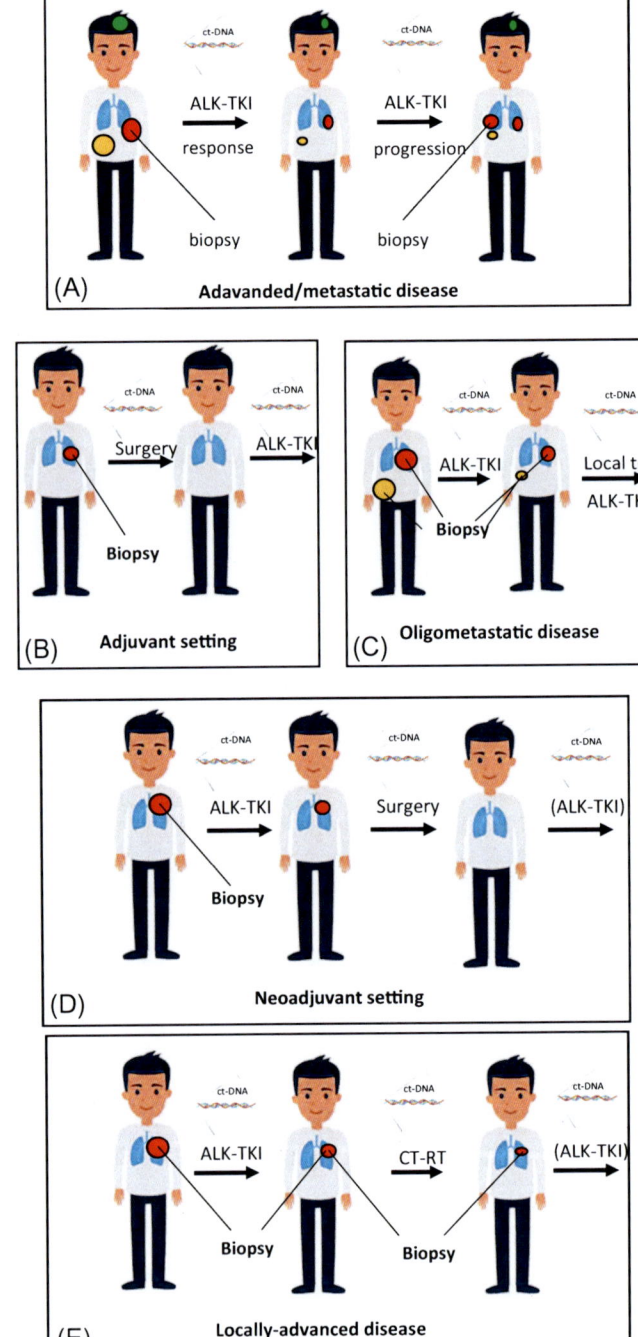

FIG. 2   Classical (A) and novel (B–E) settings of research in the field of resistance mechanisms to ALK inhibitors in non-small cell lung cancer. Biopsy at disease relapse/progresssion will be envisaged in settings B–E. *ct-DNA*, circulating tumor DNA; *ALK-TKI*, ALK-tyrosine kinase imnhibitor; *tp*, therapy; *CT-RT*, chemo-radiotherapy.

resistance mechanisms (see Chapter 4). Both for their main occurrence in pediatric patients and for their clinical features, curative intent is always the primary objective to be pursued, with a growing attention towards the reduction of long-term toxicity. The treatment backbones of the three entities are represented by "traditional" therapies: cytotoxic agents, surgery, radiotherapy and hematopoietic stem cell transplantation (HSCT). Nevertheless, as seen in previous chapters, ALK-TKIs find a role in these diseases, mainly as second-choice treatments and in combination with the mentioned treatments, providing their contributions by increasing curative rates. Moreover, compared to chemotherapy, ALK inhibitors carry an inferior burden of long-term toxicities. The lack of a wider incorporation of ALK-TKIs in these diseases can be mainly attributed to the relative success of standard therapy (70–80% of cure in pediatric ACLC) [59], to the challenging design of randomized clinical trials in such rare diseases and to the scarce interest these rare malignancies evoke in pharmaceutical industries.

Although evidence concerning resistance to ALK inhibitors in patients suffering from malignancies other than NSCLC are scarce, insights can nevertheless be driven. ACLC patients receiving ALK-TKIs inhibitors as progressing after standard chemotherapy undergo disease response, frequently complete, in the majority of the cases [60,61]. Nevertheless, once treatment with either crizotinib or ceritinib is interrupted, disease invariability recurs [62], leading to the incorporation of ALK inhibitors as bridge therapies towards HSCT. The rapid onset of disease progression after targeted treatment discontinuation suggests targeted treatment cannot eliminate every single tumor cell. Identifying the molecular features of resistant clones could potentially lead to decipher peculiar pharmacologic vulnerabilities, to be targeted together with ALK in order to avoid HSCT with its potential toxicities. In addition, almost all the cases of disease progression while on ALK-TKIs (crizotinib, ceritinib and alectinib) occurred in ACLC patients who experienced primary drug resistance [61,63,64]. This imply that ALK-positive ACLC can be dichotomized into a disease that respond or not to ALK-TKIs. Again, the molecular underpinnings of the differential clinical behavior are of major relevance to allow non-responsive diseases to be sensitive to combinatorial treatment strategies. Of note, this kind of differential responses are not recorded in ALK-rearranged NSCLC, where TP53 status and BIM polymorphisms may impact on PFS and on duration of response [65,66], but not on the initial sensitivity to ALK-TKIs.

In IMT, ALK-TKI are administered in order to reduce disease burden and allow complete tumor removal, as radical surgery is the treatment mainstay. The majority of patients respond to targeted agents and, of note, overcoming resistance with novel generation inhibitors has been reported [67–70]. Obtaining an in-depth view of the molecular events responsible for resistance to ALK-TKI in IMT is nevertheless awaited.

Differently from the cancer types approached thus far, in neuroblastoma ALK is not activated by gene fusion but by point mutations. This likely explains the unsatisfactory results of crizotinib in this disease, as the first-generation inhibitor is known to act a wild-type kinase domain and mutations occurring into kinase active sites are responsible for acquired crizotinib resistance in NSCLC [61]. Given the relatively frequency and the potential severity of this disease, the novel-generation ALK-TKIs now being tested, mainly as combinatorial strategies, will hopefully confirm ALK-mutant IMT as a suitable target of ALK inhibition.

Several other tumor types can be found to harbor ALK rearrangements (see Chapter 1) and they can benefit from ALK-TKI administration. In line with the ACLC, IMT and neuroblastoma, ALK-positive NSCLC remains the model to follow, concerning both the clinical use of targeted agents and linked resistance, with peculiarities dependent from specific tumor types and localization.

Conclusions

The wide and profound knowledge generated on the biology of resistance to ALK-TKI in ALK-positive NSCLC are the main responsible for the major clinical achievements obtained through the availability of several generations of ALK inhibitors and the development of novel treatment strategies. Data concerning molecular events entailing resistance in this setting have been mainly obtained from clinical samples of patients progressing on ALK-TKIs administered sequentially. These lessons are extremely precious now that the clinical practice is shifting towards the second- (and maybe third-) generation inhibitors given upfront, to suggest how tumors can evade direct inhibition. With the positioning of ALK inhibitors even in the early, locally advanced and oligometastatic stages of NSCLC, new areas of research will likely develop, in order to identify biologic features potentially responsible for relapse or progression, from clinical samples obtained at the moment of disease remission. The scientific methodology adopted to assess ALK resistance in lung cancer should be translated in other ALK-driven diseases, still acknowledging the biological and clinical peculiarity of the tumor type of origin.

In conclusion, insisting on dissecting ALK resistance mechanisms has provided dramatic improvements in the management of ALK-rearranged NSCLC. Both in lung cancer and in other ALK-driven diseases, chasing better survival outcomes relies on the understanding of biologic events.

References

[1] Boyd N, Dancey JE, Gilks CB, Huntsman DG. Rare cancers: a sea of opportunity. Lancet Oncol 2016;17:e52–61. https://doi.org/10.1016/S1470-2045(15)00386-1.
[2] Recondo G, Facchinetti F, Olaussen KA, Besse BFL. Making the first move in EGFR-driven or ALK-driven NSCLC: first-generation or next-generation TKI? Nat Rev Clin Oncol 2018;15(11):694–708.
[3] Lynch TJ, Bell DW, Sordella R, Gurubhagavatula S, Okimoto RA, Brannigan BW, et al. Activating mutations in the epidermal growth factor receptor underlying responsiveness of non-small-cell lung cancer to gefitinib. N Engl J Med 2004;350:2129–39. https://doi.org/10.1056/NEJMoa040938.
[4] Paez JG, Jänne PA, Lee JC, Tracy S, Greulich H, Gabriel S, et al. EGFR mutations in lung cancer: correlation with clinical response to gefitinib therapy. Science (80-) 2004;304:1497–500. https://doi.org/10.1126/science.1099314.
[5] Kwak EL, Bang Y-J, Camidge DR, Shaw AT, Solomon B, Maki RG, et al. Anaplastic lymphoma kinase inhibition in non-small-cell lung cancer. N Engl J Med 2010;363:1693–703. https://doi.org/10.1056/NEJMoa1006448.
[6] Choi YL, Soda M, Yamashita Y, Ueno T, Takashima J, Nakajima T, et al. EML4-ALK mutations in lung cancer that confer resistance to ALK inhibitors. N Engl J Med 2010;363:1734–9. https://doi.org/10.1017/CBO9781107415324.004.
[7] Katayama R, Shaw AT, Khan TM, Mino-Kenudson M, Solomon BJ, Halmos B, et al. Mechanisms of acquired crizotinib resistance in ALK-rearranged lung cancers. Sci Transl Med 2012;4:120ra17. https://doi.org/10.1126/scitranslmed.3003316.
[8] Friboulet L, Li N, Katayama R, Lee CC, Gainor JF, Crystal AS, et al. The ALK inhibitor ceritinib overcomes crizotinib resistance in non-small cell lung cancer. Cancer Discov 2014;4:662–73. https://doi.org/10.1158/2159-8290.CD-13-0846.
[9] Zou HY, Friboulet L, Kodack DP, Engstrom LD, Li Q, West M, et al. PF-06463922, an ALK/ROS1 inhibitor, overcomes resistance to first and second generation ALK inhibitors in preclinical models. Cancer Cell 2015;28:70–81. https://doi.org/10.1016/j.ccell.2015.05.010.
[10] Gainor JF, Dardaei L, Yoda S, Friboulet L, Leshchiner I, Katayama R, Dagogo-Jack I, Gadgeel S, Schultz K, Singh M, Chin E, Parks M, Lee D, DiCecca RH, Lockerman E, Huynh T, Logan J, Ritterhouse LL, Le LP, Muniappan A,

Digumarthy S, Channick C, Keyes C, Ge SA. Molecular mechanisms of resistance to first- and second-generation ALK inhibitors in ALK-rearranged lung cancer. Cancer Discov 2016;6(10):1118–33.

[11] Yoda S, Lin JJ, Lawrence MS, Burke BJ, Friboulet L, Langenbucher A, et al. Sequential ALK inhibitors can select for lorlatinib-resistant compound ALK mutations in ALK-positive lung cancer. Cancer Discov 2018;8:714–29. https://doi.org/10.1158/2159-8290.CD-17-1256.

[12] Okada K, Araki M, Sakashita T, Ma B, Kanada R, Yanagitani N, et al. Prediction of ALK mutations mediating ALK-TKIs resistance and drug re-purposing to overcome the resistance. EBioMedicine 2019;41:105–19. https://doi.org/10.1016/j.ebiom.2019.01.019.

[13] Recondo G, Mezquita L, Facchinetti F, Planchard D, Gazzah A, Bigot L, et al. Diverse resistance mechanisms to the third-generation ALK inhibitor lorlatinib in ALK-rearranged lung cancer. Clin Cancer Res 2020;26. https://doi.org/10.1158/1078-0432.CCR-19-1104.

[14] Solomon BJ, Kim DW, Wu YL, Nakagawa K, Mekhail T, Felip E, Cappuzzo F, Paolini J, Usari T, Tang Y, Wilner KD, Blackhall FMT. Final overall survival analysis from a study comparing first-line crizotinib with chemotherapy: results from PROFILE 1014. J Clin Oncol 2018;36(22):2251–8.

[15] Mok T, Camidge DR, Gadgeel SM, Rosell R, Dziadziuszko R, Kim DW, Pérol M, Ou SI, Ahn JS, Shaw AT, Bordogna W, Smoljanović V, Hilton M, Ruf T, Noé JPS. Updated overall survival and final progression-free survival data for patients with treatment-Naïve advanced ALK-positive non-small-cell lung cancer in the ALEX study. Ann Oncol 2020;31:1056–64.

[16] Solomon BJ, Mok T, Kim D-W, Wu Y-L, Nakagawa K, Mekhail T, et al. First-line Crizotinib versus chemotherapy in ALK -positive lung cancer. N Engl J Med 2014;371:2167–77. https://doi.org/10.1056/NEJMoa1408440.

[17] Mok TS, Wu YL, Thongprasert S, Yang CH, Chu DT, Saijo N, et al. Gefitinib or carboplatin-paclitaxel in pulmonary adenocarcinoma. N Engl J Med 2009;361(10):947–57. https://doi.org/10.1056/NEJMoa0810699.

[18] Fukuoka M, Wu YL, Thongprasert S, Sunpaweravong P, Leong SS, Sriuranpong V, et al. Biomarker analyses and final overall survival results from a phase III, randomized, open-label, first-line study of gefitinib versus carboplatin/paclitaxel in clinically selected patients with advanced non-small-cell lung cancer in Asia (IPASS). J Clin Oncol 2011;29(21):2866–74. https://doi.org/10.1200/JCO.2010.33.4235.

[19] Facchinetti F, Tiseo M, Di Maio M, Graziano P, Novello S. Tackling ALK in non-small cell lung cancer: the role of novel inhibitors. Transl Lung Cancer Res 2016;5:301–21. https://doi.org/10.21037/tlcr.2016.06.10.

[20] Peters S, Camidge DR, Shaw AT, Gadgeel S, Ahn JS, Kim DW, Ou SI, Pérol MDR. Alectinib versus crizotinib in untreated ALK-positive non-small-cell lung cancer. N Engl J Med 2017;377(9):829–38.

[21] Costa DB, Shaw AT, Ou SHI, Solomon BJ, Riely GJ, Ahn MJ, et al. Clinical experience with crizotinib in patients with advanced ALK-rearranged non-small-cell lung cancer and brain metastases. J Clin Oncol 2015;33 (17):1881–8. https://doi.org/10.1200/JCO.2014.59.0539.

[22] Shaw AT, Felip E, Bauer TM, Besse B, Navarro A, Postel-Vinay S, et al. Lorlatinib in non-small-cell lung cancer with ALK or ROS1 rearrangement: an international, multicentre, open-label, single-arm first-in-man phase 1 trial. Lancet Oncol 2017;18:1590–9. https://doi.org/10.1016/S1470-2045(17)30680-0.

[23] Bauer TM, Shaw AT, Johnson ML, Navarro A, Gainor JF, Thurm H, et al. Brain penetration of lorlatinib: cumulative incidences of CNS and non-CNS progression with lorlatinib in patients with previously treated ALK-positive non-small-cell lung cancer. Target Oncol 2020;15(1):55–65. https://doi.org/10.1007/s11523-020-00702-4.

[24] Sequist LV, Han J, Ahn M, Cho BC, Yu H, Kim S, et al. Articles Osimertinib plus savolitinib in patients with EGFR cancer after progression on EGFR tyrosine kinase inhibitors: interim results from a multicentre, open-label, phase 1b study. Lancet Oncol 2020;21:373–86. https://doi.org/10.1016/S1470-2045(19)30785-5.

[25] Wu YL, Cheng Y, Zhou J, Lu S, Zhang Y, Zhao J, Kim DW, Soo RA, Kim SW, Pan H, Chen YM, Chian CF, Liu X, Tan DSW, Bruns R, Straub J, Johne A, Scheele J, Park KYJ. Tepotinib plus gefitinib in patients with EGFR-mutant non-small-cell lung cancer with MET overexpression or MET amplification and acquired resistance to previous EGFR inhibitor (INSIGHT study): an open-label, phase 1b/2, multicentre, randomised trial. Lancet Respir Med 2020. https://doi.org/10.1016/S2213-2600(20)30154-5.

[26] Zhu VW, Klempner SJ, Ou SHI. Receptor tyrosine kinase fusions as an actionable resistance mechanism to EGFR TKIs in EGFR-mutant non-small-cell lung cancer. Trends Cancer 2019;5:677–92. https://doi.org/10.1016/j.trecan.2019.09.008.

[27] Dagogo-Jack I, Rooney M, Lin JJ, Nagy RJ, Yeap BY, Hubbeling H, et al. Treatment with next-generation ALK inhibitors fuels plasma ALK mutation diversity. Clin Cancer Res 2019;25. https://doi.org/10.1158/1078-0432.CCR-19-1436.

[28] Shaw AT, Friboulet L, Leshchiner I, Gainor JF, Bergqvist S, Brooun A, et al. Resensitization to crizotinib by the lorlatinib ALK resistance mutation L1198F. N Engl J Med 2015;374:54–61. https://doi.org/10.1056/NEJMoa1508887.

[29] Ramalingam SS, Vansteenkiste J, Planchard D, Cho BC, Gray JE, Ohe Y, et al. Overall survival with osimertinib in untreated, EGFR-mutated advanced NSCLC. N Engl J Med 2020;382:41–50. https://doi.org/10.1056/NEJMoa1913662.

[30] Soria JC, Tan DSW, Chiari R, Wu YL, Paz-Ares L, Wolf J, et al. First-line ceritinib versus platinum-based chemotherapy in advanced ALK-rearranged non-small-cell lung cancer (ASCEND-4): a randomised, open-label, phase 3 study. Lancet 2017;389:917–29. https://doi.org/10.1016/S0140-6736(17)30123-X.

[31] Camidge DR, Kim HR, Ahn MJ, Yang JCH, Han JY, Lee JS, et al. Brigatinib versus crizotinib in ALK-positive non-small-cell lung cancer. N Engl J Med 2018;379:2027–39. https://doi.org/10.1056/NEJMoa1810171.

[32] Gadgeel SM, Shaw AT, Govindan R, Gandhi L, Socinski MA, Camidge DR, et al. Pooled analysis of CNS response to alectinib in two studies of pretreated patients with ALK-positive non-small-cell lung cancer. J Clin Oncol 2016;34(34):4079–85. https://doi.org/10.1200/JCO.2016.68.4639.

[33] Gadgeel S, Peters S, Mok T, Shaw AT, Kim DW, Ou SI, et al. Alectinib versus crizotinib in treatment-naive anaplastic lymphoma kinase-positive (ALK+) non-small-cell lung cancer: CNS efficacy results from the ALEX study. Ann Oncol Off J Eur Soc Med Oncol 2018;29:2214–22. https://doi.org/10.1093/annonc/mdy405.

[34] Pennell NA, Neal JW, Chaft JE, Azzoli CG, Jänne PA, Govindan R, et al. Select: a phase II trial of adjuvant erlotinib in patients with resected epidermal growth factor receptor-mutant non-small-cell lung cancer. J Clin Oncol 2019;37(2):97–104. https://doi.org/10.1200/JCO.18.00131.

[35] Zhong WZ, Wang Q, Mao WM, Xu ST, Wu L, Shen Y, et al. Gefitinib versus vinorelbine plus cisplatin as adjuvant treatment for stage II–IIIA (N1 – N2) EGFR-mutant NSCLC (ADJUVANT/CTONG1104): a randomised, open-label, phase 3 study. Lancet Oncol 2018;19(1):139–48. https://doi.org/10.1016/S1470-2045(17)30729-5.

[36] Herbst RS, Tsuboi M, John T, Grohé C, Majem M, Goldman JW, Kim S-W, Marmol D, Yuri Rukazenkov Y-LW. Osimertinib as adjuvant therapy in patients with stage IB-IIIA EGFR mutation positive NSCLC after complete tumor resection: ADAURA. J Clin Oncol 2020;38:LBA5.

[37] Dumont D, Dô P, Lerouge D, Planchard G, Riffet M, Dubos-Arvis C, et al. Off-label use of crizotinib as a neoadjuvant treatment for a young patient when conventional chemotherapy gave no benefits in stage IIIA non-small cell lung cancer. Am J Case Rep 2017;18:890–3. https://doi.org/10.12659/AJCR.903528.

[38] Zhang C, lei LS, Nie Q, Dong S, Shao Y, ning YX, et al. Neoadjuvant crizotinib in resectable locally advanced non-small cell lung cancer with ALK rearrangement. J Thorac Oncol 2019;14(4):726–31. https://doi.org/10.1016/j.jtho.2018.10.161.

[39] Parikh AB, Hammons L, Gomez JE. Neoadjuvant tyrosine kinase inhibition in locally-advanced non-small cell lung cancer: two cases and a brief literature review. Anticancer Res 2019;39(2):897–902. https://doi.org/10.21873/anticanres.13191.

[40] Kilickap S, Onder S, Dizdar O, Erman M, Uner A. Short-time use of crizotinib as neoadjuvant in ALK-positive non-small cell lung carcinoma can be a chance for resectability. Cancer Chemother Pharmacol 2019;83(6):1195–6. https://doi.org/10.1007/s00280-019-03810-9.

[41] Zhang C, Yan LX, Jiang BY, Wu YLZW. Feasibility and safety of neoadjuvant alectinib in a patient with ALK-positive locally advanced NSCLC. J Thorac Oncol 2020;15:e95–9.

[42] De Koning HJ, Van Der Aalst CM, De Jong PA, Scholten ET, Nackaerts K, Heuvelmans MA, et al. Reduced lung-cancer mortality with volume CT screening in a randomized trial. N Engl J Med 2020;382:503–13. https://doi.org/10.1056/NEJMoa1911793.

[43] Antonia SJ, Villegas A, Daniel D, Vicente D, Murakami S, Hui R, et al. Overall survival with durvalumab after chemoradiotherapy in stage III NSCLC. N Engl J Med 2018;379:2342–50. https://doi.org/10.1056/nejmoa1809697.

[44] Mazières J, Drilon A, Lusque A, Mhanna L, Cortot AB, Mezquita L, Thai AA, Mascaux C, Couraud S, Veillon R, Van Den Heuvel M, Neal J, Peled N, Früh M, Ng TL, Gounant V, Popat S, Diebold J, Sabari J, Zhu VW, Rothschild SI, Bironzo P, Martinez A, Curioni-Fon GO. Immune checkpoint inhibitors for patients with advanced lung cancer and oncogenic driver alterations: results from the IMMUNOTARGET registry. Ann Oncol 2019;30(8):1321–8.

[45] Giaj-Levra N, Giaj-Levra M, Durieux V, Novello S, Besse B, Hasan B, et al. Defining synchronous oligometastatic non–small cell lung cancer: a systematic review. J Thorac Oncol 2019;14(12):2053–61. https://doi.org/10.1016/j.jtho.2019.05.037.

[46] Dingemans AMC, Hendriks LEL, Berghmans T, Levy A, Hasan B, Faivre-Finn C, et al. Definition of synchronous Oligometastatic non–small cell lung cancer—a consensus report. J Thorac Oncol 2019;14(12):2109–19. https://doi.org/10.1016/j.jtho.2019.07.025.

[47] Gomez DR, Tang C, Zhang J, Blumenschein GR, Hernandez M, Jack Lee J, et al. Local consolidative therapy vs. maintenance therapy or observation for patients with oligometastatic non–small-cell lung cancer: long-term results of a multi-institutional, phase II, randomized study. J Clin Oncol 2019;37(18):1558–65. https://doi.org/10.1200/JCO.19.00201.

[48] Palma DA, Olson R, Harrow S, Gaede S, Louie AV, Haasbeek C, Mulroy L, Lock M, Rodrigues GB, Yaremko BP, Schellenberg D, Ahmad B, Senthi S, Swaminath A, Kopek N, Liu M, Moore K, Currie S, Schlijper R, Bauman GS, Laba J, Qu XM, Warner ASS. Stereotactic ablative radiotherapy for the comprehensive treatment of oligometastatic cancers: long-term results of the SABR-COMET phase II randomized trial. J Clin Oncol 2020;38(25):2830–8.

[49] Gan GN, Weickhardt AJ, Scheier B, Doebele RC, Gaspar LE, Kavanagh BD, et al. Stereotactic radiation therapy can safely and durably control sites of extra-central nervous system oligoprogressive disease in anaplastic lymphoma kinase-positive lung cancer patients receiving crizotinib. Int J Radiat Oncol Biol Phys 2014;88:892–8. https://doi.org/10.1016/j.ijrobp.2013.11.010.

[50] Weickhardt AJ, Scheier B, Burke JM, Gan G, Lu X, Bunn PA, et al. Local ablative therapy of oligoprogressive disease prolongs disease control by tyrosine kinase inhibitors in oncogene-addicted non-small-cell lung cancer. J Thorac Oncol 2012;7:1807–14. https://doi.org/10.1097/JTO.0b013e3182745948.

[51] McCoach CE, Bivona TG, Blakely CM, Doebele RC. Neoadjuvant oncogene-targeted therapy in early stage non–small-cell lung cancer as a strategy to improve clinical outcome and identify early mechanisms of resistance. Clin Lung Cancer 2016;17(5):466–9. https://doi.org/10.1016/j.cllc.2016.05.025.

[52] Panagiotou OA, Högg LH, Hricak H, Khleif SN, Levy MA, Magnus D, Murphy MJ, Patel B, Winn RA, Nass SJ, Constanti M. Clinical application of computational methods in precision oncology. JAMA Oncol 2020;6(8):1282–6.

[53] Sharma SV, Lee DY, Li B, Quinlan MP, Takahashi F, Maheswaran S, et al. A chromatin-mediated reversible drug-tolerant state in cancer cell subpopulations. Cell 2010;141:69–80. https://doi.org/10.1016/j.cell.2010.02.027.

[54] Hata AN, Niederst MJ, Archibald HL, Gomez-Caraballo M, Siddiqui FM, Mulvey HE, et al. Tumor cells can follow distinct evolutionary paths to become resistant to epidermal growth factor receptor inhibition. Nat Med 2016;22:262–9. https://doi.org/10.1038/nm.4040.

[55] Russo M, Crisafulli G, Sogari A, Reilly NM, Arena S, Lamba S, et al. Adaptive mutability of colorectal cancers in response to targeted therapies. Science 2019;366(6472):1473–80. https://doi.org/10.1126/science.aav4474.

[56] Boumahdi S, de Sauvage FJ. The great escape: tumour cell plasticity in resistance to targeted therapy. Nat Rev Drug Discov 2020;19(1):39–56. https://doi.org/10.1038/s41573-019-0044-1.

[57] Chaudhuri AA, Chabon JJ, Lovejoy AF, Newman AM, Stehr H, Azad TD, et al. Early detection of molecular residual disease in localized lung cancer by circulating tumor DNA profiling. Cancer Discov 2017;7(12):1394–403. https://doi.org/10.1158/2159-8290.CD-17-0716.

[58] Abbosh C, Frankell A, Garnett A, Harrison T, Weichert M, Licon A, Veeriah S, Daber B, Moreau M, Chesh A, Litchfield K, Lim E, Cooke D, Puttick C, Al Bakir M, Gomes F, Akshay Patel LCS. Phylogenetic tracking and minimal residual disease detection using ctDNA in early-stage NSCLC: a lung TRACERx study. In: Proceedings: AACR annual meeting; 2020. p. CT023.

[59] Prokoph N, Larose H, Lim MS, Burke GAA, Turner SD. Treatment options for paediatric anaplastic large cell lymphoma (ALCL): current standard and beyond. Cancers (Basel) 2018;10(4):99. https://doi.org/10.3390/cancers10040099.

[60] Carlo GP, Cristina M, Pogliani EM. Crizotinib in anaplastic large-cell lymphoma. N Engl J Med 2011;64:775–6. https://doi.org/10.1056/NEJMc1013224.

[61] Mossé YP, Lim MS, Voss SD, Wilner K, Ruffner K, Laliberte J, et al. Safety and activity of crizotinib for paediatric patients with refractory solid tumours or anaplastic large-cell lymphoma: a Children's Oncology Group phase 1 consortium study. Lancet Oncol 2013;14(6):472–80. https://doi.org/10.1016/S1470-2045(13)70095-0.

[62] Gambacorti-Passerini C, Mussolin L, Brugieres L. Abrupt relapse of ALK-positive lymphoma after discontinuation of crizotinib. N Engl J Med 2016;374:95–6. https://doi.org/10.1056/NEJMc1511045.

[63] Geoerger B, Schulte J, Zwaan CM, Casanova M, Fischer M, Moreno L, et al. Phase I study of ceritinib in pediatric patients (Pts) with malignancies harboring a genetic alteration in ALK (ALK+): safety, pharmacokinetic (PK), and efficacy results. J Clin Oncol 2015;33:10005. https://doi.org/10.1200/jco.2015.33.15_suppl.10005.

[64] Sekimizu M, Fukano R, Choi I, Kada A, Saito A, Asada R, et al. Phase II trial of CH5424802 (alectinib hydrochloride) for recurrent or refractory ALK-positive anaplastic large cell lymphoma. Blood 2018;79(3):407–13. https://doi.org/10.1182/blood-2018-99-112708.

[65] Kron A, Alidousty C, Scheffler M, Merkelbach-Bruse S, Seidel D, Riedel R, et al. Impact of TP53 mutation status on systemic treatment outcome in ALK-rearranged non-small-cell lung cancer. Ann Oncol 2018;29(10):2068–75. https://doi.org/10.1093/annonc/mdy333.

[66] Zhang L, Jiang T, Li X, Wang Y, Zhao C, Zhao S, et al. Clinical features of Bim deletion polymorphism and its relation with crizotinib primary resistance in Chinese patients with ALK/ROS1 fusion-positive non–small cell lung cancer. Cancer 2017;123(15):2927–35. https://doi.org/10.1002/cncr.30677.

[67] Theilen TM, Soerensen J, Bochennek K, Becker M, Schwabe D, Rolle U, et al. Crizotinib in ALK + inflammatory myofibroblastic tumors—current experience and future perspectives. Pediatr Blood Cancer 2018;65(4). https://doi.org/10.1002/pbc.26920.

[68] Brivio E, Zwaan CM. ALK inhibition in two emblematic cases of pediatric inflammatory myofibroblastic tumor: efficacy and side effects. Pediatr Blood Cancer 2019;66(5):e27645. https://doi.org/10.1002/pbc.27645.

[69] Parker BM, Parker JV, Lymperopoulos A. A case report: pharmacology and resistance patterns of three generations of ALK inhibitors in metastatic inflammatory myofibroblastic sarcoma. J Oncol Pharm Pract 2019; 25(5):1226–30. https://doi.org/10.1177/1078155218781944.

[70] Yuan C, Ma MJ, Parker JV, Mekhail TM. Metastatic anaplastic lymphoma kinase-1 (ALK-1)-rearranged inflammatory myofibroblastic sarcoma to the brain with leptomeningeal involvement: favorable response to serial ALK inhibitors: a case report. Am J Case Rep 2017;18:799–804. https://doi.org/10.12659/AJCR.903698.

Index

Note: Page numbers followed by *f* indicate figures and *t* indicate tables.